SECOND ASCE
CONFERENCE ON

CIVIL ENGINEERING AND NUCLEAR POWER

Vol. II: Geotechnical Topics

SPONSORED BY
AMERICAN SOCIETY OF CIVIL ENGINEERS
STRUCTURAL DIVISION
NUCLEAR STRUCTURES AND MATERIALS COMMITTEE
GEOTECHNICAL DIVISION
TENNESSEE VALLEY SECTION
THE UNIVERSITY OF TENNESSEE
DEPARTMENT OF CIVIL ENGINEERING

Knoxville, Tennessee
September 15-17, 1980

Library of Congress Catalog Card No. 80-67611
ISBN 0-87262-248-7

FOREWORD

The second conference on Civil Engineering and Nuclear Power is a continuation of the conference held at the ASCE National Convention in April 1979 and earlier conferences entitled Structural Design of Nuclear Facilities held in 1972, 1973, and 1975. The objectives of this conference are to provide an opportunity for North American engineers to keep abreast of current developments relating to the design and construction of nuclear facilities and to bring together in one conference engineers who are involved in the various facets of design and construction of nuclear facilities for an exchange of information and viewpoints.

As the state of the nuclear industry changes, so do the needs for information of engineers involved in the industry. The present dearth of orders for nuclear units, issues surrounding the nuclear fuel cycle, retrofitting and reevaluation of existing facilities all dictated a change in the structure of ASCE Nuclear Structures and Materials Committee and in this conference. You will find at this conference sessions dealing with the nuclear fuel cycle, structural design and materials technologies, supports for cable trays and HVAC ductwork, geotechnical including exploration and construction problems as well as the more traditional topics of seismic analysis and impactive and impulsive loads.

The proceedings of the conference are arranged in six volumes as follows:

Volume I—Materials and Structural Design
Design and Analysis for Cable Trays and HVAC Ducts

Volume II—Geotechnical Topics

Volume III—Nuclear Fuel Cycle

Volume IV—Impactive and Impulsive Loads

Volume V—Report by the Committee on Impactive and Impulsive Loads, Nuclear Structures and Materials Committee

Volume VI—Seismic Analysis

I would like to express my appreciation to all those who contributed to putting on this conference. While it is difficult to single out a few individuals, I believe it is appropriate to mention Kenneth P. Buchert who served as Technical Program Chairman and R. Joe Hunt who served as Liaison to the Tennessee Valley Section of ASCE.

The papers in these Proceedings will be (1) eligible for ASCE prizes and awards under the rules that govern such awards; (2) open to discussion in an appropriate ASCE Journal; and (3) indexed by ASCE. However, a paper in the Proceedings will not be eligible for republication by ASCE unless it has been significantly revised, updated with new information, condensed into more concise and readable form, or otherwise made obviously and significantly more useful to the profession than the original paper.

Ronald G. Domer
Chairman of Conference and Chairman of ASCE Nuclear Structures and Materials Committee
Knoxville, Tennessee

CIVIL ENGINEERING AND NUCLEAR POWER

VOLUME II

GEOTECHNICAL TOPICS

TABLE OF CONTENTS

*Not Available for Publication

TABLE OF CONTENTS (Continued)

*Not Available for Publication

FIELD MEASUREMENT OF DYNAMIC SOIL PROPERTIES

by

Kenneth H. Stokoe, II[1], A. M. ASCE

INTRODUCTION

Soils exhibit nonlinear stress-strain behavior when loaded to failure. For problems involving dynamic soil response, nonlinear behavior in shear is usually of most concern. This nonlinear behavior is typically characterized by an initial tangent shear modulus, G_{max}, a failure shearing stress, τ_{max}, and a curve linking G_{max} and τ_{max} as shown in Fig. 1. The initial loading curve, also called the backbone curve, represents the nonlinear behavior of the soil subjected to monotonic loading to failure. When combined with various hypotheses for hysteretic behavior, this curve forms the basis for defining the shape and location of hysteresis loops generated during subsequent cyclic loading (Idriss et al, 1976 and 1978; Joyner and Chen, 1975; Pyke, 1979; and Streeter et al, 1974).

Seismic methods such as the crosshole, downhole and surface refraction methods have traditionally been employed in the field to determine initial tangent moduli (Ballard and McLean, 1975; Murphy, 1972; Wilson et al, 1978; and Woods, 1978). One of the reasons these methods have been successfully used for this purpose is the magnitude of the strains generated in the soil during their use. The variation in shear modulus with shearing strain amplitude can be evaluated from the backbone curve shown in Fig. 1. The resulting modulus-strain relationship is shown in Fig. 2. As seen in this figure, there is a threshold shearing strain of approximately 0.001 percent below which shear modulus

---

[1]Associate Prof., Dept. of Civil Engrg., Univ. of Texas at Austin

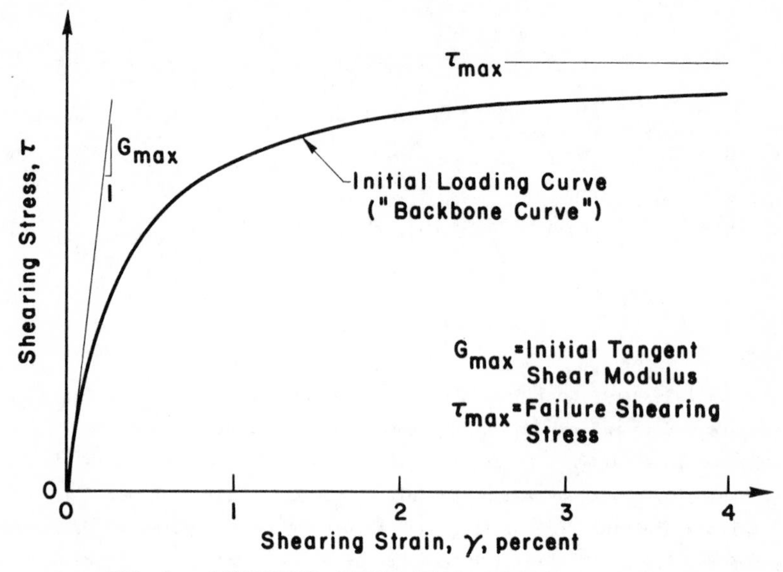

FIG. 1 - GENERALIZED STRESS-STRAIN BEHAVIOR IN SHEAR

can be considered constant and equal to the initial tangent modulus. Because field seismic methods normally operate below the threshold strain and because frequencies generated in these tests are typically less than several hundred Hertz, measurement of G_{max} compatible with earthquake engineering problems is performed, and strain amplitude and strain rate can generally be ignored.

Other seismic methods which have also been employed to determine initial tangent moduli are the up-hole, in-hole and steady-state Rayleigh wave methods. The uses of these methods and of the crosshole, downhole and surface refraction methods are discussed herein. The crosshole and downhole methods are presently the most widely used field seismic methods for evaluation of small-amplitude moduli in geotechnical earthquake engineering and much of the discussion is directed towards these two methods.

Recently, two seismic methods have been developed which permit measurement of dynamic soil behavior from the small-strain range into the nonlinear range. These methods are the *in situ impulse test* method which was developed by Shannon and Wilson and Agbabian Associates (1976)

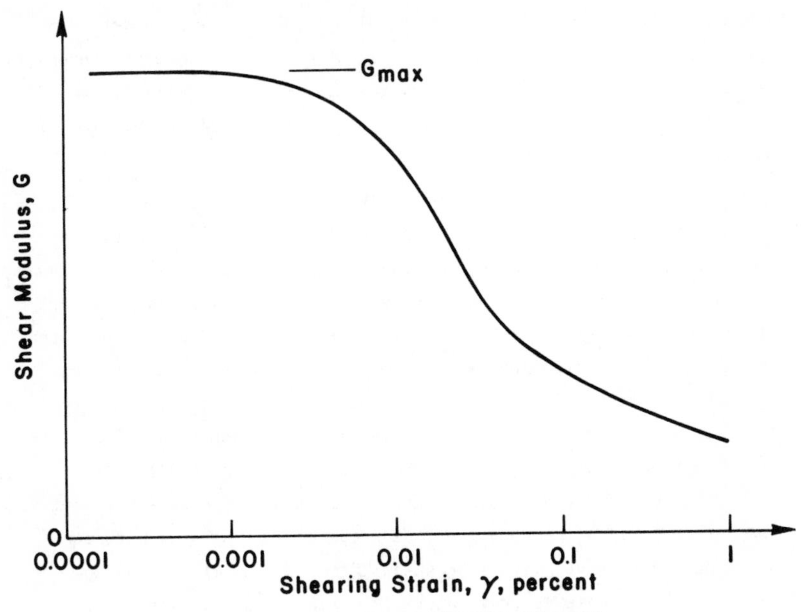

FIG. 2 - GENERALIZED VARIATION IN SHEAR MODULUS WITH SHEARING STRAIN

and the *cylindrical in situ test* method, CIST method, which was developed by the Air Force Weapons Laboratory (1977). The in situ impulse test is similar to conventional crosshole tests except that a large-impulse source mechanism is used which permits evaluation of shear modulus at strains in excess of 0.1 percent. The CIST method, which is also an adaptation of the crosshole test, permits evaluation of constrained modulus at strains in excess of several percent. These methods and their uses and limitations are discussed herein.

The state-of-practice in geotechnical earthquake engineering in terms of field measurement of dynamic soil properties is generally limited to measurement of dynamic moduli, shear and constrained moduli, at small-amplitude strains. Although the in situ impulse test and CIST methods are useable at higher strains, economics and availability have severely limited their use. No readily useable field methods are available for measurement of material damping at any strain amplitude. Therefore, dynamic laboratory tests are used to determine material damping. These laboratory tests are also typically used to determine the variation in moduli with strain amplitude at and above the threshold strain level. Field and laboratory results are then combined with engineering judgement to estimate the nonlinear dynamic properties of soil in situ.

BODY WAVES

Because the generation and measurement of body waves is such an important part in most seismic methods used to determine initial tangent moduli, an understanding of the salient characteristics and behavior of these waves is essential. Two types of body waves can be generated in the field: compression waves, P-waves, and shear waves, S-waves. These waves are termed body waves because they propogate through the interior or body of a soil mass. This is in contrast to surface waves which only propagate along the surface of an unlayered soil mass.

Compression waves generate particle motion in the soil parallel to the direction in which the waves are propagating. As a result, compression waves travel with a "push-pull" motion as illustrated in Fig. 3. Shear waves generate particle motion perpendicular to the direction of wave propagation which creates distortion in the soil as illustrated in Fig. 3. It is the particle motion created by the body waves which receivers in a soil mass monitor. Therefore, for optimum performance in any seismic test, the directions of particle motion of the P- and S-waves should be known, and receivers should be oriented along these directions.

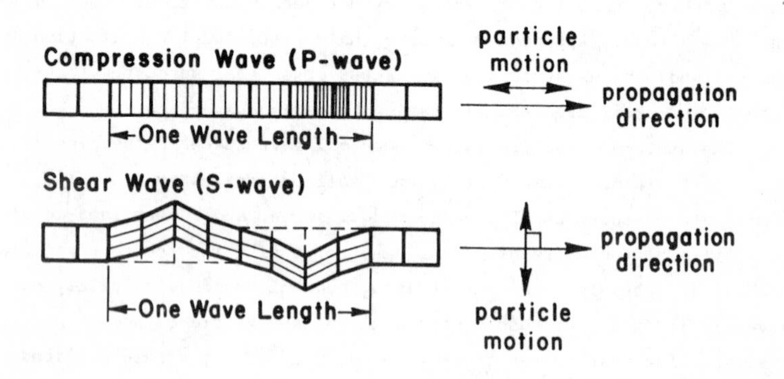

FIG. 3 - BODY WAVE MOTIONS

Seismic methods in which body waves are measured simply consist of measuring the time required for these waves to travel given distances. Once travel times and distances have been measured, wave velocities are calculated by dividing distance of travel by travel time. The initial tangent modulus related to each wave velocity can be calculated by assuming an isotropic homogeneous medium from:

Shear Modulus: $G = \rho v_s^2$ (1)

Constrained Modulus: $M = \rho v_p^2$ (2)

where v_s is shear wave velocity, v_p is compression wave velocity and ρ is the mass density of the soil ($\rho = \gamma/g$ where γ is total unit weight and g is gravitational acceleration). The constrained modulus, M, and not Young's modulus, is determined from the compression wave velocity in the soil mass because plane wave motion generally occurs and hence no lateral deformation results in a direction perpendicular to the propagation direction as shown in Fig. 3.

Additional elastic constants which can be calculated once both body wave velocities have been measured are:

Bulk Modulus: $B = \rho \left(v_p^2 - \frac{4}{3} v_s^2\right)$ (3)

Poisson's Ratio: $\nu = \dfrac{0.5(v_p/v_s)^2 - 1}{(v_p/v_s)^2 - 1}$ (4)

Young's Modulus: $E = 2\rho v_s^2 (1 + \nu)$ (5)

Other characteristics of body waves that are important in properly performing and analyzing seismic tests are the relative magnitudes of the P- and S-wave velocities and the influence of water in the soil on these velocities. Compression wave velocity is always greater than shear wave velocity for the same soil conditions. For this reason, shear wave measurements are more complicated than compression wave measurements because the shear wave is never the first arrival at the monitoring point.

Water in the soil has little effect on shear wave velocity but can have a major effect on compression wave velocity determined from first arrivals. Shear wave velocity is controlled by the stiffness in shear of the soil skeleton. The main influence which water in the soil has on S-wave velocity is manifested in the effect that it has on the effective state of stress in the soil skeleton. As effective stress increases, the stiffness of the soil skeleton increases, and thus shear wave velocity increases. If different samples of the same soil are confined at the same effective stress but have degrees of saturation varying from zero to 100 percent, the shear wave velocity will be essentially the same for each sample.

In comparison with shear wave velocity, compression wave velocity as determined from first arrivals can be significantly affected by water in the soil. For degrees of saturation less than about 98 percent, compression wave velocity is controlled by the stiffness of the soil skeleton in uniaxial loading. The main influence of water on v_p over this range in degree of saturation comes from its affect on the effective stress on the soil skeleton. However, if the degree of saturation equals 100 percent, v_p measured in the field from first arrivals no longer results from a wave propagating through the soil skeleton but results from a fluid wave propagating through the soil-water system. In the author's experience, typical values for v_p measured in saturated soils in the field range from 1370 to 1980 m/s (4500 to 6500 fps). For those cases where the degree of saturation is nearly 100 percent, the fluid wave propagating through the soil-water-air system still controls the measured compression wave velocity (Allen et al, 1980), with the value of v_p being *very* sensitive to degree of saturation or compressibility of the system. Fig. 4 shows the typical influence of degree of saturation on fluid wave velocity, hence the value of v_p measured in the field, over this small range in degree of saturation.

Because of the significant effect of water on P-wave velocity in nearly saturated soil, great care must be taken when comparing field and laboratory values of v_p. Also, P-wave velocities considerably less than 1370 to 1980 m/s (4500 to 6500 fps) should be expected in field measurements in areas of fluctuating water tables, in recently submerged compacted soils, and in soils with decomposing organic matter even though these cases are usually considered saturated for other practical problems.

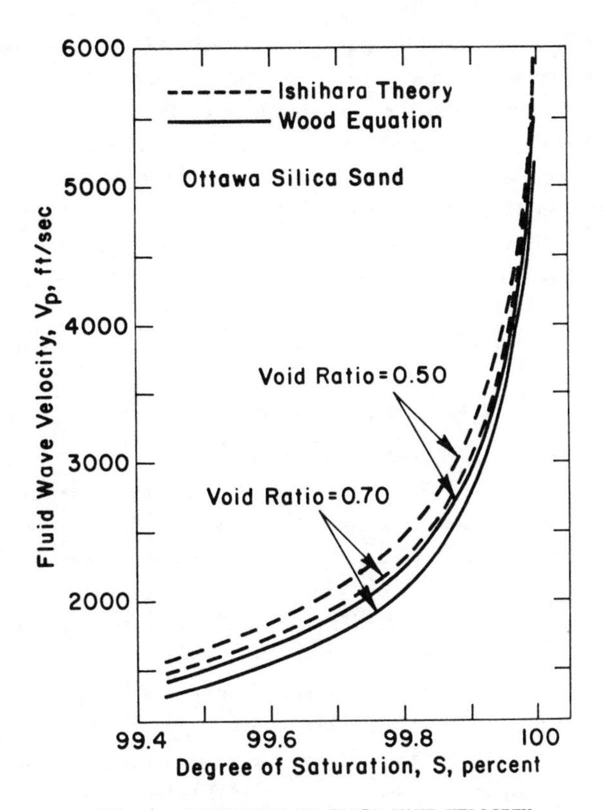

FIG. 4 - VARIATION IN FLUID WAVE VELOCITY WITH DEGREE OF SATURATION FOR AN INFINITE MEDIUM (from Allen, 1977)

In some instances, strain amplitudes and strain rates are of concern. Strain amplitudes for body waves can be calculated based on a plane wave approximation from:

$$\varepsilon = \frac{\dot{u}_p}{v_p} \tag{6}$$

$$\gamma = \frac{\dot{u}_s}{v_s} \tag{7}$$

where ε is normal strain, γ is shearing strain, u_p is P-wave particle velocity and $\dot{u}_s$ is S-wave particle velocity. Strain rates are sometimes considered when investigating shear wave velocities and shear moduli. Based on the author's experience in crosshole and downhole testing, strain rates in these tests range from about 0.0001 to 0.01 in./in./sec. This range overlaps the ranges of strain rates found in conventional static testing and in earthquake ground motion. However, the average of the range in field seismic testing generally tends to be somewhat higher than the average found in static soil testing and somewhat lower than the average found in strong ground motion shaking. Depending on soil type, these variations in strain rates can generally be ignored unless frequencies in excess of 500 Hz are used in the field.

CROSSHOLE AND DOWNHOLE SEISMIC METHODS

The two most widely used seismic methods to determine accurate and detailed compression and shear wave velocity profiles for engineering analyses are the crosshole and downhole methods. Requisite components in these methods include: mechanical sources which are strong, directional and repeatable shear wave generators; receivers with proper coupling, orientation and frequency response; recording equipment with accurate timing, proper frequency response and more than one channel; precise and consistent triggering systems; proper data collection and analysis procedures; and well-trained, conscientious field personnel.

Crosshole Seismic Method - In the crosshole seismic method, the time for compression and shear waves to travel between several points at the same depth within a soil mass is measured. With travel times, wave velocities are calculated after travel distances have been determined.

One field procedure which is successfully employed in the engineering profession today (Stokoe and Hoar, 1978) is shown schematically in Fig. 5. Two or more cased boreholes with spacings on the order of 3 to 4.6 m (10 to 15 ft) are used in a linear array. These boreholes are drilled and cased to the desired depth several days before testing. Either aluminum or plastic casing is used, and the casing is grouted in place. Standard penetration test (SPT) equipment is employed as a mechanical source. This is accomplished by positioning a drill rig at the location of the source borehole shown in Fig. 5 and advancing this borehole (usually uncased). The source borehole is

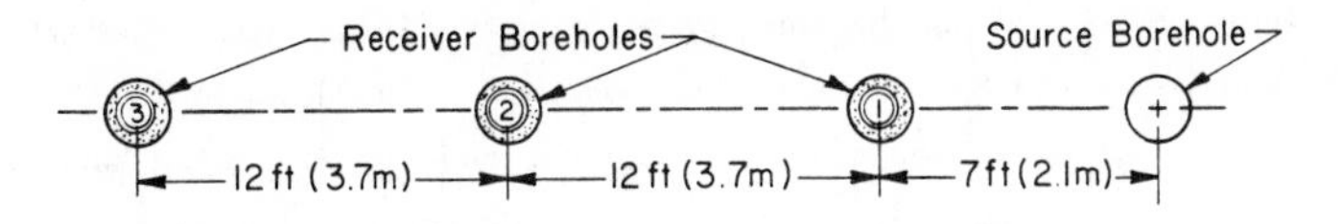

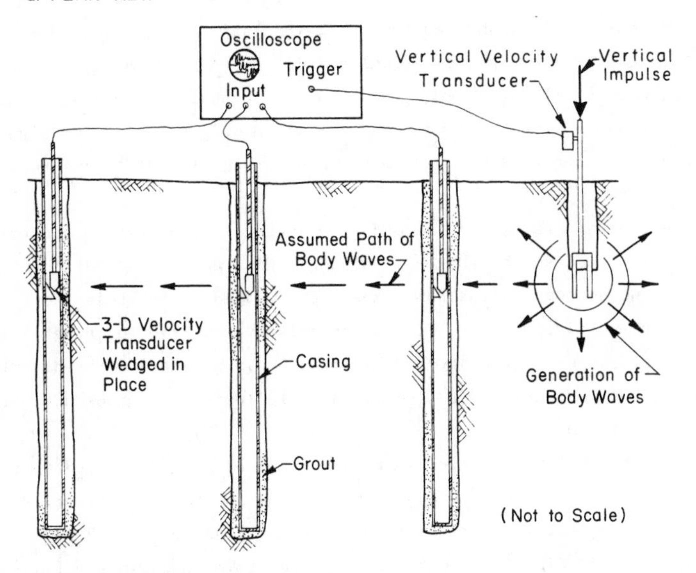

FIG. 5 - CROSSHOLE SEISMIC METHOD
(from Stokoe and Hoar, 1978)

advanced to the first depth at which travel time measurements are to be performed. SPT equipment is then used to generate vertical impulses in the bottom of this borehole. Each impulse generates compression and shear waves in the soil and simultaneously triggers a recording device, an oscilloscope. Wave arrivals are monitored by receivers wedged in the cased boreholes at the same depth. After measurements have been performed for all travel paths at this depth, the source borehole is advanced to the next depth, and the test is repeated. In this manner, testing is continued to the final depth.

Many variations of the above procedure in crosshole testing have been successfully employed. These procedures are essentially contained in reviews by Ballard and McLean (1975), Murphy (1972), Wilson et al, (1978) and Woods (1978). Some of the more widely used variations include: 1. using as few as one receiver borehole, 2. using hand-held hammers to provide the source input usually in conjunction with electrical triggers, 3. using in-hole sources which couple energy to the borehole wall at any desired depth (Ballard, 1976; Auld, 1977; and Hoar, 1980), 4. using a torsional source to generate SH-waves (Hoar and Stokoe, 1978) rather than the SV-waves generated with a vertical impusle, and 5. using signal enhancement recorders to add the signals at each receiver from several impulses or using digital oscilloscopes (Hoar, 1980).

A typical travel time record from vertical velocity transducers wedged in two cased boreholes is shown in Fig. 6a. The record consists of top and bottom traces from transducers wedged in boreholes No. 1 and No. 2, respectively. (See Fig. 5 for relative location of boreholes.) These results were generated by simultaneously monitoring the output from each transducer for one hammer drop following the procedure outlined in Fig. 5.

The initial wave arrivals at each receiver are identified on the record by P for the compression wave and S for the shear wave. Direct (t_D) and interval (t_I) travel time determinations for the shear wave are illustrated in Fig. 6b. Each direct travel time shown in the figure represents the time between triggering of the oscilloscope and arrival of the shear wave at the receiver in the borehole. The time interval between the oscilloscope triggering and generation of the body waves in the soil at the bottom of the SPT rod is also included in these direct time measurements. This time is approximately equal to the compression wave travel time in the rod *if* an instantaneous trigger is used (Hoar and Stokoe, 1978). Therefore, the rod travel times must be determined at each measurement depth and subtracted from the total travel time shown in the figure to determine the actual direct travel time of the shear wave (or compression wave) in the soil.

The interval time shown in Fig. 6b represents the S-wave travel time between the two receiver boreholes. At times there are significant advantages to using interval travel times rather than source-to-receiver travel times because interval times minimize many of the

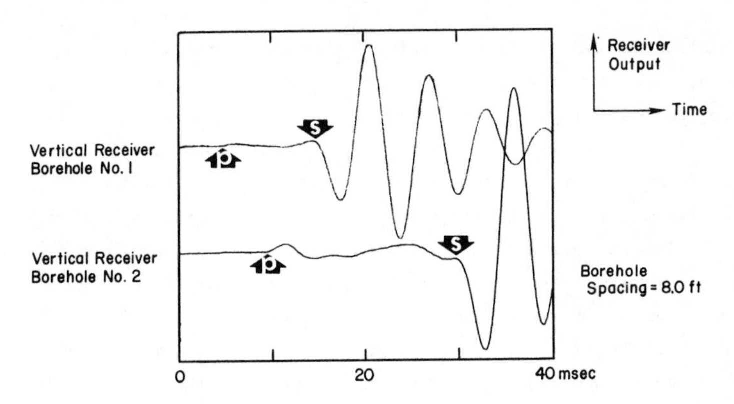

a. Record of Initial Wave Arrivals

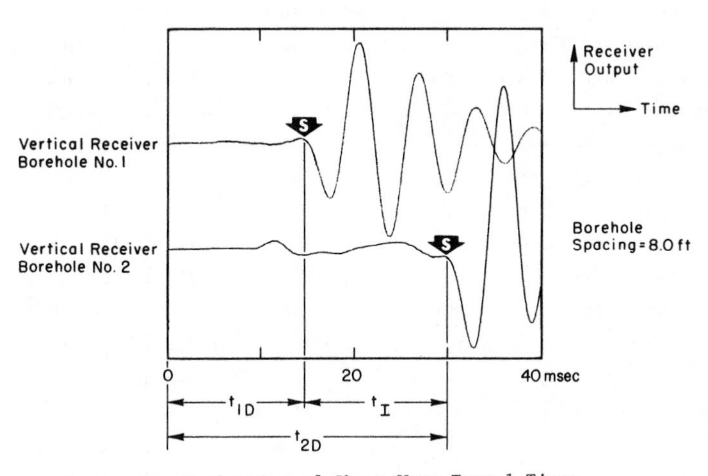

b. Evaluation of Shear Wave Travel Times

FIG. 6 - TRAVEL TIME RECORD USING TWO VERTICAL RECEIVERS IN THE CROSSHOLE SEISMIC TEST

effects which can adversely influence these measurements (Stokoe and Hoar, 1978).

Additional travel time records similar to the ones shown in Fig. 6 but at expanded sweep rates are typically obtained to improve the accuracy and resolution with which P-wave travel times are determined. Receiver orientation for records of this type can be either vertical or radial horizontal with the radial orientation generally preferred. However, the use of radial horizontal receivers requires the use of three-component geophones combined with a receiver orientation system.

Small explosive sources placed in fluid-filled boreholes are usually better suited than mechanical sources for compression wave velocity measurements because explosive sources generate sharper and larger initial P-wave arrivals (Warrick, 1974; and McLamore, et al, 1978). This is especially true for measurements made over large distances and for source-to-receiver measurements based on initial P-wave arrivals. However, mechanical sources perform adequately for interval P-wave velocity measurements over shorter travel paths which are typical of crosshole measurements (3 to 6 m (10 to 20 ft)).

Downhole Seismic Method - In the downhole method, the time for compression and shear waves to travel between a source on the surface and points within the soil mass is measured. Wave velocities are then calculated from the corresponding travel times after travel distances have been determined.

Figure 7 shows schematically one field procedure which is successfully used today (Stokoe and Hoar, 1978). One cased borehole (or more) is installed to the desired depth several days before testing. The borehole is identical to those used as receiver boreholes in the crosshole test. (In fact, the same boreholes are normally used in the crosshole and downhole tests so that redundancy exists in the measurement.) At the same time, a 0.6-m (2-ft.) cube of concrete is cast-in-place at the location shown in Fig 7. This concrete block is embedded so that the top of the block is about even with the ground surface. Two sections of angle iron are placed in the block at the time of casting as shown in Fig 7. The purpose of the angle irons is to provide a stable striking surface for the hammer impulse and to provide good contact for the electrical trigger circuit. The embedded ends of the two angle iron sections are welded together before placement in the block to prevent the angle irons from being driven through the block during use.

The test is performed by wedging a horizontal velocity transducer (usually part of a three-component (3-D) velocity transducer package) in the cased borehole at the first measurement depth. One of the angle irons in the block is struck with a 4.5-kg (10-lb.) sledge hammer.

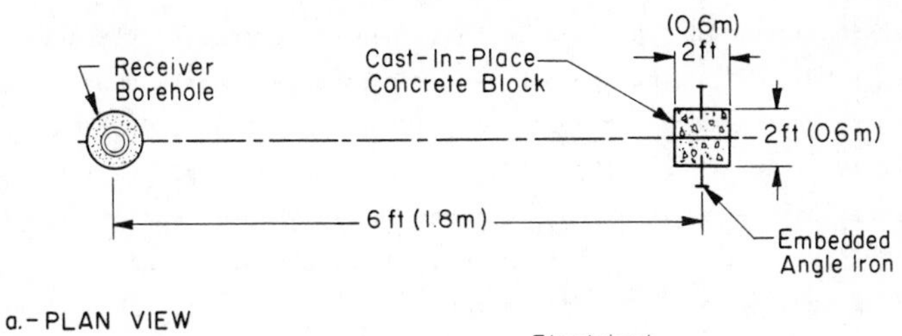

a.-PLAN VIEW

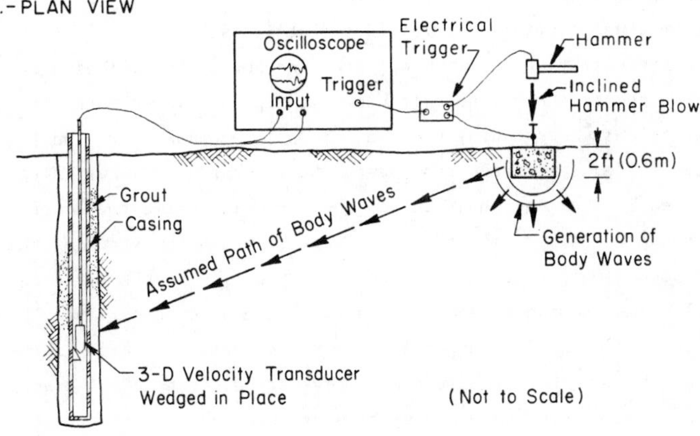

b.-CROSS-SECTIONAL VIEW

FIG. 7 - DOWNHOLE SEISMIC METHOD
(after Stokoe and Hoar, 1978)

Body waves are generated in the soil and their arrivals at the horizontal velocity transducer in the cased borehole are recorded. The test is then repeated except that the block is struck in the opposite direction. Following this procedure, testing is repeated at selected intervals to the final measurement depth.

Variations of the above approach to downhole testing which have been successfully employed include: 1. using different types of source generating mechanisms such as horizontal impulses applied to planks (Warrick, 1974; Beeston and McEvilly, 1977; Wilson et al, 1978; and Patel, 1980) or to steel plates embedded in the ends of trenches (Warrick, 1974; Auld, 1977; Hoar and Stokoe, 1980; and Patel, 1980), mechanically coupled explosives (Jolly, 1956; and Shima and Ohta, 1967).

and torsional impulses applied to cast-in-place concrete cylinders (Patel, 1980), 2. using two 3-D receiver packages in the same borehole to measure true interval travel times (Stokoe and Hoar, 1978; Hoar and Stokoe, 1980; and Patel, 1980), 3. using a receiver orientation system to assure that receiver orientation and polarity match the direction and polarity of the source impulse (Patel, 1980), and 4. using other recording devices such as oscillograph recorders, signal enhancement recorders, and digital oscilloscopes (Patel, 1980).

A typical travel time record from one horizontal velocity transducer wedged in the borehole is shown in Fig. 8. The record consists of two traces generated by reversed blows to a plank source on the surface. The upper trace represents the monitoring of wave energy generated by striking the plank in one direction while the lower trace represents the monitoring of wave energy generated by striking the plank in the opposite direction. The vertical position of the trace on the oscilloscope viewing screen was changed between reversed hammer blows.

The difference between crosshole records with a source on the borehole bottom and downhole records is very evident in the traces in Fig. 8. Compression wave energy remains coincident for reversed hammer blows over much of the time before the shear wave arrives while the shear wave energy is polarized under these conditions, and hence the initial S-wave arrival reverses (Schwarz and Musser, 1972). When analyzing downhole records similar to this one, it is generally assumed that the time between triggering the oscilloscope and coupling of the waves into the soil is negligible. Therefore, travel times determined directly from the oscilloscope trace represent direct travel times of the wave in the soil only.

A typical travel time record from two horizontal velocity transducers wedged at different depths in the same cased borehole is shown in Fig. 9a. The record consists of an upper trace from the shallower transducer and a lower trace from the deeper transducer. Vertical spacing between the transducers was controlled by attaching each transducer to the opposite ends of a steel pipe with both transducers oriented in the same direction.

Both transducers in Fig. 9a were oriented in the direction of maximum shear wave particle motion. Therefore, the initial arrival of the shear wave at each transducer is easily identifiable without

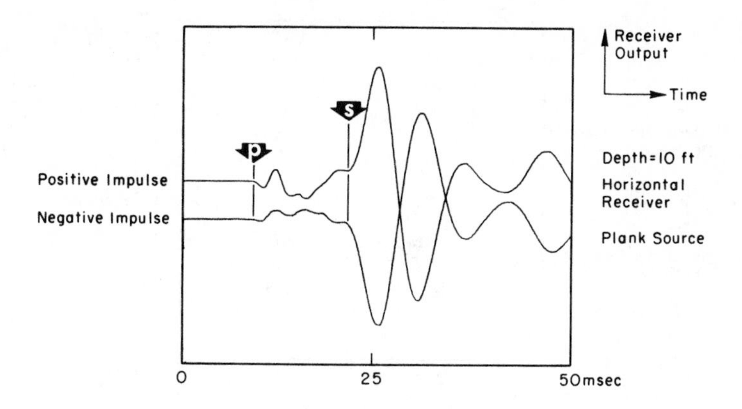

FIG. 8 - TRAVEL TIME RECORD USING REVERSED IMPULSES IN THE DOWNHOLE SEISMIC TEST

reversing the hammer blows. Direct and interval travel time determinations based on initial shear wave arrivals are illustrated in Fig. 9b. With this procedure and configuration, true interval travel times can be determined using the downhole seismic method, and errors in triggering and interval path length determination are essentially eliminated.

Other travel time records similar to the one shown in Fig. 9 but at an expanded sweep rate and using a time delay (between striking the downhole source and triggering the oscilloscope) can also be obtained to improve the resolution and accuracy with which interval downhole P- and S-wave travel times can be determined.

For the best interval downhole travel time measurements, the depth interval between receivers should be large enough to determine accurately interval travel times but small enough to obtain measurements in significant soil layers. At some sites it may be impossible to satisfy both of these requirements. Spreading of the waves causes the initial shear wave arrival to become less distinct with depth which makes it difficult to identify initial shear wave arrivals to within a sufficient tolerance compatible with the time interval between arrivals. A spacing of 1.5 to 3.0 m (5 to 10 ft.) seems to work well near the surface, but at depths greater than about 15 m (50 ft) the vertical spacing should be increased. Of course, the best spacing for each site

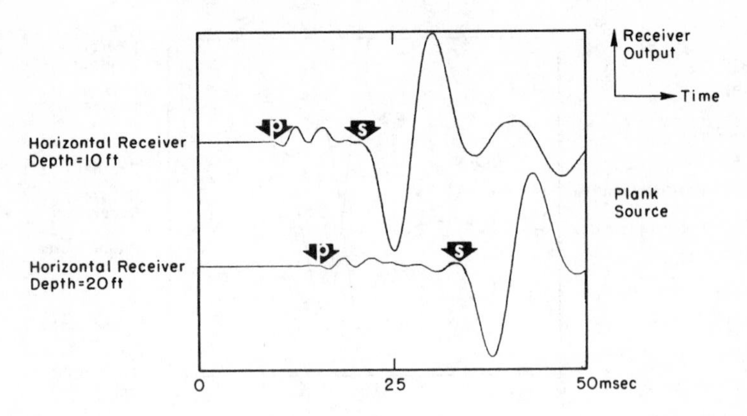

a. Record of Initial Wave Arrivals

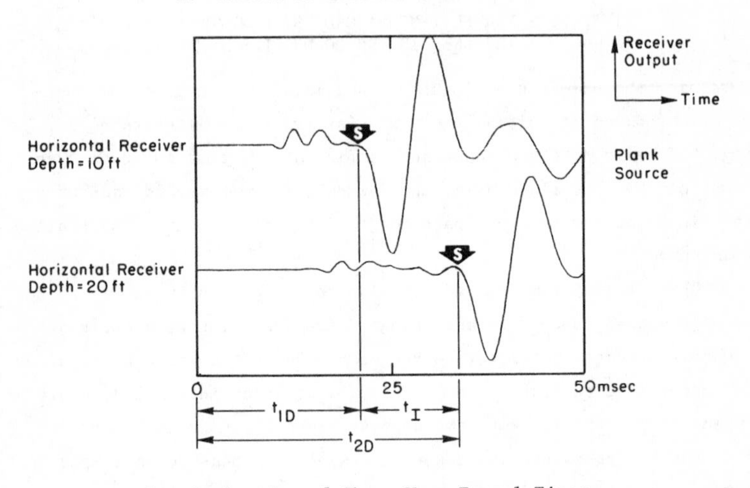

b. Evaluation of Shear Wave Travel Times

FIG. 9 - TRAVEL TIME RECORD USING TWO HORIZONTAL RECEIVERS IN THE DOWNHOLE SEISMIC TEST

will depend on the actual velocity profile. It should be noted that this effect is not a significant problem in crosshole tests because the spacing between the source and receivers is almost constant with depth so that depth does not influence the phase of the wave arriving at each borehole.

Data Collection and Analysis - The manner in which travel time data is collected and analyzed in these seismic tests can significantly affect calculated values of wave velocities. For instance, in the crosshole test, measurement of a direct wave is assumed. However, this assumption can usually be checked if three or more boreholes are used to collect the data (Murphy, 1972; and Stokoe and Abdel-razzak, 1975). For the crosshole procedure outlined in Fig. 5, an additional problem with misinterpretation of a refracted wave as a direct wave can occur as shown in Fig. 10. This can happen in P-wave and S-wave analyses. Analytical techniques can be combined with layering at the site (which should be determined during the boring phase) to identify this problem (Stokoe and Hoar, 1978).

As an example of the manner in which data collection and analysis can affect downhole seismic measurements, consider measurements performed and analyzed in the following manner. At one site, an uncased borehole was advanced by means of solid stem augers. At 1.5-m (5-ft) depth intervals, a receiver was pushed into the soil at the bottom of the borehole and downhole measurements were performed. The results of these measurements were used to determine the shear wave travel time versus depth curve presented in Fig 11 which is the conventional method of presenting these results. The typical procedure used to analyze the travel time measurements is to draw a straight line or a series of connected straight lines through the data points. The wave velocity is then taken from the slopes of these lines. If only a few straight line segments which do not pass through all the data points are used, then only average trends of the data result. These average trends are shown by the two lines in Figure 11.

If the data points in Fig. 11 are connected by a series of straight lines that intersect all data points, then the shear wave velocity profile presented in Figure 12 for pseudo-interval velocities is obtained. These velocities are not true interval velocities because the output from receivers at successive depths were recorded separately, not simultaneously. However, as a first approximation, they were analyzed as interval velocities but are referred to as pseudo-interval velocities.

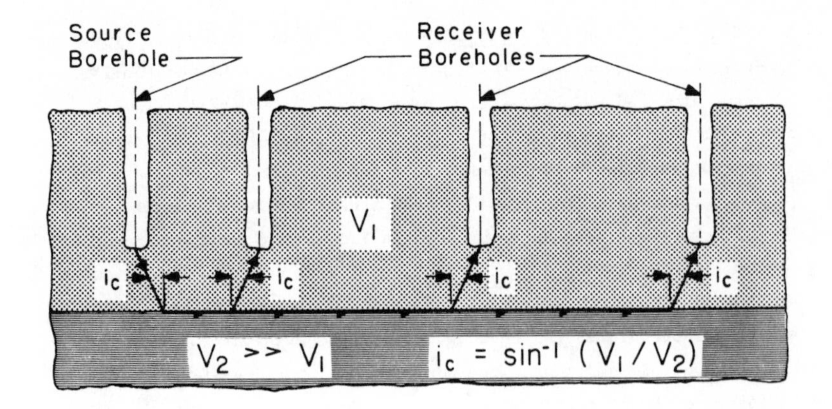

FIG. 10 - REFRACTED WAVE TRAVEL PATHS FOR CONSTANT-VELOCITY LAYERS (from Stokoe and Hoar, 1978)

True interval velocities were also determined at this site and are shown in Fig. 12. These values represent the most representative S-wave velocity profile at the site when compared with crosshole measurements.

The technique of analysis using interval velocities or pseudo-interval velocities contains much more definition of the shear wave velocity profile than the method illustrated in Fig. 11. This definition is usually necessary for most engineering analyses. In addition, measurement of true interval velocities generally eliminates much of the scatter inherent in pseudo-interval measurements.

Comparison of Crosshole and Downhole Methods - Many times the practitioner is faced with the task of selecting the most appropriate seismic method or methods to use at a site based on various requirements and constraints. A general comparison of the important features of the crosshole and downhole methods is given in Table 1 to aid in the selection process. The check marks in the table have been added to indicate the relative strengths of each method. In the author's experience, the crosshole method has given a more detailed representation of the P- and S-wave velocity profile in situ, especially at depths greater than about 15 m (50 ft). However, the crosshole method is more expensive to conduct, and thus cost considerations many times eliminate its use.

TABLE 1 - COMPARISON OF CROSSHOLE AND DOWNHOLE SEISMIC METHODS

CROSSHOLE METHOD	DOWNHOLE METHOD
Two or More Boreholes	√One Borehole
Simple Borehole Source	√Simple Surface Source
Predominantly P- and SV-Wave	Generates P- and S-Waves
but SH-Waves Also Possible	Reversible Source
Reversible Source	Travel Path Increases With
√Constant Travel Paths	Depth
√Negligible Borehole Effects	Possible Borehole Effects
Measure Borehole Verticality	√No Verticality Measurements
√Receivers Properly Aligned	Control of Receiver Alignment
For SV-Waves	Preferable
√High Signal-To-Noise Ratio	Signal-To-Noise Ratio
At All Depths	Decreases With Depth
Detect Low-Velocity Layers	Detect Low-Velocity Layers
√Detailed Profile	More Average Profile
Possible Refraction Problems	√Minimum Refraction Problems
Useable in Noisy Areas	Useable in Noisy Areas
√Workable in Limited Space	Workable in Limited Space
More Expensive	√Less Expensive

Optimum results are, of course, obtained when both methods are used together. It is interesting to note that it is possible to combine these methods so that any point within the soil mass can be excited along the three principal P-wave directions and along the six principal S-wave directions as shown in Fig. 14.

ADDITIONAL SMALL-STRAIN SEISMIC METHODS

Initial tangent moduli of soils can be evaluated by seismic methods other than the crosshole and downhole methods. Seismic methods such as the up-hole, in-hole, surface refraction and steady-state Rayleigh wave methods can be used. These methods are shown schematically in Fig. 15. They are used, however, to a lesser extent and are

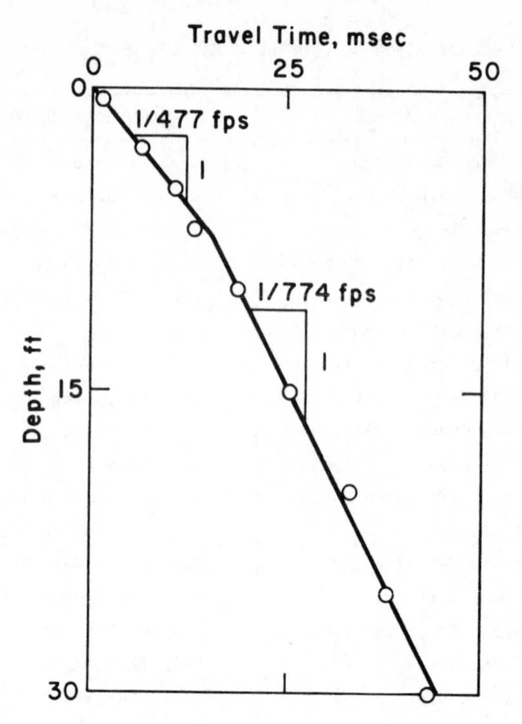

FIG. 11 - SHEAR WAVE VELOCITY PROFILE FROM DIRECT TRAVEL TIME MEASUREMENTS IN DOWNHOLE SEISMIC TEST

in general less well suited for dynamic property measurements than the crosshole and downhole methods.

The up-hole method in which the source is placed in the borehole and receivers are placed on the ground surface is most appropriately used for P-wave velocity measurements. The in-hole method in which the source and receivers are placed in the borehole is currently being developed by the OYO Corp. (1979). The method consists of lowering a probe containing a mechanical source and two receivers to the desired depth in an uncased, fluid-filled borehole. Compression and shear waves propagating in the soil along the borehole wall are then measured. This device has the advantages of requiring only one borehole and of performing measurements to great depths.

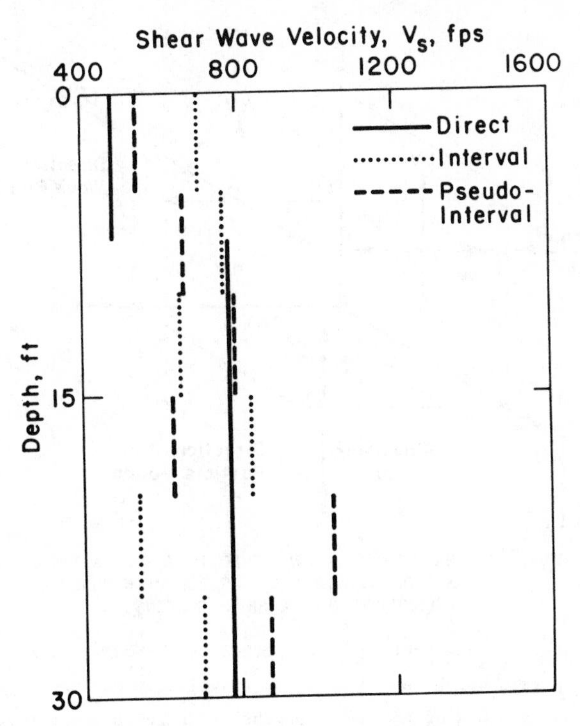

FIG. 12 - VARIATION IN SHEAR WAVE VELOCITY PROFILE WITH TYPE OF TRAVEL TIME MEASUREMENT IN DOWNHOLE SEISMIC TEST

The surface refraction and steady-state Rayleigh wave methods have the advantage of requiring no boreholes. Both the source and receivers are placed on the ground surface which represents a substantial reduction in cost. The surface refraction method also has the advantage of covering large areas quickly which is beneficial for preliminary planning and feasibility studies. However, with the surface refraction method, only the upper portion of layers are typically sampled, low velocity layers are not detected, and the depth of penetration is limited in S-wave measurements because of the significant amount of energy required for deep penetration and because of the decrease in quality of the data with depth (Wilson et al, 1978).

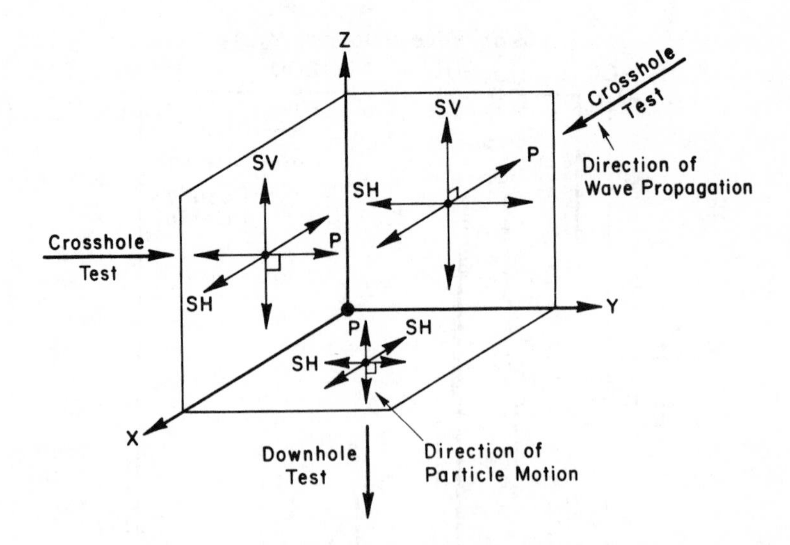

FIG. 13 - WAVE PROPAGATION DIRECTIONS AND PARTICLE MOTION DIRECTIONS POSSIBLE WITH COMBINED CROSSHOLE AND DOWNHOLE TESTING

The steady-state Rayleigh wave method is the only seismic method presented herein which is not based on travel time measurements. Rather, measurements of the wave lengths of Rayleigh waves (surface waves) generated by a steady-state vertical vibrator are performed. With wave lengths, λ_R, and excitation frequencies, f, determined

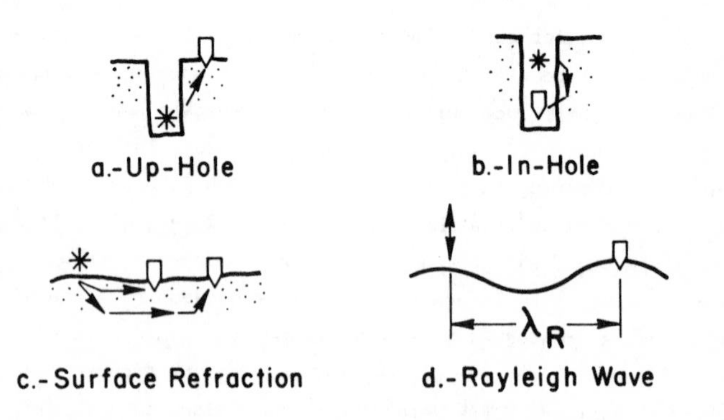

FIG. 14 - SCHEMATIC REPRESENTATION OF ADDITIONAL SMALL-STRAIN SEISMIC METHODS

Rayleigh wave velocity is calculated by multiplying λ_R times f. The effective depth of sampling is empirically related to $\frac{1}{2}\lambda_R$ (Ballard, 1964), and Rayleigh wave velocity is nearly equal to shear wave velocity. The major disadvantage with this method is the large force-generating equipment necessary to explore to significant depths (Woods, 1978).

LARGE-STRAIN SEISMIC METHODS

Two seismic methods have recently been developed for evaluating in situ moduli above the threshold shearing strain of about 0.001 percent. These methods are the in situ impulse test method (Shannon and Wilson, Inc. and Agbabian Associates, 1976) which is used for evaluating nonlinear shear moduli by the geotechnical earthquake engineering profession and the cylindrical in situ test, CIST, (Air Force Weapons Laboratory, 1977) which is used by the military to investigate nonlinear constrained moduli or stress-strain behavior in uniaxial loading. Each of these methods represents a significant advance in the state-of-the-art in field measurement of dynamic soil properties and in the solution to the problem of relating field and laboratory measurements. However, their wide-spread use has been severely limited by economics, feasibility and technical requirements.

In Situ Impulse Test - The in situ impulse test method is a modification of the crosshole seismic method presently used to determine initial tangent moduli. The main differences between these methods are the magnitude of the vertical impulse used to generate the SV-waves, the spacing between the source and first receiver borehole, and the data reduction techniques.

The in situ impulse test method is shown in Fig. 15. Three receiver boreholes and a source borehole are used. A linear borehole array is employed with spacings of 1.2, 2.4 and 4.9 m (4, 8 and 16 ft) between the source and receiver boreholes numbers 1, 2 and 3, respectively. An in-hole source is used which is composed of three plates that are hydraulically forced against the sides of the uncased source borehole. The vertical impulse is generated by dropping a 68-kg (150-lb) hammer about 30 cm (1 ft) onto a Belleville spring, with the

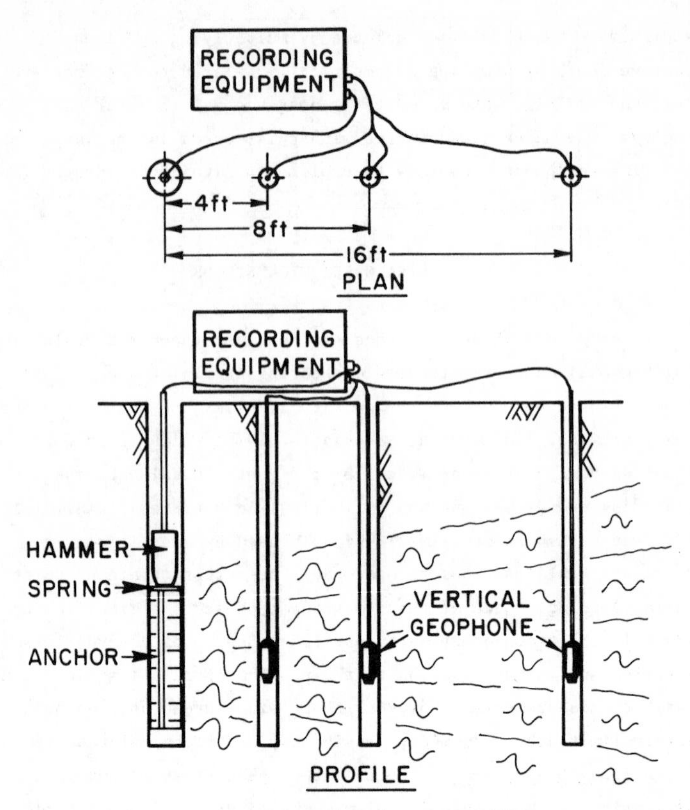

FIG. 15 - IN SITU IMPULSE TEST METHOD (from Wilson et al, 1978)

spring being used to shape the impulse applied to the soil (Wilson et al, 1978).

A typical velocity-time record monitored at the anchor position and by the three vertical receivers for one impulse is shown in Fig. 16. As can be seen in the record, essentially all of the energy goes into generating shear wave motion. Travel times of the zero crossing points as well as peak particle velocities are determined at each receiver position as shown in Fig. 16. The velocity-time history at the anchor is also analyzed with special procedures to account for factors such as slippage and radial jack stresses (Wilson et al, 1978).

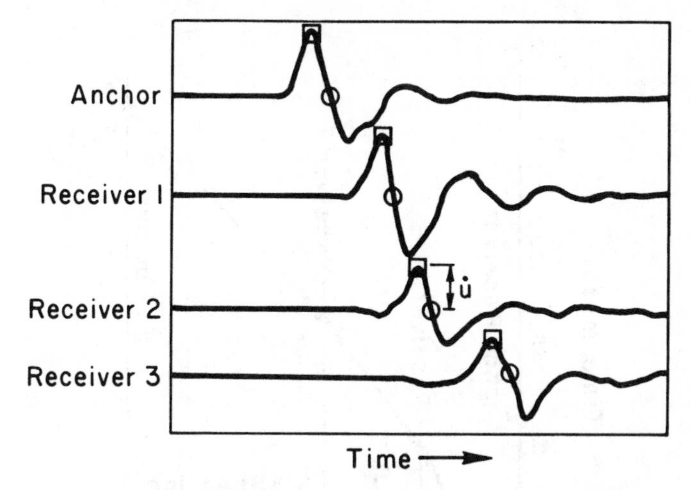

FIG. 16 - TYPICAL VELOCITY - TIME RECORD FROM IN SITU IMPULSE TEST (from Wilson et al, 1978)

Shear wave travel time versus distance from the source is then plotted as shown in Fig. 17, and shear wave velocity is determined from the slope of the curve at each monitoring location. Shearing strain amplitude is calculated from Eq. 7 using particle velocity and propagation velocity at each monitoring location. Velocity-strain and modulus-strain curves can then be plotted at each measurement depth.

The main limitations of this method are the cost, the ability to maintain the spacing between the boreholes within tolerable limits at significant depths, disturbance of the soil between the source and first receiver borehole, and the use of the more complicated data reduction procedures to analyze the anchor motion from which the highest-strain data is developed.

Cylindrical In Situ Test. - The cylindrical in situ test (CIST) method is an adaptation of the crosshole method in which the impulse is applied over most of the length of the source borehole. This is accomplished by filling a 61-cm (24-in.) diameter borehole with 1.5 m

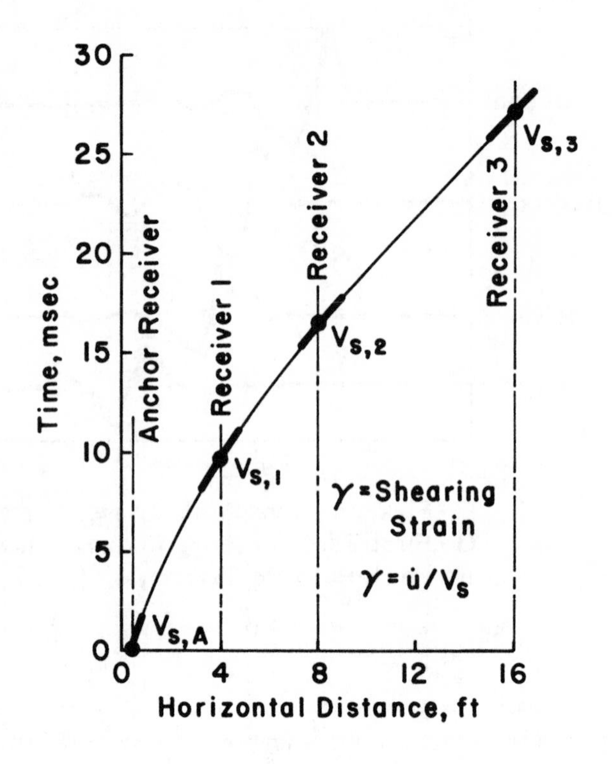

FIG. 17 - TYPICAL TIME-DISTANCE CURVE FROM IN SITU IMPULSE TEST (from Wilson et al, 1978)

(5-ft) long racks of primacord explosives as shown in Fig. 18. Accelerometers are placed in a closely spaced pattern around the source borehole as shown in the figure. The test is performed by detonating the racks of explosive simultaneously which generates a cylindrical P-wave front propagating away from the source. The passage of this wave front past the accelerometers is recorded and these records are integrated to yield velocity-time histories at each monitoring point.

An iterative procedure is then followed whereby measured and predicted velocity-time histories are compared (Bratton and Higgins, 1978). One- and two-dimensional finite difference models of wave propagation in the test are used with assumed material models. When

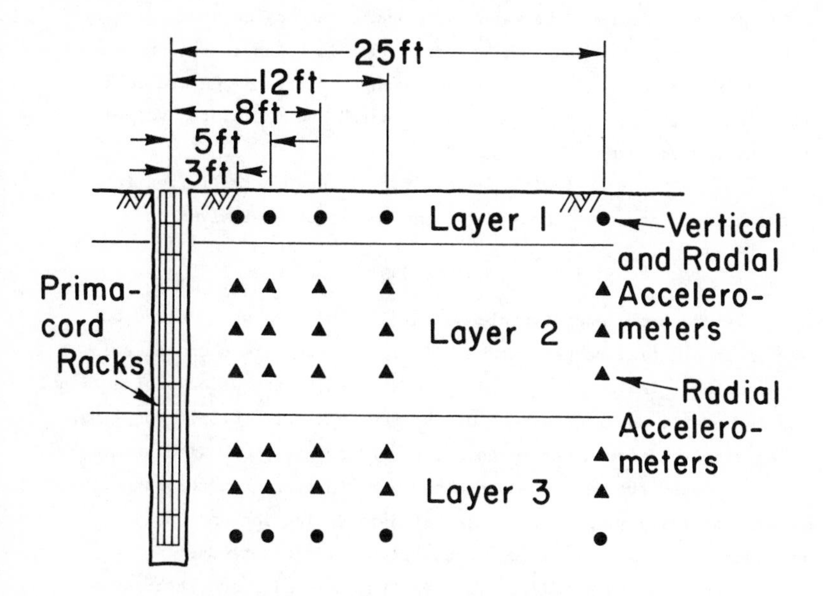

FIG. 18 - CYLINDRICAL IN SITU TEST (CIST) METHOD
(from Bratton and Higgins, 1978)

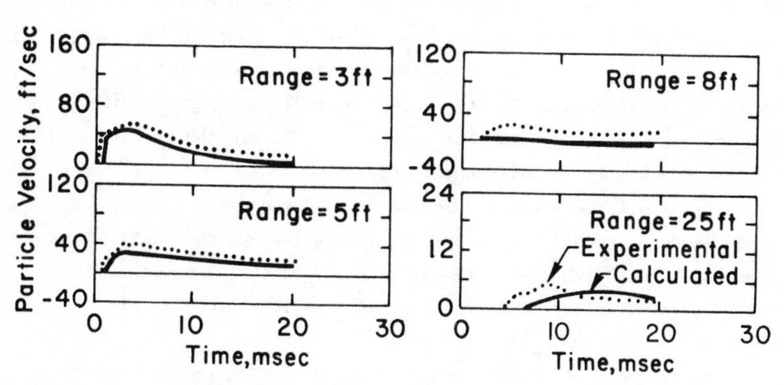

FIG. 19 - COMPARISON OF CALCULATED AND MEASURED VELOCITY-TIME HISTORIES IN CIST
(from Bratton and Higgins, 1978)

good agreement is found between the measured and calculated velocity-time histories, the constitutive soil model is assumed to represent correctly the material over the large range in strains generated in the field. A typical comparison of measured and calculated velocity-time histories is shown in Fig. 19.

The CIST method in its present form is obviously limited in application because of the significant amount of explosives used.

CONCLUSIONS

The crosshole and downhole seismic methods are well suited for evaluating initial tangent moduli in situ. Requisite components in these methods include: mechanical sources which are strong, directional and repeatable shear wave generators; receivers with proper coupling, orientation and frequency response; recording equipment with accurate timing, proper frequency response and more than one channel; precise and consistent triggering systems; proper data collection and analysis procedures; and well-trained, conscientious field personnel.

Accurate and detailed profiles of compression and shear wave velocities with depth can be measured with either the crosshole or downhole method. A thorough understanding of the characteristics and behavior of the P- and S-waves is essential in conducting and analyzing these tests. The influence of water in soil on wave velocities is particularly important. The technique of analysis using interval wave velocities can be used with these methods to improve the quality of the results, especially in the downhole test.

Other seismic methods which have also been used to evaluate initial tangent moduli are the up-hole, in-hole, surface refraction, and steady-state Rayleigh wave methods. The use of these methods is presently very limited in terms of field measurement of dynamic soil properties.

Two seismic methods have been developed which permit measurement of dynamic soil behavior from the small-strain range into the nonlinear range. These methods are the in situ impulse test method which is used for evaluating nonlinear shear moduli and the cylindrical in situ test (CIST) method which is used to investigate nonlinear constrained moduli or stress-strain behavior in uniaxial loading. These methods have had only limited use because of economics, feasibility and technical requirements.

ACKNOWLEDGEMENTS

The writer wishes to acknowledge the many helpful and fruitful discussions over the past years on the topic of seismic measurements with D. G. Anderson, R. F. Ballard, Jr., J. R. Hall, Jr., V. R. McLamore, F. E. Richart, Jr., R. E. Warrick, and R. D. Woods. Sincere appreciation is also expressed to the following graduate students: E. J. Arnold, R. J. Hoar, L. G. Long, L. D. Olson and N. S. Patel with whom the writer has had the pleasure of working and who in fact did most of the work.

REFERENCES

1. Air Force Weapons Laboratory, (1977), "Cylindrical In Situ Test at Selected Nuclear and High-Explosive Test Sites," AFWL-TR-76-709, Kirtland Air Force Base, N. M., February.

2. Allen, N. F., (1977), "Fluid Wave Propagation in Saturated and Nearly Saturated Sands," Ph.D. dissertation, University of Michigan, Ann Arbor, 278 pg.

3. Allen, N. F., Richart, F. E., Jr., and Woods, R. D., (1980), "Fluid Wave Propagation in Saturated and Nearly Saturated Sand," Journal of the Geotechnical Engineering Division, ASCE, Vol. 106, No. GT3, pp. 235-254.

4. Auld, B., (1977), "Cross-Hole and Down-Hole v_s by Mechanical Impulse." Journal of the Geotechnical Engineering Division, ASCE, Vol. 103, GT12, pp. 1381-1398.

5. Ballard, R. F., Jr., (1964), "Determination of Soil Shear Moduli at Depths by In Situ Vibratory Techniques," Miscellaneous Paper No. 4-691, U.S. Army Engineer Waterways Experiment Station, P. O. Box 631, Vicksburg, MS 39180

6. Ballard, R. F., Jr., (1976), "Method for Crosshole Seismic Testing," Journal of the Geotechnical Engineering Division, ASCE, Vol. 102, No. GT2, pp. 1261-1273.

7. Ballard, R. F., Jr., and McLean, F. G., (1975), "Seismic Field Methods for In Situ Moduli," Proceedings of the Conference on In Situ Measurement of Soil Properties, ASCE, Vol. 1, Raleigh, N. C., pp. 121-150.

8. Beeston, H. E., and McEvilly, T. V., (1977), "Shear Wave Velocities From Down-Hole Measurements," Earthquake Engineering and Structural Dynamics, Vol. 5, No. 2, pp. 181-190.

9. Bratton, J. L., and Higgins, C. J., (1978), "Measuring Dynamic In Situ Geotechnical Properties," Proceedings of the Conference on Earthquake Engineering and Soil Dynamics, ASCE, Vol. 1, Pasadena, California, pp. 272-289.

10. Hoar, R. J., (1980), "In Situ Seismic Velocity and Attenuation Measurements for Dynamic Analyses," Ph.D. dissertation, University of Texas at Austin.

11. Hoar, R. J. and Stokoe, K. H., II, (1978), "Generation and Measurement of Shear Waves In Situ," Dynamic Geotechnical Testing, ASTM STP 654, American Society for Testing and Materials, pp. 3-29.

12. Hoar, R. J., and Stokoe, K. H. II, (1980), "In Situ Shear Wave Velocity Measurements, Lynn Ary Park, Anchorage, Alaska," report submitted to the United States Geological Survey, Brach of Engineering Geology, Denver, Colorado.

13. Idriss, I. J., Dobry, R., Doyle, E. H., and Singh, R. D., (1976), "Behavior of Soft Clays under Earthquake Loading Conditions," Proceedings, Offshore Technology Conference, OTC 2671, Houston, Texas.

14. Idriss, I. M., Dobry, R., and Singh, R. D., (1978), "Nonlinear Behavior of Soft Clays During Cyclic Loading," Journal of the Geotechnical Engineering Division, ASCE, Vol. 104, No. GT12, Dec., pp. 1427-1447.

15. Jolly, R. N., (1956), "Investigation of Shear Waves," Geophysics, Vol. XXI, No. 4, pp. 905-938.

16. Joyner, W. B., and Chen, A. T. F., (1975), "Calculation of Nonlinear Ground Responses in Earthquakes," Bulletin of the Seismological Society of America, Vol. 65, No. 5, Oct., pp. 1315-1336.

17. McLamore, V. R., Anderson, D. G., and Espana, C., (1978), "Crosshole Testing Using Explosive and Mechanical Energy Sources," Dynamic Geotechnical Testing, ASTM STP 654, American Society for Testing and Materials, pp. 30-55.

18, Murphy, V. J., (1972), "Geophysical Engineering Investigation Techniques for Microzonation," Proceedings of the International Conference on Microzonation for Safer Construction, Research and Application, Vol. 1, Seattle, Washington, pp. 131-159.

19. Ogura, K., (1979), "Development of a Suspension Type S-Wave Log System," Technical Note of OYO Corp., RP-4105, Nov., 23 pg.

20. Patel, N. S., (1980), "Generation and Attenuation of Seismic Waves in Downhole Testing," M.S. thesis, The University of Texas at Austin, Texas.

21. Pyke, R., (1979), "Nonlinear Soil Models for Irregular Cyclic Loadings," Journal of the Geotechnical Engineering Divison, ASCE, Vol. 105, No. GT6, June, pp. 715-726.

22. Schwarz, S. C., and Musser, J. M., (1972), "Various Techniques for Making In Situ Shear Wave Velocity Measurements--A Description and Evaluation," Proceedings of the International Conference on Microzonation for Safer Construction, Research and Application, Vol. 2, Seattle, Washington, pp. 593-608.

23. Shannon and Wilson, Inc., and Agbabian Associates, (1976), "In Situ Impulse Test: An Experimental and Analytical Evaluation of Data Interpretation Procedures," Report No. NUREG-0028, Nuclear Regulatory Commission, 264 pp.

24. Shima, E., and Ohta, Y., (1967), "Experimental Study on Generation and Propagation of S-Waves: I., Designing of SH-Wave generator and its Field Tests," Bulletin of the Earthquake Research Institute, Vol. 45, pp. 19-31.

25. Stokoe, K. H., II, and Abdel-razzak, K. H., (1975), "Shear Moduli of Two Compacted Fills," Proceedings of the Conference on In Situ Measurement of Soil Properties, ASCE, Vol. 1, Raleigh, N.C., pp. 442-449.

26. Stokoe, K. H., II, and Hoar, R. J., (1978), "Variables Affecting In Situ Seismic Measurements," Proceedings of the Conference on Earthquake Engineering and Soil Dynamics, ASCE Geotechnical Engineering Division, Vol. II, pp. 919-939.

27. Streeter, V. L., Wylie, E. B., and Richart, F. E., Jr., (1974), "Soil Motion Computations by Characteristics Method," Journal of the Geotechnical Engineering Division, ASCE, Vol. 100, No. GT3, March, pp. 247-263.

28. Warrick, R. E., (1974), "Seismic Investigation of a San Francisco Bay Mud Site," Bulletin of the Seismological Society of America, Vol. 64, No. 2, pp. 375-385.

29. Wilson, S. D., Brown, F. R., Jr., and Schwarz, S. D., (1978), "In Situ Determination of Dynamic Soil Properties," Dynamic Geotechnical Testing, ASTM STP 654, American Society for Testing and Materials, pp. 295-317.

30. Woods, R. D., (1978), "Measurement of Dynamic Soil Properties," Proceedings of the Conference on Earthquake Engineering and Soil Dynamics, ASCE, Geotechnical Engineering Division, Vol. I, pp. 91-178.

EFFECT OF ANISOTROPIC CONSOLIDATION ON DYNAMIC PROPERTIES OF COHESIVE SOILS

INTRODUCTION

In the dynamic stability analysis of an earth embankment, shear modulus and damping ratio are required to determine the stress-strain response of a soil under various dynamic loading conditions. These parameters are established in the laboratory by the widely accepted resonant column technique. To simulate field conditions, the laboratory specimen is consolidated at a predetermined confining pressure equivalent to the effective overburden pressure. However, in order to simplify test procedures, the resonant column test is usually performed on a specimen consolidated isotropically which, in fact, is not representative of stress conditions of a soil element in an earth embankment.

The initial stress, defined as the stress existing in a soil element prior to being subjected to an external load, may be isotropic or anisotropic, depending upon the slope of the surface overlying the soil element. In most cases the anisotropic condition prevails due to the shearing stresses existing in an embankment. This study was undertaken to investigate the effect of anisotropic consolidation on the shear modulus and damping ratio of normally consolidated and overconsolidated soils. Also considered were the effects of the overconsolidation ratio and the confining pressure.

LITERATURE REVIEW

Investigations reported in the literature are related to the influence of anisotropic consolidation on static and dynamic strength of soils which can be categorized as follows: dynamic strength of anisotropically consolidated soils, undrained shear strength of anisotropically consolidated clay, and drained and undrained shear strengths of anisotropically consolidated sand.

Dynamic Strength of Anisotropically Consolidated Soils

Lee and Seed (1967) studied three types of soils under cyclic loading conditions, including a fine sand, a medium sand, and a compacted silt. To ensure a fully saturated specimen, soils were boiled under a vacuum. Under anisotropic conditions the consolidation stress ratio was defined as the ratio of the major principal stress, σ_{1c}, to the minor principal stress, σ_{3c}, and by the symbol $K_c = \sigma_{1c}/\sigma_{3c}$. Test results, shown in figure 1, indicate the strength under cyclic loading increases considerably with the increasing value of the consolidation stress ratio (K_c) over a wide range of densities and confining pressures.

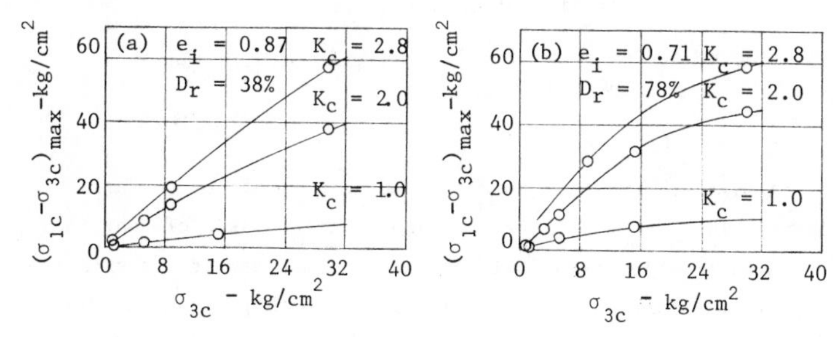

Figure 1. Dynamic Strength at Various Anisotropic-Consolidation Stresses

Undrained Shear Strength of an Anisotropically Consolidated Clay

Lowe and Karafiath (1960) investigated the effect of anisotropic consolidation on the undrained shear strength of compacted clays and correlated the undrained shear strength with the normal stress at failure, shown in figure 2. The lower limiting line in this graph covers $K_c = 1$, which represents the conventional consolidated-undrained triaxial compression test. The upper limiting line, $K_c = K_f$, represents the conventional consolidated-drained triaxial compression test. The lines in between represent the undrained shear strength of soils at various consolidation stress ratios. It is apparent the undrained shear strength increases with increasing consolidation stress ratio. As a result, the need for applying anisotropic-consolidated shear strength in the stability analysis is emphasized.

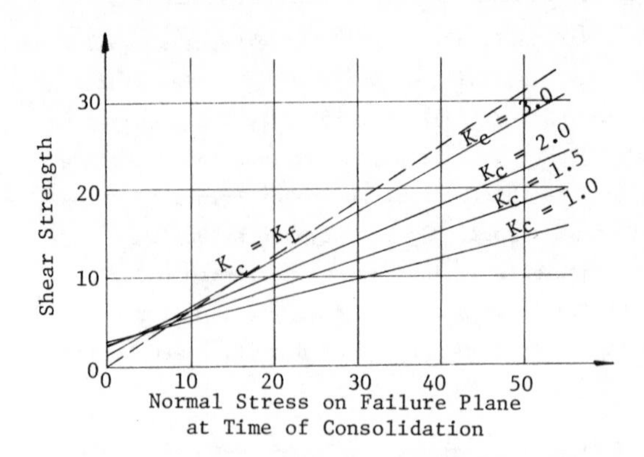

Figure 2. Anisotropically Consolidated Undrained Shear Strength

A similar study was conducted by Ladd (1965) on normally consolidated clays. His test results indicate the ratio of undrained shear strength to major principal consolidation stress is practically unchanged ($\pm$15 percent) at K_o (coefficient of earth pressure at rest). However, the variation of undrained shear strength with K_o was apparent.

Lee and Morrison (1970) performed a series of consolidated-undrained triaxial compression tests with pore pressure measurements to investigate the stress-strain characteristic, the pore pressure behavior, and the shear strength of compacted clay under anisotropic stress conditions. It was concluded the stress-strain and pore pressure behavior is dependent on the consolidation stress condition. However, it was not considered a practical or reliable method to obtain results for anisotropically consolidated-undrained conditions from isotropically consolidated-undrained tests.

Hollow cylindrical samples of a remolded saturated kaolinite clay were studied by Broms and Ratnam (1963). Test results revealed the shear strength parameters in terms of effective stress, ϕ' and c', were affected by the consolidation conditions, being higher under anisotropic consolidation for both normally consolidated and over-consolidated samples.

Duncan and Seed (1965) investigated the effect of anisotropy and reorientation of principal stresses on the shear strength of saturated clay and evaluated their significance in short-term stability analyses. "Perfect sampling tests" were used in the investigation in which plane strain tests were performed where samples were anisotropically consolidated simulating in-situ testing. It is stated that neglecting the effect of anisotropy and reorientation of principal stresses tends to increase the computed factor of safety in the "ϕ = o method" in a stability analysis.

Drained and Undrained Shear Strength of an Anisotropically Consolidated Sand

Lee and Seed (1970) conducted an investigation to determine the drained and undrained strengths of saturated sand after consolidation under anisotropic stress conditions and to compare these results with those determined for the same sand after consolidation under isotropic stress conditions. No significant difference in shear strength was found in the drained tests. In the undrained tests, however, the difference is highly dependent on the magnitude of the confining pressure; the higher the confining pressure the greater the difference in shear strength.

Based on the literature review presented above, investigators pointed out the need to determine static and dynamic soil properties under anisotropic stress conditions if field conditions are to be simulated in the laboratory.

TESTING APPARATUS

A general view of the resonant column apparatus used in this study is shown in figure 3. Two triaxial chambers, each containing an oscillator and a specimen, are shown. In the chamber the bottom end of the specimen is sealed to the pedestal and is considered the fixed end; the top end of the specimen is attached rigidly to the center ring of the oscillator and is the free end or driving end. During apparatus assembly the weight of the oscillator is supported by a pneumatic jack. During consolidation and testing it is counterbalanced by a dead weight located outside the chamber. The dead weight is also used to control anisotropic consolidation. Torque is

Figure 3. A General View of the Resonant Column Apparatus

applied to the top of the specimen, and its response is measured at the top by an accelerometer displayed on the oscilloscope and recorded by a digital voltmeter.

MATERIAL

The two types of soils, a sandy silt (ML) and a fat clay (CH), were included in this study. The particle-size distribution curves are shown in figure 4. Index properties are as follows: ML - specific gravity = 2.70; liquid limit = 30 percent; plastic limit = 22.7 percent; CH - specific gravity = 2.75; liquid limit = 61.8 percent; plastic limit = 22.4 percent.

SPECIMEN PREPARATION

The soils were wetted to a moisture content of 23.0 percent for sandy silt and 18.2 percent for the fat clay and stored individually in sealed plastic containers to allow moisture equalization. To avoid clumping of the soil and to assure uniform densities of the specimens, the wetted soil was passed through a No. 10 sieve. An amount of wetted soil necessary to achieve a desired dry density was weighed and molded in a 1.4-inch-diameter cylinder. By compacting both ends of the mold with different lengths

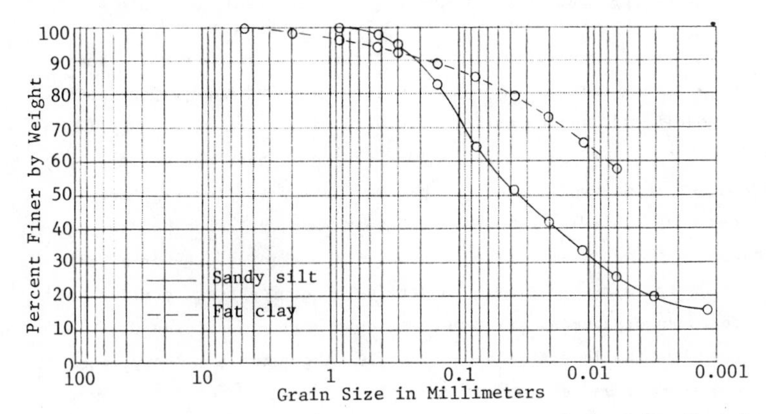

Figure 4. Particle-Size Distribution Curves of the Soils Tested

of steel rod, a specimen of 2.8 inches in height was formed. After removal from the mold, the specimen was weighed to obtain its initial weight. Care was taken to avoid specimen disturbance while setting up the specimen in the apparatus.

TEST PROCEDURE

To achieve a normally consolidated or overconsolidated specimen under an isotropic- or anisotropic-consolidation condition, the following procedures were followed:

1. Isotropically consolidated--The desired confining pressure was applied to the specimen and the drainage valve opened to allow consolidation. Specimen deformation was measured to determine the completion of consolidation, usually requiring 20 hours.

2. Anisotropically consolidated--Initially an isotropically consolidated specimen was prepared in accordance with step 1. Then an additional axial load in increments of one-fifth of the total load was applied to achieve the desired consolidation stress ratio ($K_c = \sigma_1/\sigma_3$). Deformation was checked to ensure that primary consolidation for each loading was completed prior to applying the next load.

3. <u>Overconsolidated</u>--A higher confining pressure was applied to the specimen, then reduced to a lower level to obtain a desired overconsolidation ratio designated as OCR.

After completion of consolidation the drainage valve was closed and measurements taken of the shear modulus and damping ratio. Shear moduli and damping ratios were determined at a strain amplitude ranging from 1.0×10^{-5} to 1.0×10^{-4} in./in. Since the maximum strain amplitude applied on the specimen during the test was not greater than 1.0×10^{-4} in./in., the test was considered nondestructive. The same specimen was tested under various confining pressures and consolidation stress ratios. To verify the nondestructive nature of the test, an experiment was conducted on a soil classifying CL. The specimen was remolded at a density of 101.1 pcf and a moisture content of 20.8 percent and subjected to confining pressures of 1.0 tsf and 3.0 tsf. Two measurements of shear moduli and damping ratios were made at each confining pressure about every 2 hours. Test results shown in figure 5 indicate essentially no difference between the two subsequent measurements. This confirmed that the maximum strain amplitude of 1.0×10^{-4} in./in. in fact is nondestructive.

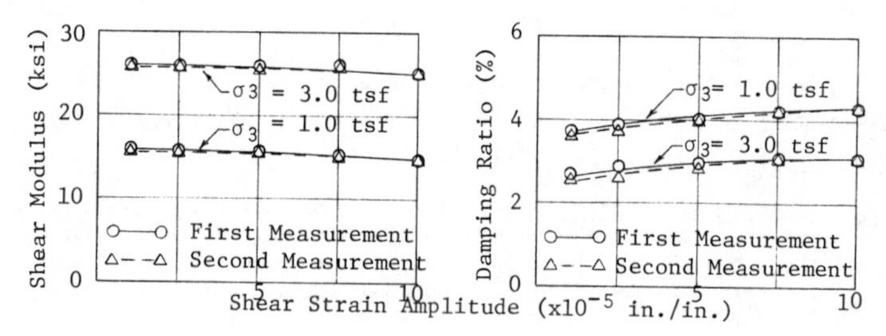

Figure 5. Shear Modulus and Damping Ratio of a Lean Clay

DISCUSSION OF TEST RESULTS

Shear moduli generally are decreasing and damping ratios increasing with increasing strain amplitude. To establish relationships between shear moduli, damping ratios, and the strain amplitude, five amplitudes were selected and applied at specific test conditions on the same specimen.

Effect of the Confining Pressure on Dynamic Soil Properties

Test results of specimens composed of sandy silt and fat clay respectively and consolidated isotropically and anisotropically at various confining pressures are shown in figures 6 and 7. The upper graphs show results of specimens consolidated at a stress ratio equal to 1 under isotropic conditions; the lower graphs illustrate performance of specimens consolidated at a stress ratio equal to 2 under anisotropic conditions. Both soils indicate shear moduli at a given confining pressure are higher under anisotropic than isotropic conditions. The effect of confining pressure on the shear modulus is much greater with the sandy silt than the fat clay. For example, the shear modulus at 2.0 tsf confining pressure is about three times as high as the shear modulus at 0.5 tsf confining pressure for the sandy silt and only less than one and one-half times for the fat clay. The rate of decrease in shear modulus with increasing strain amplitude is greater with the sandy silt than the fat clay under anisotropic conditions, as indicated by a steeper slope of the curves shown in figure 6 for $K_c = 2$. However, the rate of decrease in shear modulus is about the same for both soils under isotropic conditions. The same effect of confining pressure on the shear modulus is also indicated for both overconsolidated soils at a ratio equal to 6 as shown in figure 8.

The damping ratio generally decreases with increasing confining pressure. No significant difference was found between the two types of soils under the isotropic or anisotropic test condition and the normally consolidated or overconsolidated soil condition.

Figure 9 shows the relationship between maximum shear modulus and effective normal stress. On a log-log scale a linear relationship between these two parameters exists for both soil types. The slope of the line for the sandy silt is steeper than that of the fat clay. The confining pressure has a greater effect on the shear modulus of this soil type than on the fat clay.

Effect of the Consolidation Stress Ratio on Dynamic Soil Properties

Figures 10 and 11 show the shear modulus and damping ratio of a sample of anisotropically consolidated sandy silt and fat clay at a consolidation stress ratio ranging up to 2.5 for the sandy silt and to 3.0 for the fat clay. An attempt was made to consolidate the sandy silt at stress ratio of 3.0 but failed because the specimen

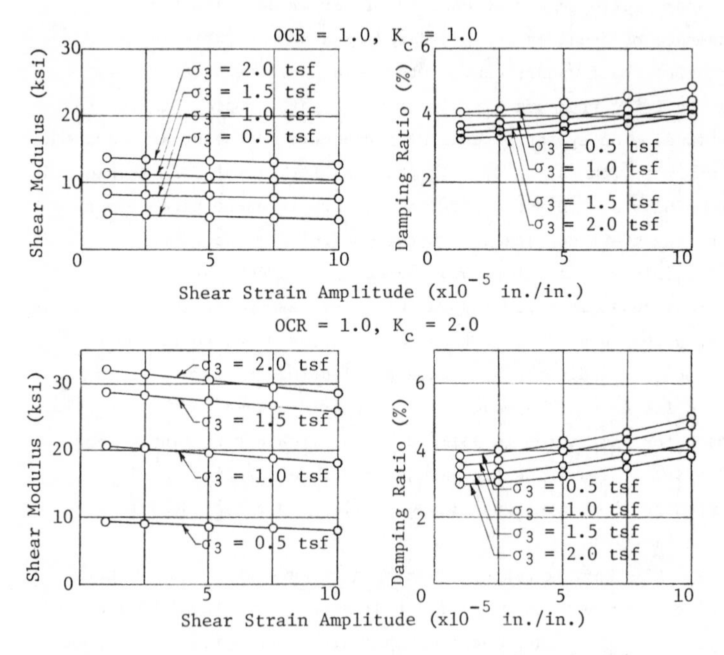

Figure 6. Shear Modulus and Damping Ratio of Sandy Silt Normally Consolidated at Various Confining Pressures Under Both Isotropic and Anisotropic Conditions

could not sustain the additional axial load applied to achieve this stress ratio. Both graphs indicate the shear modulus increases and the damping ratio decreases with increasing consolidation stress ratio. The effect of this ratio on the shear modulus is more pronounced at a higher confining pressure. When the consolidation stress ratio increases from 1.0 to 3.0, the shear modulus at a strain amplitude of 1.0 x 10^{-5} in./in. increases from 17,000 psi to 20,200 psi at 0.5 tsf confining pressure and from 23,100 psi to 32,250 psi at a confining pressure of 2.0 tsf. In other words, there is an 18.8 percent increase in the shear modulus resulting from a change in the consolidation stress ratio from 1.0 to 3.0 for 0.5 tsf

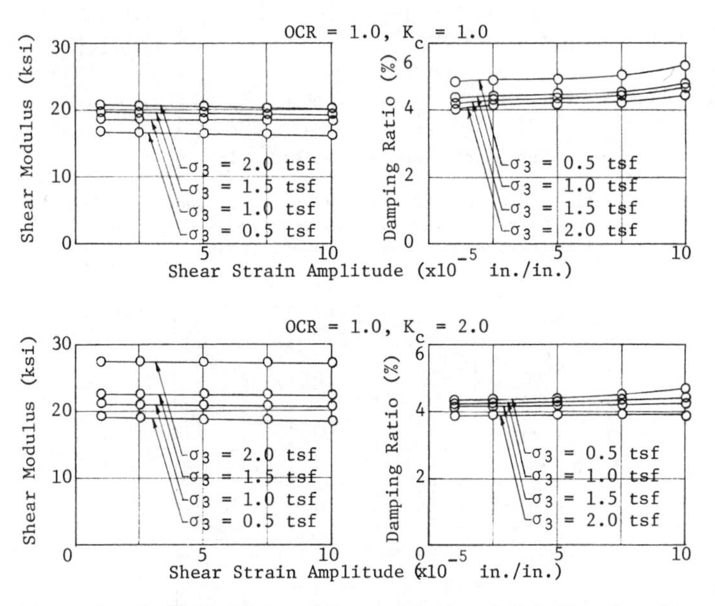

Figure 7. Shear Modulus and Damping Ratio of Fat Clay Normally Consolidated at Various Confining Pressures Under Both Isotropic and Anisotropic Conditions

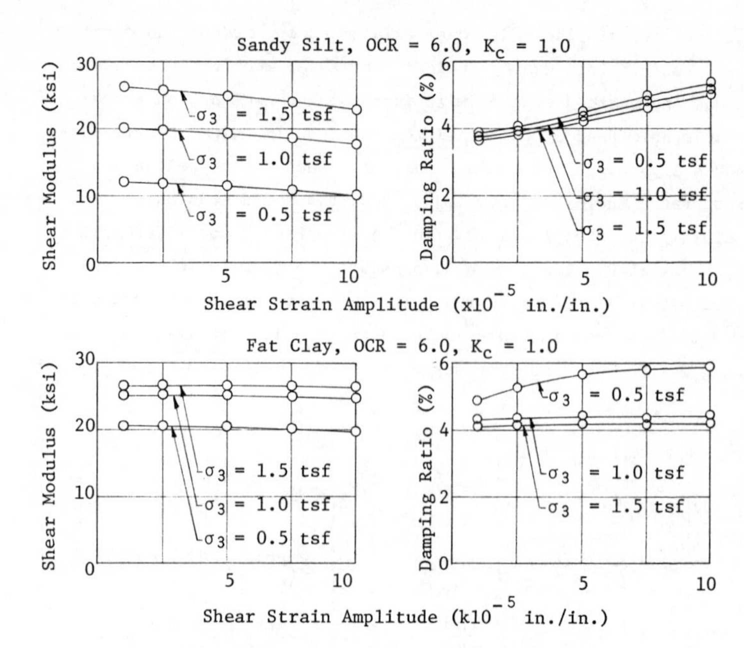

Figure 8. Shear Modulus and Damping Ratio of Sandy Silt and Fat Clay Overconsolidated at Various Confining Pressures Under Isotropic Conditions

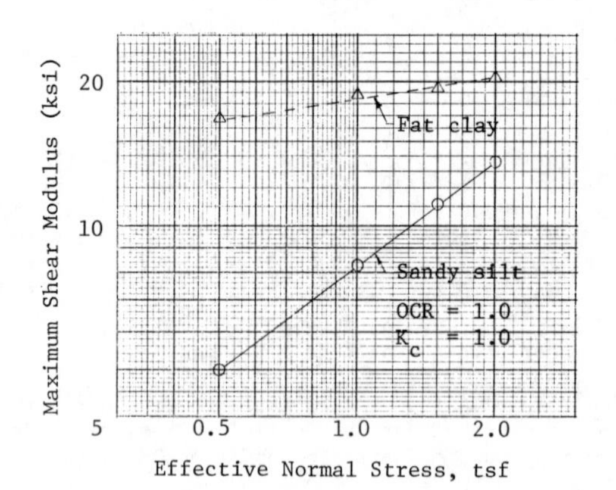

Figure 9. The Relationship Between Maximum Shear Modulus and Effective Normal Stress

confining pressure and a 40.0 percent increase for the higher confining pressure. This example clearly points to the significance of the effect of anisotropic consolidation on the shear modulus of a soil. Also, the rate of increase in the shear modulus with the increasing consolidation stress ratio is greater as the value of the consolidation stress ratio increases. See figures 10 and 11. Comparing the effect of different soil types on the shear modulus under identical test conditions, the sandy silt has a higher shear modulus than the fat clay. The influence of the consolidation stress ratio on the damping ratio is less significant than that on the shear modulus and less noticeable at a higher confining pressure.

Effect of the Overconsolidation Ratio on Dynamic Soil Properties

In a natural deposit or an embankment the soil is consolidated to some degree, depending on its overburden and

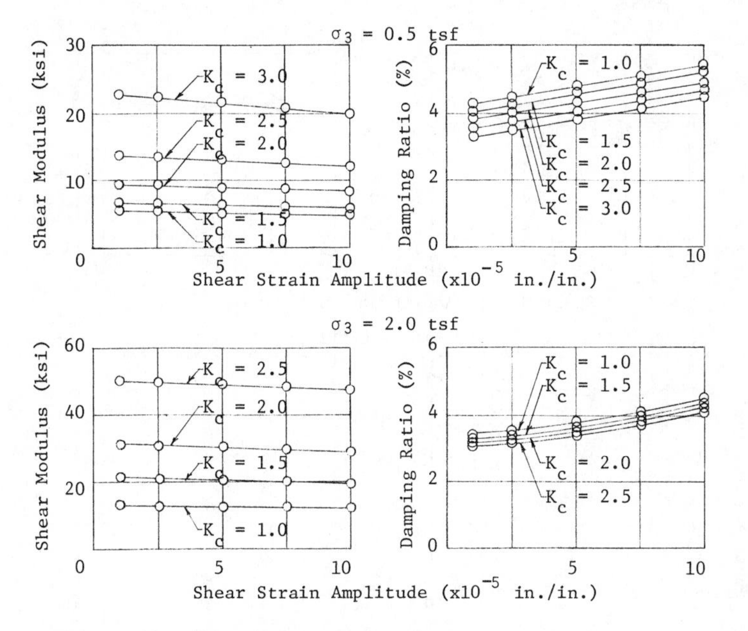

Figure 10. Shear Modulus and Damping Ratio of Sandy Silt Under Anisotropic Condition

preconsolidation pressure. In this study an overconsolidation ratio (OCR) ranging up to 14 was investigated: The case of OCR = 1.0 represents a normally consolidated condition and OCR > 1.0 represents an overconsolidated condition. Confining pressures of 0.5 tsf and 1.5 tsf were applied, and tests were performed on a specimen consolidated at ratios up to 14 at the 0.5 tsf confining pressure and up to 6 for the 1.5 tsf confining pressure. Test results of shear modulus and damping ratio are presented in figures 12 and 13. The overconsolidation ratio has a similar effect on shear modulus and damping ratio as the consolidation stress ratio and confining pressure discussed previously. The effect of overconsolidation ratio on the shear modulus is significant. As illustrated in figure 12, for 0.5 tsf confining pressure the shear modulus at a strain amplitude of 1×10^{-5} in./in. increases from 5200 psi under a normally consolidated

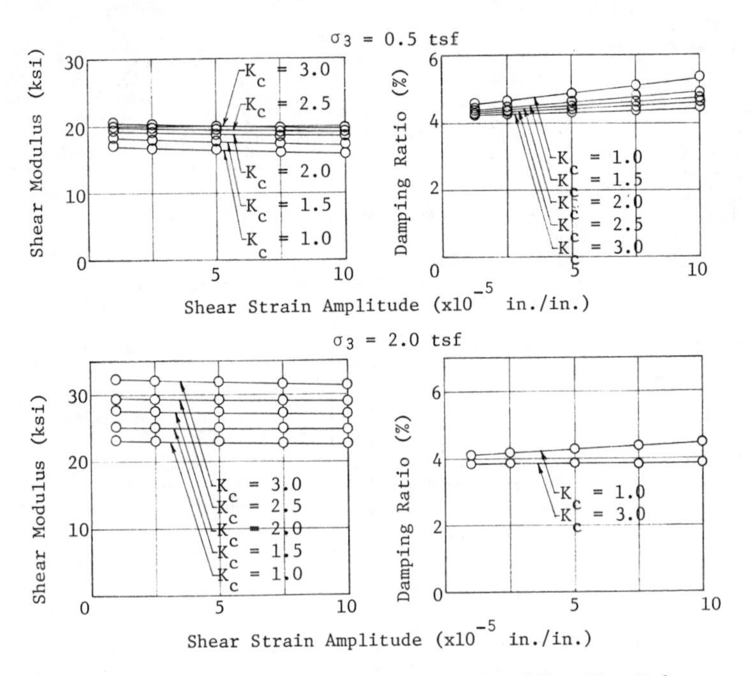

Figure 11. Shear Modulus and Damping Ratio of Fat Clay Under Anisotropic Condition

condition (OCR = 1) to 22,800 psi under an overconsolidated condition (OCR = 14). This effect (increase in shear modulus) seems to be even greater as the confining pressure increases as shown in the lower graphs, figures 12 and 13.

Figure 14 shows the plot of the shear modulus at a strain amplitude of 1 x 10^{-5} in./in. versus the overconsolidation ratio on a log-log scale. For both soils the relationship seems to be about linear at a ratio of less than 10. The line becomes concave upward for the ratio greater than 10. This indicates the shear modulus sharply increases for a highly overconsolidated soil (OCR > 10). Comparing the effect of overconsolidation on the shear modulus of different soil types, it seems the effect of the sandy silt is much greater than that of the fat clay as indicated by a steeper slope shown in figure 14.

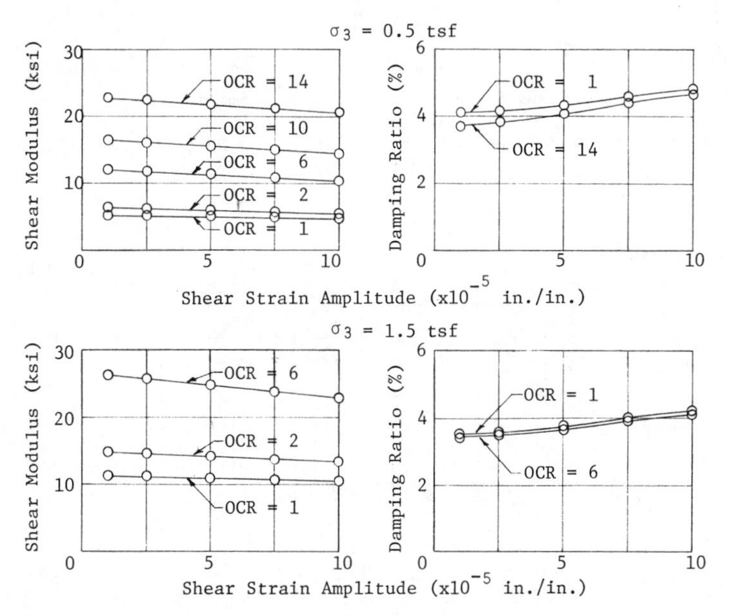

Figure 12. Shear Modulus and Damping Ratio of a Normally Consolidated and Overconsolidated Sandy Silt

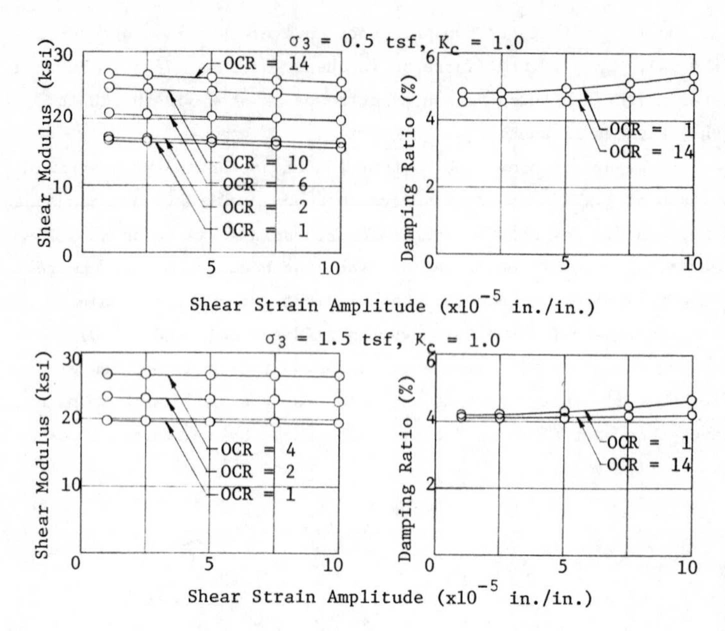

Figure 13. Shear Modulus and Damping Ratio of a Normally Consolidated and Overconsolidated Fat Clay

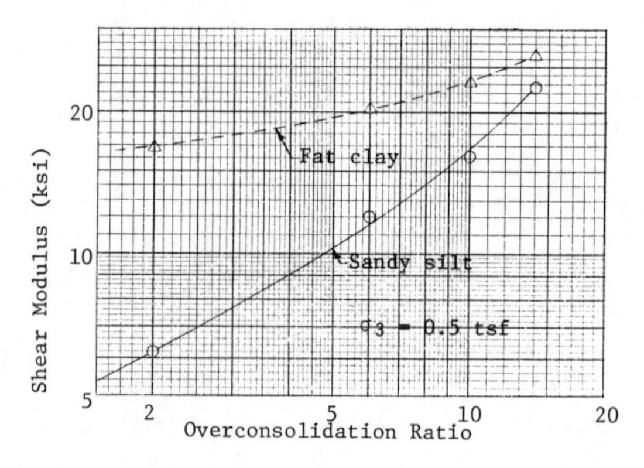

Figure 14. The Relationship Between Shear Modulus and Overconsolidation Ratio

Joint Effects of the Consolidation Stress Ratio and the Overconsolidation Ratio on Dynamic Soil Properties

The effects of the consolidation stress ratio and the overconsolidation ratio on the shear modulus and damping ratio of soils have been discussed earlier. However, each of these parameters do necessarily exist in combination under in-situ stress conditions. For example, a soil element in place may be subjected to some shearing force ($K_c > 1$) as well as overconsolidation (OCR > 1). Thus tests were conducted considering combined anisotropic and overconsolidated conditions. As earlier established the overconsolidation ratio has a negligible effect on the damping ratio. Thus the combined consolidation stress ratio and overconsolidation ratio produces a similar effect on the damping ratio as the consolidation stress ratio.

As indicated from the study by Lee and Morrison (1970), typical values for the consolidation stress ratio (K_c) in an embankment range from 1.5 to 2.5 with $K_c = 2.0$ being representative. Thus test results for $K_c = 2.0$ were selected for comparison. Figures 15 and 16 show shear moduli at various confining pressures and overconsolidations. The rate of decrease in the shear modulus with increasing strain amplitude is greater for the sandy silt than the fat clay, and the higher the confining pressure the higher the rate. In comparing figures 12 and 15, the shear modulus is slightly higher at $K_c = 2.0$ than $K_c = 1.0$ at the same overconsolidation ratio. It can be concluded that the combined effect of the consolidation stress ratio and the overconsolidation ratio on the shear modulus is greater than the effect of either, and that the effect is cumulative. The same applies to the overconsolidated fat clay as shown in figures 13 and 16. The following tabulation lists shear moduli for isotropically normally consolidated ($K_c = 1.0$ and OCR = 1.0) and anisotropically overconsolidated ($K_c = 2.0$ and OCR = 6.0) soils.

Soil Type	Confining Pressure tsf	Shear Modulus (psi) at Various Test Conditions	
		$K_c = 1.0$, OCR = 1.0	$K_c = 2.0$, OCR = 6.0
Sandy silt	0.5	5,200	16,200
	1.5	11,200	28,700
Fat clay	0.5	16,700	22,200
	1.5	19,600	27,600

In comparing the two sets of shear moduli for the same soil type and confining pressure, a significant difference in test results between the two test conditions exists. The higher values of the shear moduli of anisotropically overconsolidated soils are not only more realistic with regard to in-situ conditions but allow for a more economical design application.

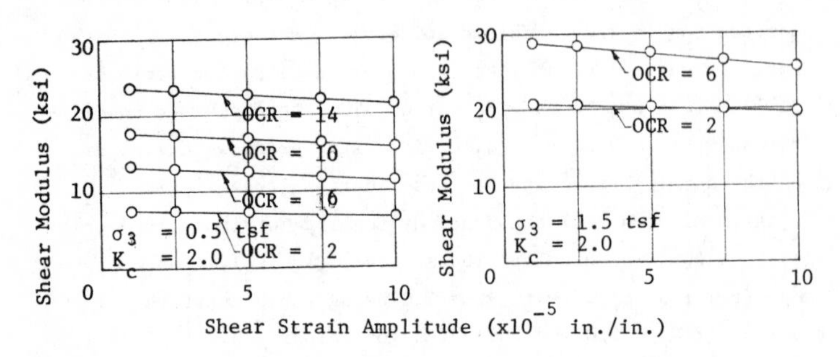

Figure 15. Shear Modulus of Anisotropically Overconsolidated Sandy Silt

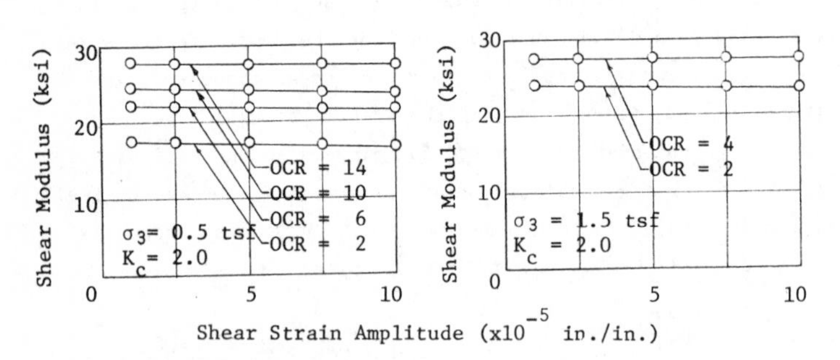

Figure 16. Shear Modulus of Anisotropically Overconsolidated Fat Clay

CONCLUSION

A soil element within an earth embankment is usually subjected to shearing forces, which depend upon its location in the embankment and the slope of the overlying surface. Consequently, to simulate those in-place conditions in the laboratory, specimens should be tested anisotropically consolidated rather than isotropically.

In this study various consolidation stress ratios were investigated representing anisotropic consolidation. Resonant column tests were performed on a sandy silt and a fat clay to determine their shear moduli and damping ratios under various test conditions. The evaluation of the test results allows the following conclusions:

1. The effect of the consolidation stress ratio on the shear modulus of a soil is significant as the modulus is higher under anisotropical condition than under isotropical condition, but its effect on the damping ratio is minor.
2. The overconsolidation ratio has a similar effect on the shear modulus and damping ratio as the consolidation stress ratio, showing both ratios increasing with increasing shear modulus and with decreasing damping ratio.
3. The effect of the combined anisotropic consolidation and overconsolidation on the shear modulus of a soil is greater than the effect of either of these parameters and the combined effect is cumulative. It is therefore recommended to perform the soil dynamic testing on anisotropically overconsolidated specimens and to consolidate specimens to the in-situ stress condition.

REFERENCES

1. Broms, B. and M. V. Ratnam, "Shear Strength of Anisotropically Consolidated Clay," Journal of Soil Mechanics and Foundations Division, ASCE, vol. 89, No. SM6, November 1963, pp. 1-26.

2. Duncan, J. M. and H. B. Seed, "The Effect of Anisotropy and Reorientation of Principal Stresses on the Shear Strength of Saturated Clay," Soil Mechanics and Bituminous Materials Laboratory Research Report, University of California, November 1965.

3. Ladd, C. C., "Stress-Strain Behavior of Anisotropically Consolidated Clays During Undrained Shear," Proceedings of Sixth International Conference of Soil Mechanics and Foundation Engineering, vol. 1, 1965, pp. 282-286.

4. Lee, K. C. and R. A. Morrison, "Strength of Anisotropically Consolidated Compacted Clay," Journal of Soil Mechanics and Foundation Division, ASCE, vol. 96, No. SM6, November 1970, pp. 2025-2043.

5. Lee, K. L. and H. B. Seed, "Dynamic Strength of Anisotropically Consolidated Sand," Journal of Soil Mechanics and Foundation Division, ASCE, vol. 93, No. SM5, September 1967, pp. 169-189.

6. Lee, K. L. and H. B. Seed, "Undrained Strength of Anisotropically Consolidated Sand," Journal of Soil Mechanics and Foundation Division, ASCE, vol. 96, No. SM2, March 1970, pp. 411-428.

7. Lowe III, J. and L. Karafiath, "Effect of Anisotropic Consolidation on the Undrained Strength of Cohesive Soils," Research Conference on Shear Strength of Cohesive Soils, Boulder, Colorado, 1960, pp. 837-858.

EVALUATION OF FREE-FREE AND FREE-FIXED RESONANT COLUMN TESTING DEVICES

INTRODUCTION

Shear modulus and damping ratio are two soil properties required in a stability analysis under dynamic conditions. These two parameters are determined in the laboratory using the resonant column method. The two types of resonant column devices available are the free-fixed and the free-free type. By comparison, in the free-fixed boundary type resonant column test, the soil response to vibrations is measured at the top of the specimen (torque driven end). In the free-free boundary type resonant column test, the measurement of soil response to vibrations is taken at the top of the specimen with the torque driven end at the bottom.

The purpose of this comparative study is to evaluate the relative performance of the two types of resonant column devices. Reconstituted specimens of Monterey No. 0 sand were tested at various confining pressures and the shear modulus and damping ratio determined at various strain amplitudes for each confining pressure. The same test specimen was subjected to ascending confining pressures up to 43.5 psi followed by a sequence of descending pressure. Tests performed under identical conditions in the two devices were compared and test reproducibility established for each device individually. The comparison addresses itself to the correlation of the performance of the two devices based on the outcome of the test series. Also included in the discussion are the results of destructive and non-destructive phases in the free-fixed resonant column test with respect to the magnitude of strain amplitude applied.

MATERIAL

The material used in this investigation was a clean, uniform sand from Monterey, California, known as Monterey No. 0 sand. The sand was classified poorly graded sand, SP, with a specific gravity of 2.65. The particle-size distribution curve is shown in figure 1.

Reconstituted specimens were used of 1.4-inch diameter by 2.8 inches long for the free-fixed resonant column device and 2.8-inch diameter by 5.6 inches long for the free-free resonant column device.

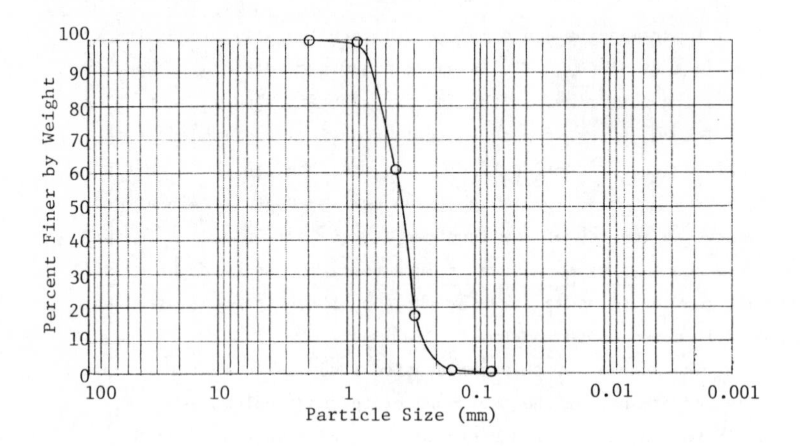

Figure 1. Particle-Size Distribution Curve of Monterey No. 0 Sand

APPARATUS

A general view of the free-fixed apparatus which was developed by Bobby Hardin of the University of Kentucky is shown in figure 2. The schematic diagram is shown in figure 3. In this system, the bottom end of the specimen is virtually motionless and is considered the fixed end. The other end of the specimen is attached to top cap which in turn is rigidly connected to the oscillator and is considered the free end or vibrating end. Two pairs of magnets and coils supported by brackets in the oscillator constitute the torsional driving system which generates sinusoidal torque applied to the specimen when sinusoidal voltage is provided to the system by a generator. An accelerometer is also attached to the oscillator bracket

Figure 2. A General View of Free-Fixed Resonant Column Device

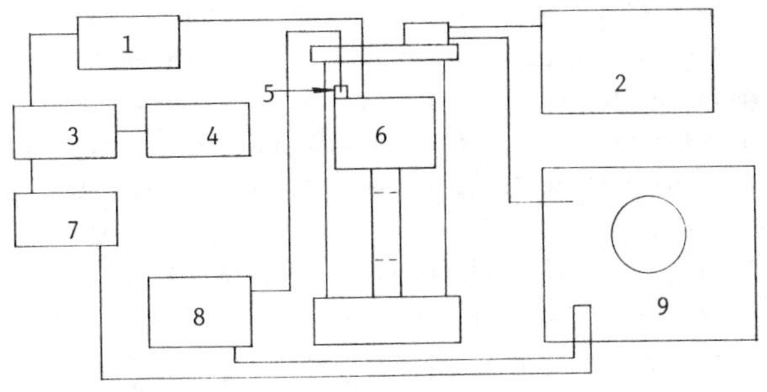

1. Bridge Amplifier
2. Audio (Sine Wave) Oscillator
3. Switch Box
4. Digital DC Voltmeter
5. Accelerometer
6. Oscillator
7. AC Meter Converter, DC Output
8. Charge Amplifier
9. Oscilloscope

Figure 3. Schematic Diagram of Free-Fixed Resonant Column Apparatus

to monitor the soil specimen response during vibration. An oscilloscope is used to display the level of torque applied and to aid in determining the system resonant frequency. The voltage drop across a precision resistor in series with the driving coils is displayed on X-axis and the output of the accelerometer is displayed on Y-axis. A resonant frequency is determined when the display on the X-Y oscilloscope produces a vertical axis ellipse.

A general view of the free-free resonant column device, developed by Vincent Drnevich of the University of Kentucky, is shown in figure 4 and its schematic diagram is shown in figure 5. The power capacity of this unit is greater than the first device, thus a higher strain amplitude can be obtained. It is also capable of testing specimen up to 6 inches in diameter. The two beams seen in figure 4 are the active beam attached to the chamber base and the passive beam bolted to the floor. At each end of the beams, a pair of magnet and coil is attached. The active velocity transducers are located beneath the chamber attachment plate of the active beam and the passive velocity transducers on the top platen. The active transducers monitor the voltage level of a driving force which is displayed on the X-axis of the oscilloscope. The passive transducers measure the specimen response during vibrations which is displayed on the Y-axis. A power amplifier boosts the output of the oscillator. As the specimen is vibrated at a specific amplitude, the resonant frequency is determined by varying frequencies until the X-Y oscilloscope trace forms a single slope.

For a more detailed description of the devices and a discussion of the theoretical analyses and calculations, see Drnevich (1978) and Hardin and Music (1965).

SPECIMEN PREPARATION

The preparation of reconstituted specimens of uniform density and minimum segregation required the following procedure suggested by Drnevich (1978).

1. Determining of the average thickness of the membrane and the inside diameter and height of the mold.
2. Assembling of the membrane and mold on the bottom platen and applying a vacuum between the membrane and mold.

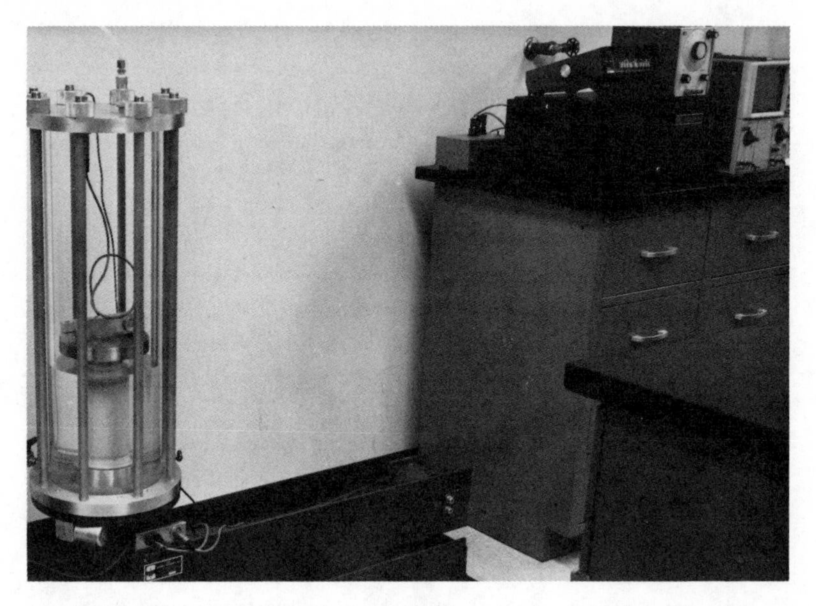

Figure 4. A General View of Free-Free Resonant Column Device

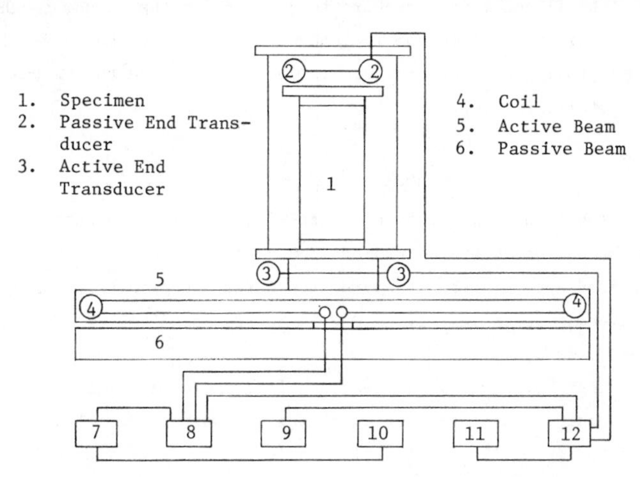

7. Power Amplifier
8. Control Box
9. Oscilloscope
10. Sine Wave Oscillator
11. AC Voltmeter
12. Switch Box

Figure 5. Schematic Diagram of Free-Free Resonant Column Apparatus

3. Weighing out the exact amount of dry sand to achieve the specified density.
4. Placing a special cylinder with a No. 10 sieve at the bottom into the mold and filling the cylinder with the weighed dry sand.
5. The cylinder is raised slowly and the mold tapped as the sand flows out.
6. The top surface is flattened with the total material placed in the mold and a platen placed on top.
7. The membrane is sealed to the top platen with an O-ring.
8. A vacuum of 10 inches of mercury is applied to the specimen and the mold removed.
9. Specimen dimensions are measured in three places and averaged.
10. Specimen dimensions and the thickness of the membrane are measured to calculate the initial dry density of the specimen.
11. The assembly of the apparatus is completed.
12. A small confining pressure is applied and the vacuum to the specimen reduced from 10 inches to 5 inches of mercury. The initial confining pressure is then applied and the vacuum is totally removed.

TEST PROCEDURE

Two identical series of tests were performed in both devices, standardized by the use of identical dry specimens, which were prepared according to standard procedures. Confining pressures were applied in seven levels: 7.25, 14.50, 21.75, 43.50, 21.75, 14.50, and 7.25 psi.

Test No. 1 was subjected to a full range of amplitudes during ascending and descending confining pressures. Test No. 2 was subjected to low amplitudes (strain amplitude less than 1×10^{-3} percent) during ascending pressures (except 43.5 psi), then followed by the application of the full range of amplitudes during descending pressures including 43.5 psi confining pressure.

Two additional tests designated 3 and 4 were performed in the free-fixed resonant column device. These test procedures were exactly the same as tests 1 and 2 except that the maximum strain amplitude applied to the specimen at varying confining pressures was less than 1×10^{-2} percent. The purpose of these two tests was to investigate the extent of the strain amplitude which defines a non-destructive or destructive test in the resonant column.

Details of the testing procedure for each pressure level are as follows.

1. Stabilization time of 15 minutes after applying confining pressure.
2. For tests at low strain amplitudes, an amplitude of less than 1×10^{-3} percent was applied and the frequency was adjusted to the resonant condition.
3. For tests over the full range of strain amplitudes, vibration was started at the lowest amplitude and gradually increased by 50 percent of the previous amplitude until the maximum amplitude of the apparatus was reached. At each step, the frequency was adjusted to the resonant condition.
4. The magnification factor method of measuring the damping at the steady state vibration was used throughout the tests.

DISCUSSION OF TEST RESULTS

A total of six specimens, including two specimens each for both devices and two additional specimens for the free-fixed resonant column device, were tested at various confining pressures, from 7.25 psi to 43.50 psi in ascending and descending pressure conditions. According to the measured dimensions of the specimen after the mold was removed, initial densities of all specimens were determined as shown in table 1.

Table 1, Initial Density of Specimens

Type of Device	Free-Free		Free-Fixed			
Test No.	1	2	1	2	3	4
Density (pcf)	98.1	98.5	100.3	99.3	98.7	98.8

Results of damping ratio versus strain amplitude for the six individual tests are shown in figures 14 through 19.

1. Test Reproducibility

Plots of shear modulus as a function of strain amplitude are shown in figures 6 and 7 for the free-free resonant column device and figures 8 and 9 for the free-fixed resonant column device. For both devices, shear moduli were determined at a strain amplitude of less than 1×10^{-3} percent during ascending pressure condition and only two data points were determined. These figures show that for both devices, the shear moduli of test No. 1 are slightly higher than those from test No. 2 at a corresponding strain amplitude for every confining pressure. The amount of variation in shear moduli between the two tests is approximately the same throughout the range of strain amplitude as indicated by the two solid lines being almost parallel.

As the shear modulus is a function of the strain amplitude at a given confining pressure, the strain amplitude must be specified to allow a comparison of the shear modulus. Table 2 presents values of shear modulus at a strain amplitude of 1×10^{-3} percent for the free-free resonant column device. For some tests for which no data point was available at 1×10^{-3} percent strain amplitude, the value for the shear modulus was extrapolated. Test reproducibility was evaluated by the equation:

$$EOR\ (\%) = \frac{G_1 - G_2}{G_1} \times 100$$

where:

EOR = Error of reproducibility in percent

G_1 = Shear modulus of test No. 1 in ksi

G_2 = Shear modulus of test No. 2 in ksi

As the error of reproducibility is used to express the degree of test duplication, a low value of EOR is indicative of higher reproducibility. Tables 2 and 3 show the calculated values for EOR of the free-free resonant column device tests ranging from 0.8 to 3.9. For the free-fixed resonant column device the EOR values range from 1.1 to 8.1. The value of EOR is random and seemingly independent of the confining pressure and strain amplitude. Indications are tests can be better reproduced at ascending pressure conditions than at descending conditions. The overall average of EOR value of free-free resonant column device is about one half of that of free-fixed resonant column device. This may be partly caused by the difficulty of

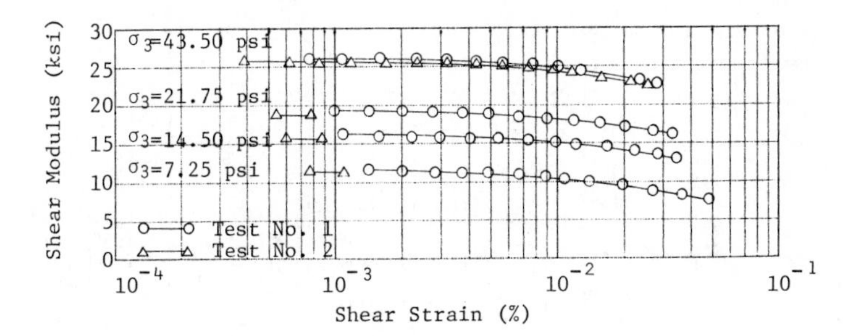

Figure 6. Shear Modulus from Test Nos. 1 and 2 During Ascending Pressures for Free-Free Resonant Column Device

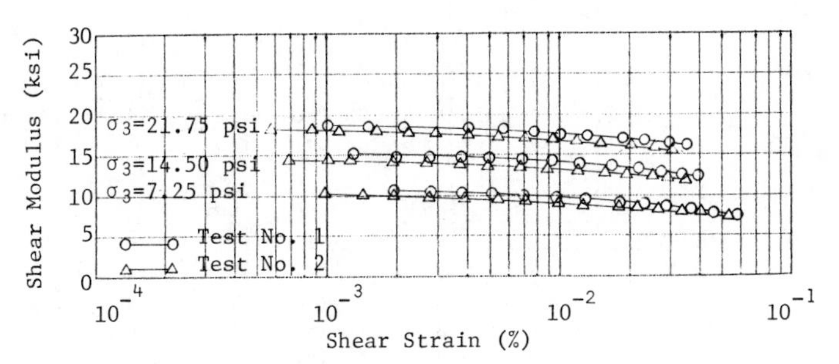

Figure 7. Shear Modulus from Test Nos. 1 and 2 During Descending Pressures for Free-Free Resonant Column Device

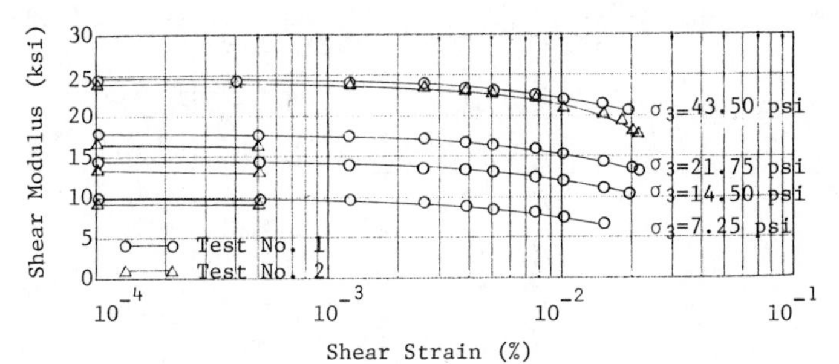

Figure 8. Shear Modulus from Test Nos. 1 and 2 During Ascending Pressures for Free-Fixed Resonant Column Device

assembling the free-fixed resonant column device possibly effecting some specimen disturbance.

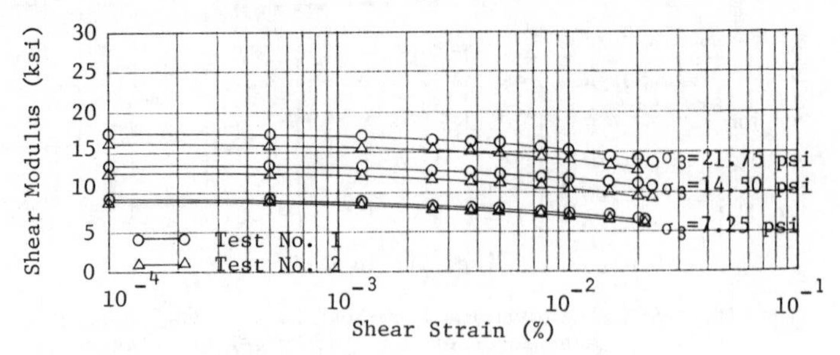

Figure 9. Shear Modulus from Test Nos. 1 and 2 During Descending Pressures for Free-Fixed Resonant Column Device

Table 2, Values of Shear Modulus and EOR from Test Nos. 1 and 2 for Free-Free Resonant Column Device

Pressure	Ascending			Descending		
Test No.	1	2	EOR	1	2	EOR
σ_3 (psi)	(ksi)	(ksi)	(%)	(ksi)	(ksi)	(%)
7.25	11.6	11.5	0.9	10.6	10.4	1.9
14.50	16.4	15.8	3.7	15.1	14.5	3.9
21.75	19.5	18.8	3.6	18.8	18.3	2.7
43.50	26.2	26.0	0.8			
Avg			2.3			2.8

Table 3, Values of Shear Modulus and EOR from Test Nos. 1 and 2 for Free-Fixed Resonant Column Device

Pressure	Ascending			Descending		
Test No.	1	2	EOR	1	2	EOR
σ_3 (psi)	(ksi)	(ksi)	(%)	(ksi)	(ksi)	(%)
7.25	9.7	9.3	4.1	8.9	8.8	1.1
14.50	14.3	13.3	7.0	13.3	12.3	7.5
21.75	17.6	16.6	5.7	17.2	15.8	8.1
43.50	24.5	24.2	1.2			
Avg			4.5			5.6

In general, results obtained from test Nos. 1 and 2 for both devices are very similar (within about 2.5 percent for free-free resonant column device and 5.0 percent for free-fixed resonant column device) and of acceptable reproducibility.

Test results further indicate an increase in damping ratio with increasing strain amplitudes. At the range of strain amplitude of less than 1×10^{-2} percent, values of damping ratio are very small but increase sharply above this amplitude. Generally, damping ratios obtained from test Nos. 1 and 2 for both devices are in agreement.

2. Comparison of Results form the Free-Free and Free-Fixed Devices

As pointed out earlier in the text, the magnitude of confining pressure and its applied sequence were identical for test Nos. 1 and 2. However, the range of strain amplitude applied at ascending pressures differed from that at descending pressures which was suspected to have an effect on the test results as reported by Hardin and Drnevich (1972). Therefore, in order to eliminate this factor a further comparison was undertaken by evaluating tests performed in the two devices under identical test conditions.

a. Shear Modulus

In the free-fixed resonant column device test, the low amplitude applied at 1×10^{-4} percent strain reached a maximum at about 2×10^{-2} percent strain. In the free-free resonant column device test, the amplitude applied ranges from 1×10^{-3} percent to a maximum at 3×10^{-2} percent strain. Figures 10 and 11 indicate the shear modulus decreases with increased strain amplitude while the rate of decrease in the shear modulus is greater at a higher strain amplitude.

For a given confining pressure and strain amplitude, the shear modulus obtained in the free-free device is higher than that attained in the free-fixed device. Variations in shear modulus between these two devices are slightly greater when the strain amplitude increases.

In comparing the shear modulus at ascending pressures with descending pressures, the differences between the two devices is somewhat lower at descending pressures. In figures 12 and 13 shear moduli are plotted versus strain amplitude obtained in test No. 2 at ascending and descending pressures for both devices. During the application of ascending confining pressures, the shear modulus was measured at strain amplitudes of less than 1×10^{-3} percent, providing two data points for each pressure at 7.25 psi, 14.50 psi, and 21.75 psi as shown in figure 12.

Results were similar in test No. 2 in that the shear moduli are higher in the free-free device than in the free-fixed device. Tables 4 and 5 list values of shear modulus at a strain amplitude of 1×10^{-3} percent of tests 1 and 2 obtained from these two devices. It is noted that variations in the values of shear moduli are similar regardless of the magnitude of the confining pressure. The overall average in the difference in shear moduli established in these two devices is approximately 2 ksi. This difference is consistent throughout and may be inherent in the design of the devices.

Specific factors possibly responsible for the deviations are:

1. Specimen boundary conditions
2. Specimen size
3. Contact conditions between the top cap and the specimen
4. Degree of specimen disturbance during assembly of apparatus
5. Location of applying vibration
6. Calibration of apparatus constants

b. Damping Ratio

Values of damping ratio obtained at all confining pressures for both devices are very small at strain amplitudes of less than 1×10^{-3} percent. See figures 14 through 19. No significant variation in the damping ratios was found between the two testing devices. However, variations between testing devices become more apparent at higher strain amplitudes with the free-fixed device showing larger damping ratios. Possible explanations for the difference in the ratios are the same as for the shear moduli discussed in the preceding chapter.

3. Nondestructive and Destructive Resonant Column Tests

The threshold between nondestructive and destructive resonant column testing is suggested by Drnevich (1978) to be the strain amplitude of 1×10^{-4} in./in. To verify this postulate, two additional tests were performed following the exact outline of the program for

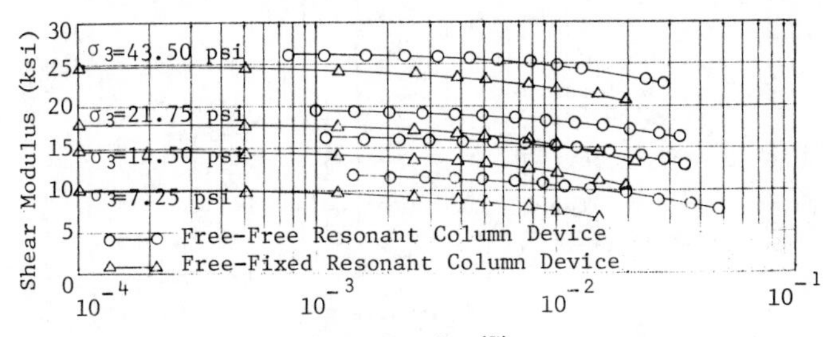

Figure 10. Shear Modulus from Test No. 1 During Ascending Pressures for Free-Free and Free-Fixed Resonant Column Device

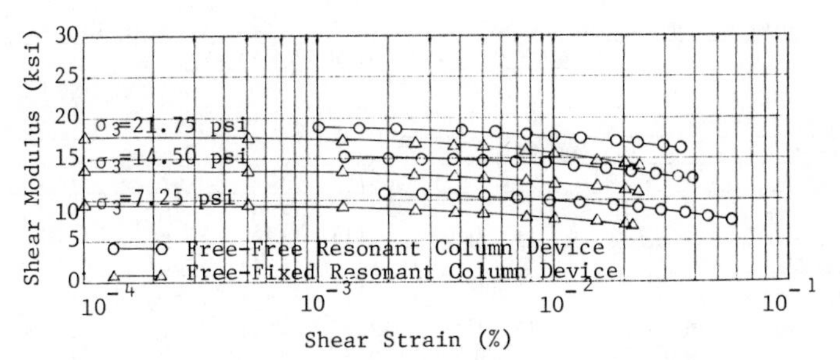

Figure 11. Shear Modulus from Test No. 1 During Descending Pressures for Free-Free and Free-Fixed Resonant Column Device

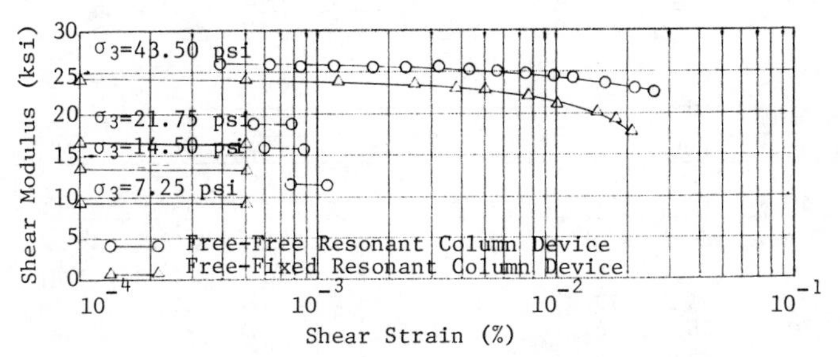

Figure 12. Shear Modulus from Test No. 2 During Ascending Pressures for Free-Free and Free-Fixed Resonant Column Device

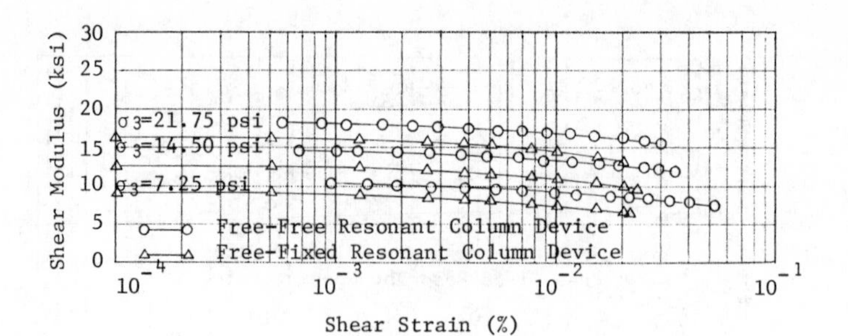

Figure 13. Shear Modulus from Test No. 2 During Descending Pressures for Free-Free and Free-Fixed Resonant Column Devices

Table 4, Values of Shear Modulus (Strain Amplitude at 1×10^{-3} %) from Test No. 1 for Both Devices

Pressure	Ascending				Descending			
Column	(1)	(2)	(3)	(4)	(5)	(6)	(7)	(8)
Type of Device	Free-Free	Free-Fixed	(1)-(2)	(3)/(1)	Free-Free	Free-Fixed	(5)-(6)	(7)/(5)
σ_3 (psi)	(ksi)	(ksi)	(ksi)	(%)	(ksi)	(ksi)	(ksi)	(%)
7.25	11.7	9.7	2.0	17.1	10.6	8.9	1.7	16.0
14.50	16.4	14.3	2.1	12.8	15.1	13.4	1.7	11.3
21.75	19.6	17.6	2.0	10.2	18.8	17.2	1.6	8.5
43.50	26.3	24.4	1.9	7.2				
Avg			2.0				1.7	

Table 5, Values of Shear Modulus (Strain Amplitude at 1×10^{-3} %) from Test No. 2 for Both Devices

Pressure	Ascending				Descending			
Column	(1)	(2)	(3)	(4)	(5)	(6)	(7)	(8)
Type of Device	Free-Free	Free-Fixed	(1)-(2)	(3)/(1)	Free-Free	Free-Fixed	(5)-(6)	(7)/(5)
σ_3 (psi)	(ksi)	(ksi)	(ksi)	(%)	(ksi)	(ksi)	(ksi)	(%)
7.25	11.5	9.4	2.1	18.3	10.4	8.8	1.6	15.4
14.50	15.8	13.5	2.3	14.6	14.4	12.3	2.1	14.6
21.75	18.7	16.5	2.2	11.8	18.2	15.8	2.4	13.2
43.50	25.8	24.1	1.7	6.6				
Avg			2.1				2.0	

test Nos. 1 and 2. The exception was that the maximum strain amplitude did not exceed 1×10^{-2} percent.

The shear modulus was determined at each confining pressure, both ascending and descending. A comparison of the shear moduli thus obtained was to show the effect of the high strain amplitude on the specimen destruction. The following equation was applied to determine the degree of destruction (DOD):

$$DOD\ (\%) = \frac{G_A - G_D}{G_A} \times 100$$

where:

G_A = Shear modulus at ascending pressure condition

G_D = Shear modulus at descending pressure condition

A value for DOD equals zero resulted when the test specimen remained intact. On the other hand, a greater DOD value indicated a greater destruction.

Tables 6, 7, and 8 list values of shear modulus at 1×10^{-2} percent strain amplitude, DOD values for the confining pressures applied and average DOD values for each test. Note that the average value of DOD of test Nos. 1 and 2 are either identical or similar indicating that consistent results can be obtained from both devices. Since the values of DOD were determined on the basis of the shear moduli at the same strain amplitude, the low average DOD values of test Nos. 3 and 4 indicate negligible destruction. By the same token the high destruction was reflected in DOD values of test Nos. 1 and 2.

Other observations based on the results in tables 6 through 8 and figures 14 through 19 of interest are;

1. Test Nos. 1 and 2 show a consistent trend in shear moduli in the ascending order being higher than those of the descending order. Test Nos. 3 and 4 do not show a definite trend and shear moduli are approximately the same. The decreases in shear moduli in test Nos. 1 and 2 are explained by the destruction which results in lower descending values.
2. It is also noted test Nos. 1 and 2 in the free-free resonant column device produced the highest values of DOD. Test Nos. 1 and 2 in the free-fixed device yielded medium values and test Nos. 3 and 4 in the free-fixed device showed the lowest values. This trend coincides with the magnitude of strain amplitude

applied. The results clearly indicate that the higher the strain amplitude applied, the greater the destruction of the specimen and consequently the greater the reduction in shear moduli.

Table 6. Values of Shear Modulus and DOD (The Maximum Strain Amplitude > 1×10^{-2}%) from Test Nos. 1 and 2 for Free-Free Resonant Column Device

Test No.	1			2		
Pressure σ_3 (psi)	Ascending (ksi)	Descending (ksi)	DOD (%)	Ascending (ksi)	Descending (ksi)	DOD (%)
7.25	11.7	10.6	9.4	11.5	10.4	9.6
14.50	16.4	15.1	7.9	15.8	14.4	8.9
21.75	19.6	18.8	4.1	18.7	18.2	2.7
43.50	26.3			25.8		
Avg			7.1			7.1

Table 7. Values of Shear Modulus and DOD (The Maximum Strain Amplitude > 1×10^{-2} %) from Test Nos. 1 and 2 for Free-Fixed Resonant Column Device

Test No.	1			2		
Pressure σ_3 (psi)	Ascending (ksi)	Descending (ksi)	DOD (%)	Ascending (ksi)	Descending (ksi)	DOD (%)
7.25	9.7	8.9	8.2	9.4	8.8	6.4
14.50	14.3	13.4	6.3	13.5	12.3	8.9
21.75	17.6	17.2	2.3	16.5	15.8	4.2
43.50	24.4			24.1		
Avg			5.6			6.5

Table 8. Values of Shear Modulus and DOD (The Maximum Strain Amplitude < 1×10^{-2} %) from Test Nos. 3 and 4 for Free-Fixed Resonant Column Device

Test No.	3			4		
Pressure σ_3 (psi)	Ascending (ksi)	Descending (ksi)	DOD (%)	Ascending (ksi)	Descending (ksi)	DOD (%)
7.25	10.1	9.8	3.0	9.2	9.0	2.2
14.50	14.2	14.2	0.0	13.2	13.4	-1.5
21.75	17.2	17.3	-0.6	16.6	16.6	0
43.50	24.2					
Avg			0.8			0.2

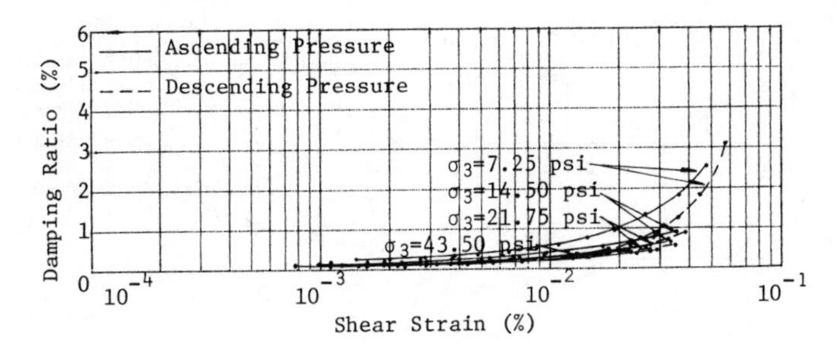

Figure 14. Damping Ratio Versus Strain Amplitude for Free-Free Resonant Column Device. Test No. 1

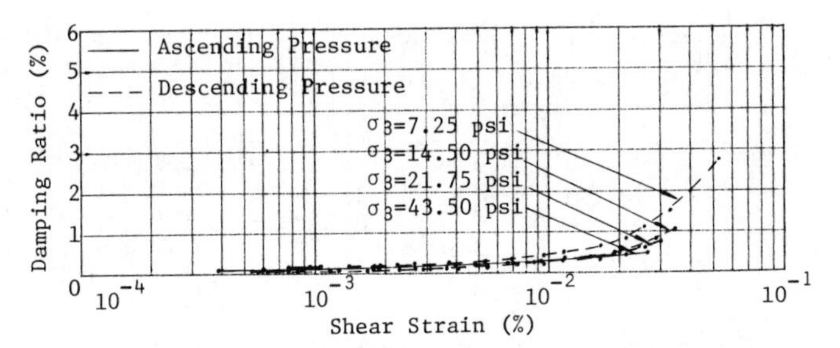

Figure 15. Damping Ratio Versus Strain Amplitude for Free-Free Resonant Column Device. Test No. 2

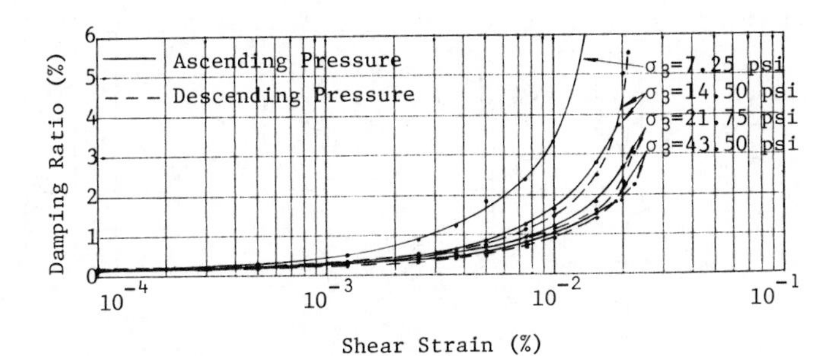

Figure 16. Damping Ratio Versus Strain Amplitude for Free-Fixed Resonant Column Device. Test No. 1

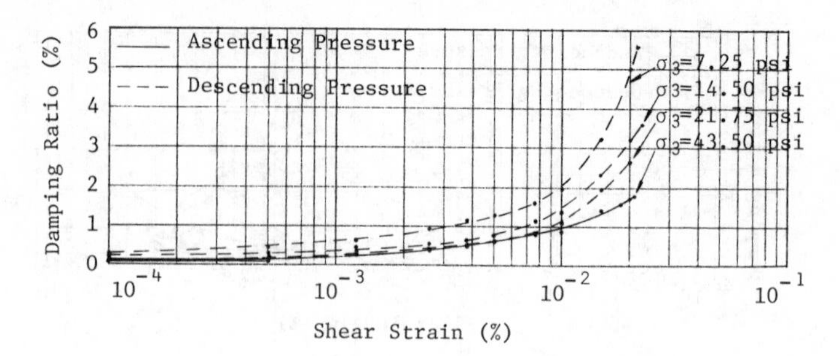

Figure 17. Damping Ratio Versus Strain Amplitude for Free-Fixed Resonant Column Device. Test No. 2

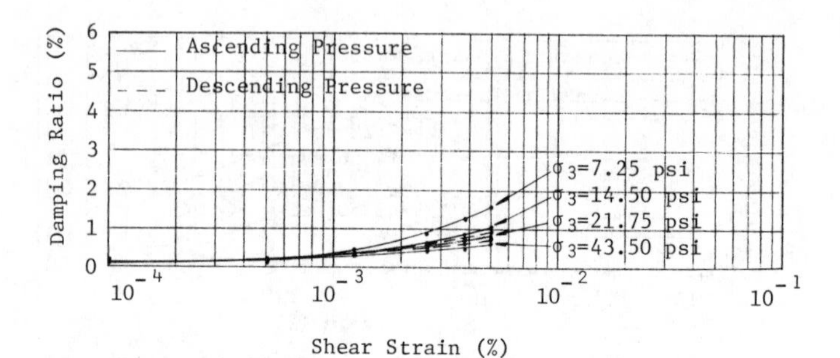

Figure 18. Damping Ratio Versus Strain Amplitude for Free-Fixed Resonant Column Device. Test No. 3

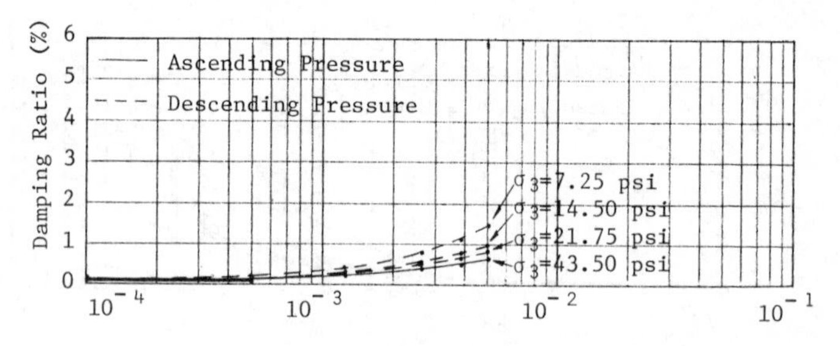

Figure 19. Damping Ratio Versus Strain Amplitude for Free-Fixed Resonant Column Device. Test No. 4

CONCLUSION

A comparison of the performance of the two types of resonant column devices, free-free and free-fixed, was based on six tests of reconstituted specimens of Monterey No. 0 sand tested at various confining pressures. At each confining pressure the shear modulus and damping ratio were determined at strain amplitudes ranging from 1×10^{-4} percent to 3×10^{-2} percent.

Conclusions are as follows:

1. Test reproducibility of both test devices, being about 5 percent between the two tests, is considered acceptable. The free-free resonant column device showed greater reproducibility than the free-fixed resonant column.
2. The shear modulus attained in the free-free resonant column device is consistently higher than that in the free-fixed resonant column, the variation being independent of the confining pressure. Damping ratios in the free-fixed test are higher than those in the free-free test. Difference in test results between the two devices are attributed to the basic differences in their design.
3. As disclosed from the test results, a strain amplitude of 1×10^{-2} percent sufficiently defines the boundary condition of nondestructive and destructive resonant column testing. As the strain amplitude exceeds this boundary, the higher the strain amplitude applied, the greater the specimen disturbance and the greater the reduction in shear modulus.

REFERENCES

1. Drnevich, V. P. "Resonant Column Test," U.S. Army Engineering Waterways Experiment Station, Miscellaneous Paper S-78-6, Vicksburg, Mississippi, July 1978.

2. Drnevich, V. P. "The Outline of the Round-Robin Resonant Column Testing Program," 1978.

3. Hardin, B. O. And V. P. Drnevich. "Shear Modulus and Damping in Soils Measurement and Parameter Effects," Journal of the Soil Mechanics and Foundations Division, ASCE, Vol. 98, No. SM6, Proc Paper 8977, June 1972.

4. Hardin, B. O. and J. Music. "Apparatus for Vibration of Soil Specimens During the Triaxial Test," Instruments and Apparatus for Soil and Rock Mechanics, ASTM STP 392, American Society for Testing and Materials, 1965.

ESTABLISHING ROCK MASS PROPERTIES FOR POWER PLANT DESIGN IN WEATHERED VOLCANIC ROCKS

J L Ehasz[I], K H Liu[II], and I H Wong[III]

SUMMARY

The engineering characteristics of weathered pyroclastic rocks at a site in Southeast Asia have been investigated and parameters necessary for nuclear power plant design were established. The degree of weathering was the most significant factor affecting the engineering properties. The materials were classified as soil-like or rock-like. Detailed static and dynamic laboratory testing was performed on the various degrees of weathered rock in order to establish the strength, compressibility and elastic properties. A field geophysical program was correlated with a laboratory sonic velocity program and correction factors were established to better appreciate the in-situ properties of the foundation rocks. The engineering properties of these weathered volcanic rocks are compared to other rocks as well as sedimentary soils.

INTRODUCTION

A description of the approach and general philosophy of establishing realistic design parameters for the nuclear power plant being constructed in a complex weathered rock region is presented. While many investigators have made detailed studies of the static strength characteristics of residual soils as well as weathered rocks, not much has been published on highly weathered pyroclastic volcanic materials. Although some field geophysical results were presented by Deere and Patton (1971), the information regarding the detailed dynamic characteristics required for nuclear plant dynamic design in these types of weathered volcanic rocks was generally unavailable. A better understanding of the dynamic characteristics of the foundation rock materials was therefore essential.

GEOLOGIC CONDITIONS AND SUBSURFACE INFORMATION

The site is located in Southeast Asia and is situated on a projecting volcanic headland flank of a volcanic complex, which is part of an island arc system. The volcanic complex is termed inactive; no historical eruption has been recorded and shallow seismic activity is lacking.

The site is underlain by late Tertiary-Quaternary indurated pyroclastic laharic and sedimentary rocks, initially derived from a volcanic complex. The total thickness of volcanic rocks beneath the site is unknown, but drilling to depth of more than 300 feet

---

I, II, III respectively. Chief Civil Engineer, Senior Civil Engineer and Consulting Engineer, Ebasco Services Incorporated, New York, New York.

below sea level has failed to reach the underlying basement rocks. Regional geology, however, suggests that basement lies several thousand feet beneath the site.

The site was extensively blanketed by vegetation and an approximate ten-foot thick mantle of reddish-brown residual soils. The tropical climate and high humidty and rainfall has significantly affected the chemical and physical engineering properties of the underlying soils sand rocks. The plant location is underlain basically by a weathering profile of the pyroclastic rocks. These pyroclastic rocks are differentiated according to clast size and consist of interlayers of volcanic breccia, lapilli tuffs and tuffs. Volcanic breccia, the dominant rock type recovered in drill core, is composed of angular to subrounded clasts greater than 32 mm in diameter within a tuffaceous matrix. Lapilli tuffs are composed of angular to round clasts 4 mm to 32 mm in diameter within a tuffaceous matrix. Tuffs are composed of volcanic fragments less than 4 mm in diameter. Individual units average more than 15 feet thick and limited lateral extent. For engineering purposes, the subsurface materials were classified, based on the degree of weathering, as residual soils (I), soil-like saprolite (IIa), rock-like saprolite (IIb), moderately weathered rock (III), and slightly weathered to fresh rock (IV). Figure 1 presents a generalized weathering profile of the subsurface materials.

In order to aid in field classification, the material was classified as soil-like saprolite if it could be sampled using a split-spoon sampler and rock-like saprolite if it could not. The moderately weathered and slightly weathered rocks were core drilled and were classified on the basis of the degree of weathering and Rock Quality Designation (RQD).

Soil-like saprolite was completely weathered from the volcanic rocks into a friable condition with soil-like characteristics, but the rock texture and structure of parent rocks are preserved. Rock-like saprolite is a highly weathered rock and the weathering extends throughout the rock mass; the rock mass is partly friable and of low hardness. Moderately weathered rock is of moderate to high strength. It can be visually distinguished from rock-like saprolite by the extent of weathering. For slightly weathered rock, weathering is only developed on the joint surface and signs of weathering are less apparent.

MECHANICS OF CHEMICAL WEATHERING

A good understanding of the chemical weathering process helps comprehend the complexity of this weathered rock system. Peltier (1950) has established the relationship between mean annual temperature and mean annual rainfall in determining the intensity of chemical weathering. Based on the site climatic conditions, Peltier's diagram (Figure 2) suggests a strong chemical weathering as the prime agent for soil formation process. The intense chemical weathering of the rock mass is caused mainly by percolation of surface water and groundwater through the rocks. The rate of chemical weathering is accelerated by the tropical heat, humidity and the decay of vegetation.

Pyroclastic rocks are formed at high pressure and temperature from a gas-charged, explosive origin. Upon cooling and achieving stability, hairline cracks and joints develop and become subject to water penetration. Subsequently, the percolation of surface water and flow of groundwater leads to oxidation and leaching processes, which alter the rock chemistry. If the chemical alteration results in an increase in volume, there is usually a subsequent disintegration of rocks. Slickensides and hairline cracks were observed in the core samples from the project site. These hairline cracks turned into fissures and became more visible in the soil-like saprolite formation. The effect of groundwater on chemical weathering was evident from a comparison of saprolite (IIa and IIb) thickness and the groundwater contours which revealed considerable similarity between central groundwater channel and zone of maximum saprolite thickness.

PROPERTIES OF WEATHERED ROCKS

An extensive laboratory testing program was formulated to investigate the static and dynamic strength characteristics of the weathered volcanic rocks. It is of particular interest to point out that substantial difficulty was encountered using conventional tube sampling techniques to recover undisturbed samples from the soil-like saprolite formation. The relict rock structure or rock fragments often caused a "refusal" condition, preventing the penetration of the thin wall tube, or shearing off the soils sample by the grinding action from the cutting edge, as in the case of Denison double tube sampler. Consequently, trenching proved to be the best method to recover "undistributed" samples of the soil-like saprolite. Rock samples were easily obtained from a NX double tube core barrels.

Static Properties – Soil-Like Materials

Soil-like materials were subjected to index and property tests, namely, unconfined compression, triaxial compression and consolidation tests. It was noted that both soil-like and rock-like saprolite had an unusually high water content and a relatively low dry unit weight. This same situation also appeared in both moderately weathered and slightly weathered rock consisting of the tuffs, and also lapilli tuffs, but to a lesser extent. A summary of all index properties of the weathered pyroclastic rocks is presented in Table 1. Further clay mineral information was established when the saprolite samples were subjected to an x-ray diffraction test. Test results indicated a very high content of halloysite clay minerals in the saprolite materials, which bore a great similarity to the Sasumua clay reported by Terzaghi (1953). Despite the high water content, varying from 30 to 68 percent, these materials have proved to be satisfactory construction materials for embankments and other engineering applications.

Due to the existence of the relict rock fragments, the soil-like saprolite exhibited an extremely heterogeneous nature. Upon unconfined compression (UC) strength testing, the results ranged from 0.1 to 14 tsf. It was evident that the inherent relict rock fragments could not be totally ignored, and yet the soil-like characteristics prevailed. Results of unconsolidated undrained (UU) tests indicated a cohesion of 1.2 ksf and a friction angle of 12° (Figure 3). This cohesion is probably the combined result of capillary tension in the voids and the interparticle cementing.

Another interesting phenomenon was also observed with soil-like saprolite during consolidated undrained triaxial compression tests with pore pressure measurements ($\overline{CIU}$). These completely weathered rock materials exhibited a very high shear strength at relatively low confining stress levels. This shear strength remained constant until reaching an effective confining stress of approximately 16 ksf. An additional frictional strength was then experienced with increasing confining pressure (Figure 4). This phenomenon might be accounted for by the existence of a strong cohesion or cementation bonding at the halloy-site clay mineral interparticle contact. This cementing or bonding force is believed to be strong enough to develop shear resistance or apparent cohesion. As the stress level was increased and approached the apparent cohesion level, the cementing agent collapsed, the micro-fissures closed, and direct intergranular or interparticle contact resulted in a buildup of frictional strength.

Reconstituted materials were also investigated and it was found that the cementation characteristic described above completely disappeared once the soil-like saprolite was reconstituted by a remolding process. The materials essentially reverted to the characteristics of a normal clayey soil. However, a high cohesion strength of 3.8 ksf was still apparent from a $\overline{CIU}$ test on a reconstituted soil-like saprolite (Figure 5).

Static Properties – Rock-Like Materials

Rock-like saprolite and moderately weathered rock resemble soil-like saprolite with respect to static strength characteristics except that they have a higher strength. Their strength characteristics are presented in Figure 6.

The compressibility of the weathered volcanic rock is particularly unique when compared to other rock groups. The degree of weathering and the cementation are responsible for the shape of the e-logP curve. The extent of chemical alteration dictated the break point or "apparent preconsolidation pressure" of the two linear portions of the e-logP curves for the weathered volcanic rocks. These break points were considered unrelated to a conventional preconsolidation effect on sedimentary soils which may originate from different mechanisms, such as glaciation or desication. The dramatic change experienced during consolidation of the weathered volcanic rocks is related to chemical alteration. This alteration process which weakens a strongly bonded rock material has been described by Sowers (1953) as "changing the physicochemical bonds between the clay particles or introducing stress by expansion or contraction of the grains during the lateration process." The break experienced in the consolidation curve was so dramatic that it was termed "threshold pressure" – a governing pressure for each individual weathered volcanic rock unit that was used to establish the imposed foundation load. If the threshold load was exceeded, significant deformation or settlement would occur. A typical e-logP curve for rock-like saprolite is presented in Figure 7. An average threshold pressure for each weathered volcanic rock is presented in Table 2.

Dynamic Properties – Soil-Like Materials

Soil-like saprolite samples obtained from test pits are subjected to cyclic triaxial and resonant column tests. The test results for a wide range of confining stresses are presented in Figure 8. It is noted that the maximum shear modulus of soil-like saprolite varied from 3000 to 5000 ksf, which are comparatively higher than ordinary sedimentary clayey and sandy soils. This high shear moduli could be accounted for the existence of the relict-rock structure in soil-like saprolite.

Figure 8, also presents the shear moduli of reconstituted soil-like saprolite. The remolding process has restructured the original skeleton of the soil-like saprolite, and destroyed the relict rock fragments and cementing agents, and subsequently resulted in a decrease in shear moduli. It is interesting to note that the exponential factor of confining stress for reconstituted soil-like saprolite, increased to a value to 1/2, a value typical of ordinary sedimentary soils.

Dynamic Properties – Rock-Like Materials

Rock-like samples obtained from coring were subjected to cyclic triaxial and resonant column tests. Laboratory sonic velocity tests were also conducted to measure both the shear and compressional wave velocities. The actual results for all strain range from cyclic triaxial and resonant column tests are graphically presented in Figure 8 for a wide range of confining stresses. These shear moduli were than normalized with respect to the confining stress used in the cyclic tests (Figure 9). It is of interest to note that shear moduli of the more highly weathered volcanic rocks are not particularly sensitive to confining stress with the increasing degree of weathering. This pheonomenon is evidenced by the low values of the exponential factor (n) of the confining stress. This low degree of sensitivity to confining stress may also be explained by considering the chemical alteration process that the rock-like materials had experienced. Essentially, the sensitivity is related to the degree of bonding or apparent cohesion. The exponential factor for each weathered volcanic rock, along with their established parameters, is presented in Table 3.

FIELD GEOPHYSICAL MEASUREMENTS

Field geophysical methods including seismic refraction, seismic reflection, cross-hole, uphold and down-hole methods were utilized to delineate the site geologic formations and to establish engineering properties of the subsurface materials.

Over 6500 linear feet of seismic refraction profiling, which traversed the site in a strategic pattern, was conducted in the proposed plant area. The compressional wave velocities obtained from the refraction work were correlated with actual borings and rock types to ensure reasonableness and compatibility. The compressional wave velocities for weathered volcanic rocks is presented in Table 4.

The stratigraphy of the site is variable both laterally and vertically. In addition, the engineering properties are highly variable since they are affected by chemical weathering, which is a highly complex function of original rock chemical content, jointing, ground water movement and topography. As described previously, based on the degree of weathering, the subsurface materials were classified in four categories (I, II, III & IV) in order of decreasing degree of chemical alteration. Thin layers and zones of increased weathering were frequently encountered in depth within the Unit III and IV. These were lenticular, of short lateral extent and generally not correlative between bore holes. This reversed weathering occurrence posed an extreme difficulty for geophysical work. The extreme heterogenity of the weathered rock formations deviated from the normal and basic assumptions of slowly varying material properties in the vertical direction.

In order to better appreciate the actual in-situ characteristics of the large rock-mass associated with the extensive nuclear plant structures, consideration was also given to all the refraction compressional wave velocity information established from over 6500 linear feet of seismic refraction profiling. The compressional wave data was averaged to establish velocity profiles to depths exceeding 300 feet, see Figure 10. The shear wave velocity profiles were established from both field and laboratory determinations and were extended to depths of over 300 feet by knowledge of the maximum compressional wave velocities. These maximums were confirmed by field measurements in fresh rock formations as well as regional studies of the average maximum crustal compressional wave velocities from the actual earthquake travel times of both macro and micro seismic motions.

ESTABLISHING ROCK MASS ENGINEERING PROPERTIES

It has been well recognized that "undisturbed" soil or rock samples suffer a certain degree of disturbance due to the stress-relief swelling, sampling technique, sample handling, etc.; which affects the micro-structure of the materials and tends to soften or reduce the elasticity in the small samples. This disturbance effect coupled with the complexity of weathering created an interesting challenge for establishing rock mass engineering properties. A summary of all engineering properties are presented in Table 5. Figure 11 presents all the UC strength test results for the weathered volcanic rocks in a vertical profile. The mean values or the averaged compressive strength along the vertical profile can be established by drawing a curve bisecting the shaded area, which envelopes the whole range of compressive strength for the weathered volcanic rocks. This average compressive strength curve correlates directly with the range of threshold pressures experienced for each weathered rock group. This indicated that the apparent preconsolidation pressures were approximately equal to the unconfined compressive strengths. It was then reasoned, that the average unconfined compressive strength curve could provide for the subject site, a reasonable index for establishing maximum bearing pressures for plant structures within variable weathered profile. The strength characteristics obtained from the other detailed laboratory tests can then be utilized to confirm and establish other aspects of design relating to bearing capacity, compressibility, foundation behavior under different loading conditions, stability of excavated slopes, etc.

A realistic appraisal or estimate of the rock mass dynamic properties is of paramount importance in the seismic design of Nuclear Power Plants; Ehasz (1979) presented various cases of the effect of foundation stiffness on the structural response under seismic loading conditions. Two (2) typical cases are presented in Figure 12 and Figure 13 to illustrate the significance and impact on the selection of foundation moduli. By varying the foundation moduli or stiffness, the resulting responses such as shear, moment and actual structural response can be very significantly affected.

In establishing the realistic stiffness to be used for seismic design various factors were considered. The sample "disturbance" was particularly significant in the weathered volcanic rocks. Thus, in order to incorporate the actual stiffness to be experienced in the rockmass, emphasis was placed on the field geophysical work, which was considered most intimately related to the moduli of the rockmass. The generalized compressional and shear wave velocity vs depth relationships as presented in Figure 10 were employed as a basis of deriving the in-situ rock mass stiffness. By comparing the field shear wave and laboratory shear wave measurements, correction factors were developed to adequately adjust the moduli obtained from laboratory to the field. The decaying effect of the shear moduli with increasing strain curve determined from laboratory testing was believed to remain unchanged. The correction factors for each of the individual weathered rocks are presented in Figure 14. The wide range of shear moduli for each material as presented in Figure 8 were then normalized individually according to their corresponding effective stress. The normalized shear modulus, presented in Figure 9 represents the rock mass dynamic charactertistics with varing degrees of weathering, confining pressures, and shear strain. The sensitivity of the shear modulus with confining pressure is of particular importance since it enables the realistic incorporation of the foundation stiffness in the soil-structure interaction analysis.

COMPARISON OF WEATHERED PYROCLASTIC ROCKS WITH OTHER WEATHERED ROCK AND SEDIMENTARY SOILS

Engineering properties of various other types of weathered rocks were used to compare with the weathered pyroclastic rocks studied. Table 6 presents the summary of engineering properties of these different types of weathered rocks. In general, this particular weathered volcanic rock group appears to be a "soft" rock and with a high cohesion when compared with those from metamorphic and igneous rocks. This may be the combined results of the different weathering process and the minerology of the original rocks.

Comparison was made between the dynamic properties of weathered volcanic rocks and sedimentary soils since sedimentary soils were considered well understood and documented (Seed and Idriss 1970, Hardin and Drnevich 1972). Special attention was given on the decaying effect of shear moduli and the increase of damping ratio with high shear strain levels. The comparison between weathered volcanic rocks and sedimentary soils for clay, and sand material are presented in Figure 15. It can be concluded from this comparison that the reduction in shear moduli and the increase in damping ratio with increasing shear strain is less for weathered rock than it is for sedimentary soils. This effect is more apparent with the less weathered volcanic rocks.

CONCLUSION

The static and dynamic strength characteristics of weathered volcanic rocks has been investigated with respect to the weathering process through the field and laboratory testing, and finally establishing of rock mass engineering properties. Emphasis was placed on the correct and adequate assessment of the rock mass engineering properties, which are significant to the seismic design of a power plant. It was established that the in-situ moduli of weathered rocks are significantly higher than laboratory determined moduli. Correction factors were applied to adjust the values of the shear modulus determined in the laboratory to field conditions. The compressibility of the weathered volcanics is unique for a rock and is somewhere similar to a highly overconsolidated lean clay. The dynamic properties of the weathered volcanic rocks were compared with the sedimentary soils, and their distinct difference in the decaying effect of shear modulus and the rate of increase of damping ratio with increasing shear strain were noted. Comparison made among weathered volcanic rocks and other weathered rocks from different origins indicates that they are generally a much softer rock.

ACKNOWLEDGMENT

Special thanks are extended to M Pavone of Ebasco Services, Inc. for the data preparation presented herein and to R. Ladd of Woodward Clyde Consultants who took extreme care with the dynamic testing of these unsual samples and added much to the interpretation of the results.

REFERENCES

Deere, D. U., and Patton, E. D. (1971) "Soil Stability in Residual Soils," Proc. 4th Pan Amer. Conf. on Soil Mech. and Found. Eng., Puerto Rico, pp. 87-170.

Dearman, W. R., Baynes F. J. and Irfan, T. Y. (1978), "Engineering Grading of Weathered Granite," Engineering Geology, Dec. pp. 345-374.

Ehasz, J. L., (1979), "Nuclear Power Plant Foundation Design and Experience," European Nuclear Conference, Hamburg, W. Germany, May.

Link, H. (1969), "The Sliding Stability of Dams," Water Power, May 1969.

Lee, K. L., and Seed, H. B. (1967), "Dynamic Strength of Anitsotropically Consolidated Sand," Journal of the Soil Mech. and Found. Div., ASCE, Vol. 93, pp. 169-190.

Peltier, L. (1950), "The Geographic Cycle in Periglacial Regions As It Is Related To Climate Geomorphology," Ann Assoc. Am. Georg. 40, pp. 214-236.

Seed, H. B. and Idris, I. M., (1970), "Soil Moduli and Damping Factors for Dynamic Response Analysis," Report No. EERC 70-10, Univ. of California, Berkeley, Calif., Dec.

Sowers, G. F., (1963), "Engineering Properties of Residual Soils Derived from Igneous and Metamorphic Rocks," Proc. 2nd Pan Amer. Conf. Soil Mech, and Found. Eng., Brazil, 1, pp. 39-61.

Sowers, G. F. (1953), "Introductory Soil Mechanics and Foundation" 2nd Edition, The Macmillan Company, New York.

Stagg, K. G., and Zienkiewicz, O. C. (1969), "Rock Mechanics in Engineering Practice," J. Wiley and Sons, New York.

Terzaghi, K. (1958), "Design and Performance of Sasumua Dam," Proc. Instn. Civil Engrs., Vol. 9, pp. 369-394.

Underwood, L. B. and Dixon, N. A., "Dam on Rock Foundations," Rock Engineering for Foundations and Slopes, Proc. of a Specialty Conf., ASCE, 1976, pp. 125-146.

Vargas, M. Silver, F. P. and Tubio, M. (1965), "Residual Clay Dams in the State of Sao Paulo, Brazil," Proc. Int. Conf. Soil Mech. Found. Eng., 6th Montreal, II, pp. 578-582.

Table 1
INDEX PROPERTIES OF WEATHERED VOLCANIC ROCKS

	Total Unit Weight γtPCF	Dry Unit Weight γtPCF	Water Content W%	Liquid Limit LL%	Plastic Limit PL%	% Passing #200 Sieve	Porosity	Core Recovery %	RQD %
Soil-Like Saprolite	80–110	53–75	30–68	34–81	21–42	34–99			
Rock-Like Saprolite	85–110	50–69	33–88				0.48–0.67	43–89	15–61
Moderately Weathered Rock	98–107	70–123	3.9–25				0.25–0.51	86–100	45–81
Slightly Weathered to Fresh Rock	98–154	76–149	3.5–29				0.10–0.55	75–100	40–89

Table 2
AVERAGE THRESHOLD PRESSURE OF WEATHERED VOLCANIC ROCKS

	Average Initial Void Ratio, eo	Average Final Void Ratio, ef	Average Threshold Pressure TSF
Soil-Like Saprolite	1.76	1.33	3.5 (335)
Rock-Like Saprolite	1.95	1.48	11.0 (1054)
Moderately Weathered Rock	1.15	0.95	38 (3640)
Slightly Weathered to Fresh Rock	–	–	>50 (4790)

( ) in KN/m^2

Table 3
RELATIONSHIP BETWEEN SHEAR MODULUS AND OTHER PARAMETERS

	A (KSF)	n	CF
Soil-Like Saprolite	2.95×10^3	1/4	1.75
Rock-Like Saprolite	6×10^3	1/6	2.5
Moderately Weathered Rock	8.4×10^3	1/4	3.0
Slightly Weathered to Fresh Rock	14×10^3	1/2.5	4.0
Reconstitute Soil-Like Saprolite	0.84×10^3	1/2	–

***As appeared in equation $G_{in\text{-}situ} = KA(\overline{\sigma}_C)^n(CF)$**
Where K=G/Gmax can be obtained from Figure 9

Table 4
SUMMARY OF COMPRESSIONAL WAVE VELOCITIES OF WEATHERED VOLCANIC ROCKS

	Compressional Wave Velocity (FPS)
Soil-Like Saprolite	1160 – 6400
Rock-Like Saprolite	5200 – 8500
Moderately Weathered Rock	6500 – 11,500
Slightly Weathered to Fresh Rock	8100 – 19,000

Table 5
SUMMARY OF ROCK MASS ENGINEERING PROPERTIES OF WEATHERED VOLCANIC ROCKS

	Coefficient of Permeability K cm/sec	Unconfined Compression Strength TSF	Unconsolidated Undrained Compression Strength	Consolidated Undrained Triaxial Compression*	Shear Wave Velocity, FPS
Soil-like Saprolite	$2 \times 10^{-4} – 8 \times 10^{-6}$	0.1–14	$C = 1.2^{ksf}, \emptyset = 17^{o}$	$\bar{C} = 1.6^{ksf}, \emptyset = 24^{o}$	470–2500
Rock-like Saprolite	$1.8 \times 10^{-3} – 5 \times 10^{-4}$	2.6–12.5	$C = 15^{ksf}, \emptyset = 17^{o}$		2000–4600
Moderately Weather Rock	8×10^{-4}	3.5–35.5		$\bar{C} = 5^{ksf}, \emptyset = 17^{o}$	2600–4600
Slightly Weathered to Fresh Rock	$1.3 \times 10^{-3} – 2.2 \times 10^{-4}$	20–390			3200–7680

*From Effective Stress Mohr's Circle

Table 6
COMPARISON OF WEATHERED VOLCANIC ROCKS WITH OTHER ROCK GROUPS

Degree of Weathering & Rock Types	Completely Weathered Rock				Highly Weathered Rock				Moderately Weathered Rocks			Slightly Weathered to Fresh Rock			
Engineering Property	Pyroclastic Rocks	Metamorphic Rocks	Igneous Rocks	Sedimentary Rocks	Pyroclastic Rocks	Metamorphic Rocks	Igneous Rocks	Sedimentary Rocks	Pyroclastic Rocks	Metamorphic Rocks	Sedimentary Rocks	Pyroclastic Rocks	Metamorphic Rocks	Igneous Rocks	Sedimentary Rocks
Foundation Conditions	Unsuitable	Conditionally Suitable	Conditional Suitable	Conditionally Suitable	Suitable Special Care Required	Suitable	Suitable	Suitable	Satisfactory	Satisfactory	Satisfactory	Favorable	Favorable	Favorable	Favorable
Excavatability	Scraping	Scraping	Scraping	Scraping	Scraping/ Ripping	Scraping/ Ripping	Scraping/ Ripping	Scraping/ Ripping	Ripping, Line Drill Prespliting	Ripping/ Blasting	Rippling/ Blasting	Blasting	Blasting	Blasting	Blasting
% Core Recovery	–	–	–	–	43 – 89	15 – 70	10 – 90		86 – 100	90–100		84 – 100	90 – 100	90 – 100	85 – 100
RQD %	–	–	–	–	15 – 61	0 – 50	0 – 50		45 – 81	50–75		40 – 89	75 – 90	75 – 90	90 – 100
Unconfined Compressive Strength (TSF)	0.1 – 14				4 – 35				15 – 35		25 – 57	34 – 390	900 – 2300	445 – 2000	21 – 880
Shear Strength	C′ = 1.6 = 24° Ø = 24°			C = 2 Ø = 10°	C = 15 Ø = 17°	Ø = 20-37°	Ø = 31°	C′=0.1-2.6 Ø =25-32	C′=5 Ø =17°	Ø=35°-44°	Ø=32-42	–	Ø=48°-73°	Ø=48°-73°	Ø=23°-62°
Compressional Wave Velocity FT/Sec	980 – 2800	1500 – 3500	1500 – 3500	1785 – 3000	3000–6800	3300–6500	3300–6500		7000–9500	5000–10,000		10,000 – 16,500	10,000 – 18,000	10,000 – 18,000	
Shear Wave Velocity (FT/Sec)	500 – 1100			1000 – 1380	1200–2600				2700–3600			3,800 – 6,500			

*C = Cohesion, KSF
C′ = Effective Cohesion, Ksf
Ø = Angle of Internal, Friction, Degree
Ø̄ = Effective Angle of Internal Friction, Degree

BASED ON W E DEARMAN (1978), L B UNDERWOOD (1976), HAROLD LINK (1969) & VARGAS (1965)

Figure 1
GENERAL GEOLOGIC FORMATION AND WEATHERING PROFILE

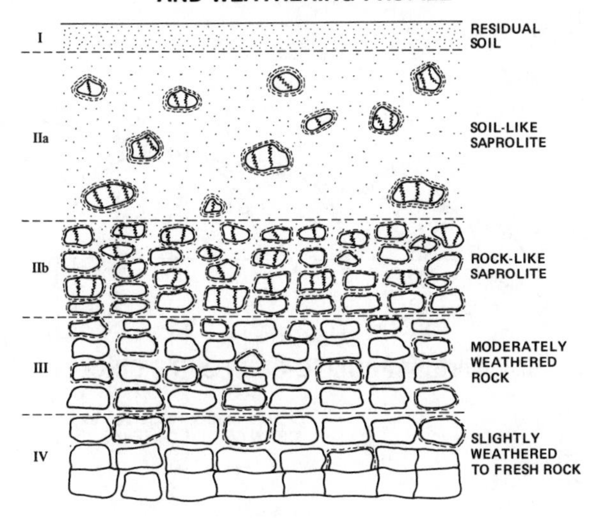

Figure 2
PELTIER'S DIAGRAM

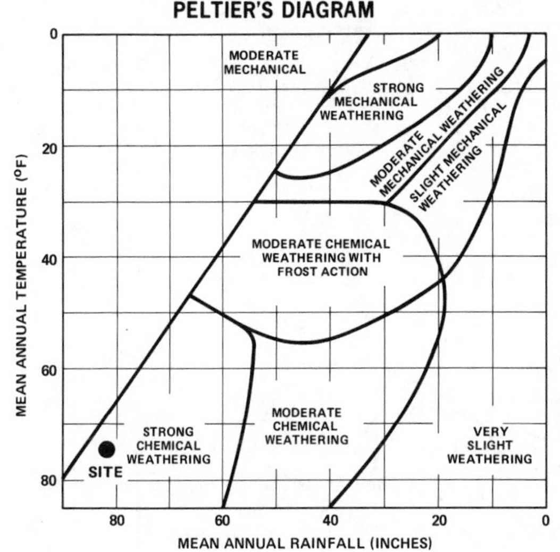

Figure 3
SOIL-LIKE SAPROLITE-UU MOHR'S CIRCLE

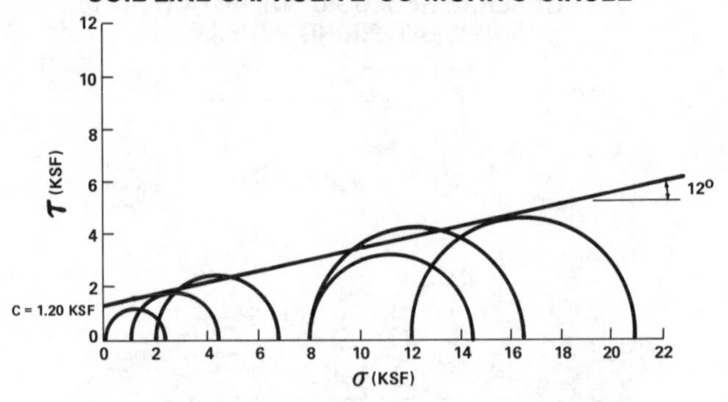

Figure 4
SOIL-LIKE SAPROLITE-$\overline{\text{CIU}}$ MOHR'S CIRCLE

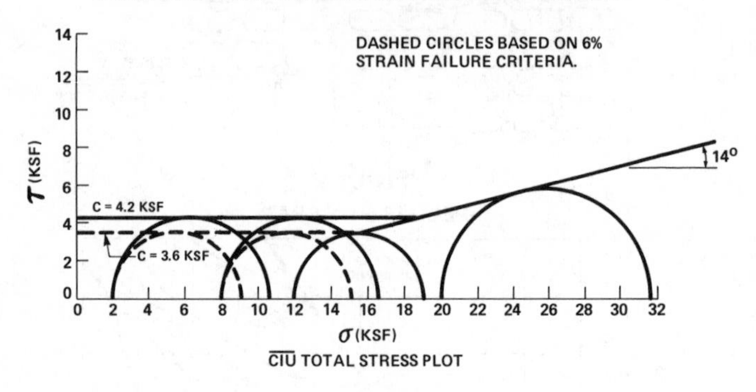

$\overline{\text{CIU}}$ TOTAL STRESS PLOT

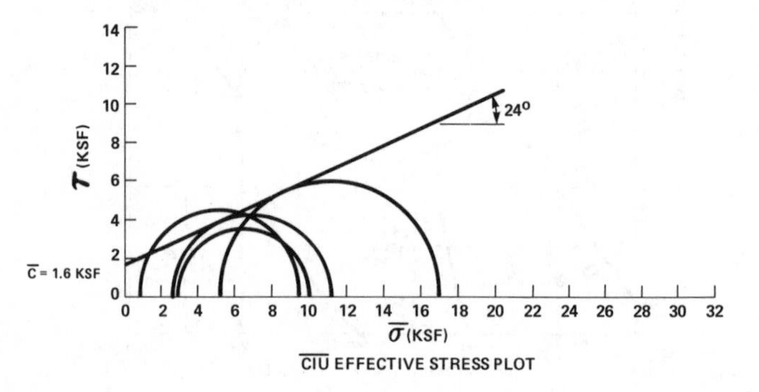

$\overline{\text{CIU}}$ EFFECTIVE STRESS PLOT

Figure 5
RECONSTITUTED SOIL-LIKE SAPROLITE-$\overline{\text{CIU}}$ MOHR'S CIRCLE

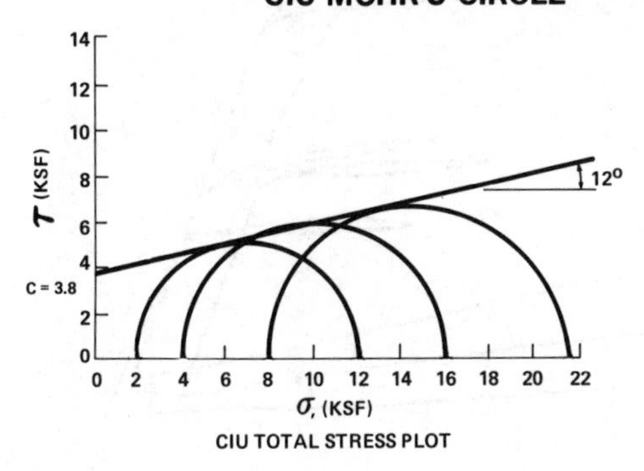

CIU TOTAL STRESS PLOT

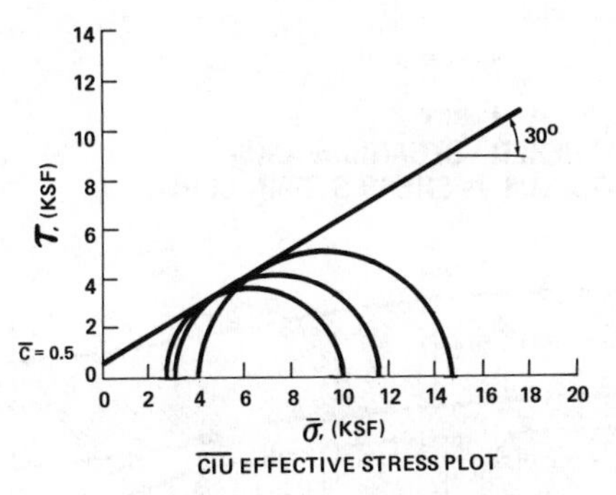

$\overline{\text{CIU}}$ EFFECTIVE STRESS PLOT

Figure 6
ROCK-LIKE SAPROLITE-UU MOHR'S CIRCLE

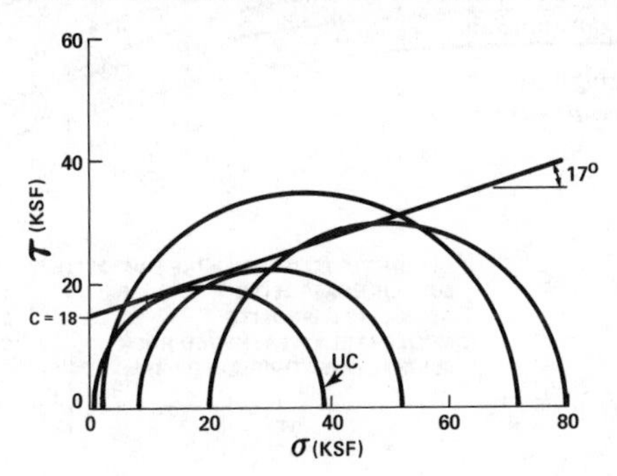

Figure 7
ROCK-LIKE SAPROLITE-CONSOLIDATION CURVE

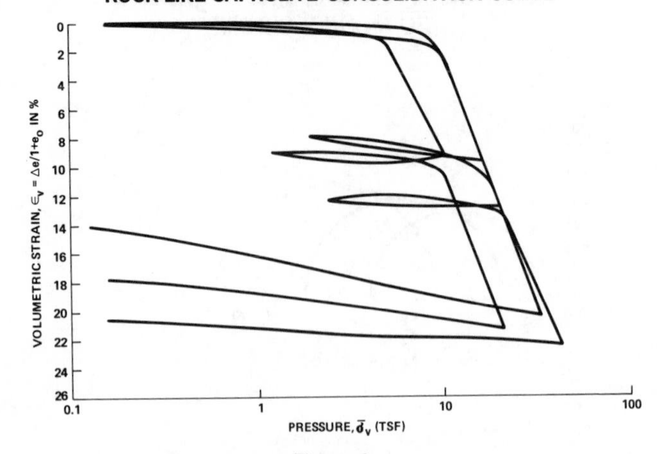

Figure 8
WEATHERED VOLCANIC ROCKS–
SHEAR MODULUS Vs SHEAR STRAIN CURVE

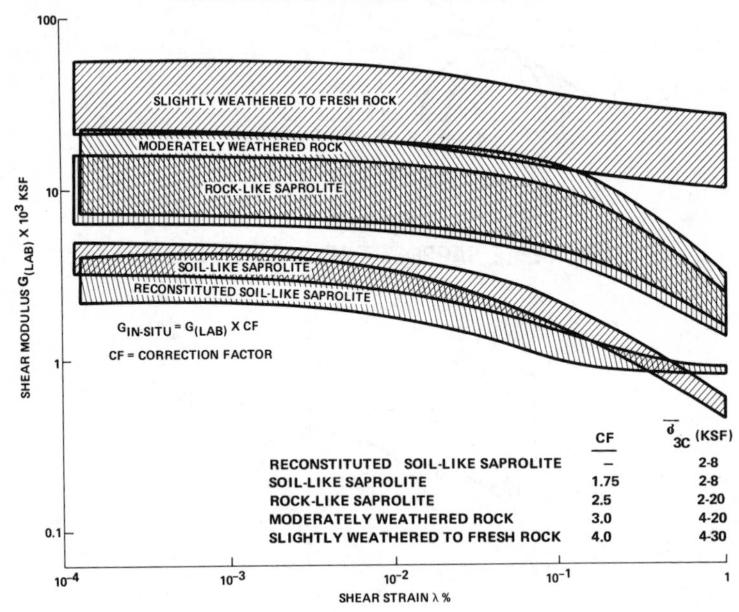

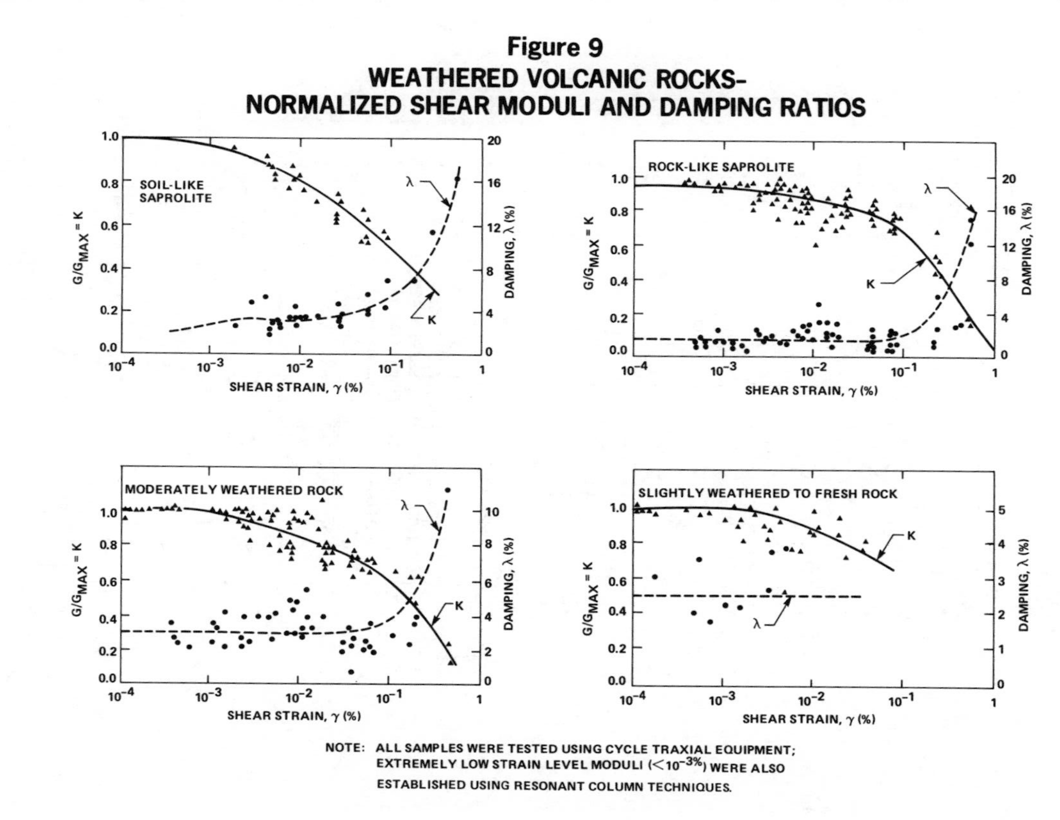

Figure 9
WEATHERED VOLCANIC ROCKS- NORMALIZED SHEAR MODULI AND DAMPING RATIOS

NOTE: ALL SAMPLES WERE TESTED USING CYCLE TRAXIAL EQUIPMENT; EXTREMELY LOW STRAIN LEVEL MODULI (<10^{-3}%) WERE ALSO ESTABLISHED USING RESONANT COLUMN TECHNIQUES.

Figure 10
FIELD SEISMIC VELOCITY PROFILE

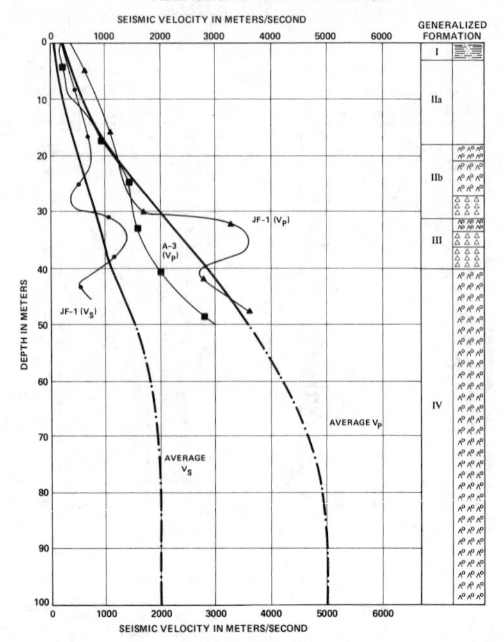

Figure 11
WEATHERED VOLCANIC ROCKS-UNCONFINED COMPRESSION STRENGTH ALONG WEATHERING PROFILE

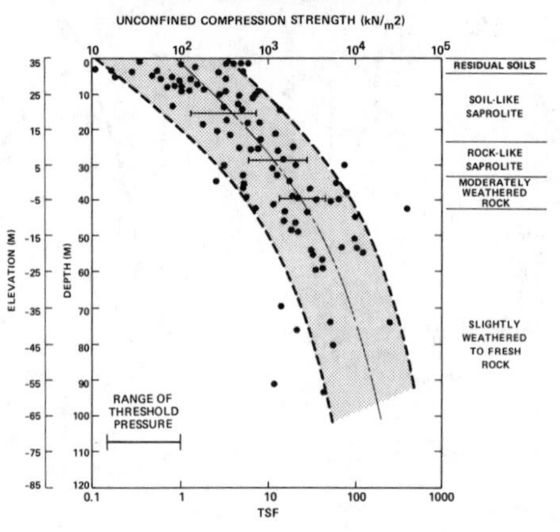

Figure 12
EFFECTS OF FOUNDATION STIFFNESS ON STRUCTURE MOMENTS AND SHEARS

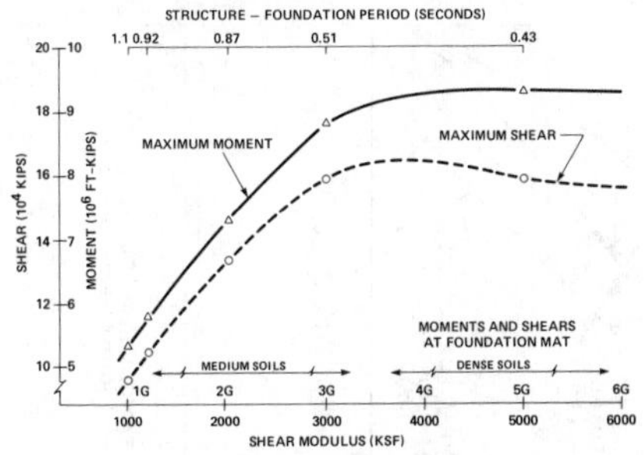

Figure 13
EFFECTS OF FOUNDATION STIFFNESS ON STRUCTURAL RESPONSE

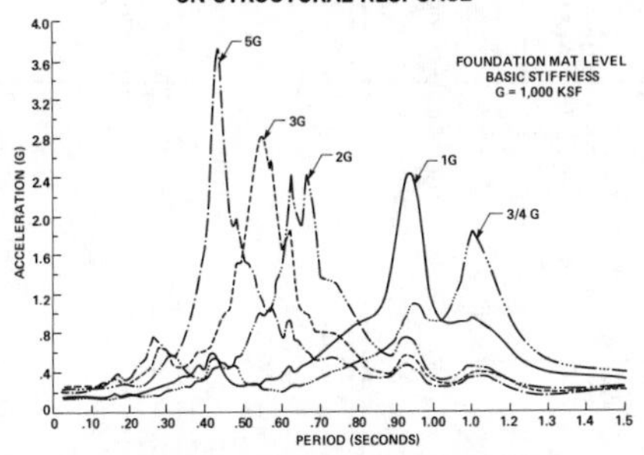

Figure 14
SHEAR WAVE VELOCITIES AND CORRECTION FACTORS FOR MODULUS

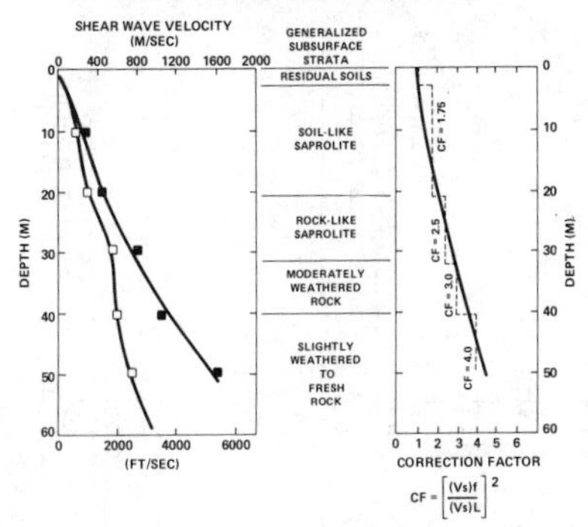

Figure 15
COMPARISON OF MODULI AND DAMPING BETWEEN WEATHERED ROCKS AND SEDIMENTARY SOILS

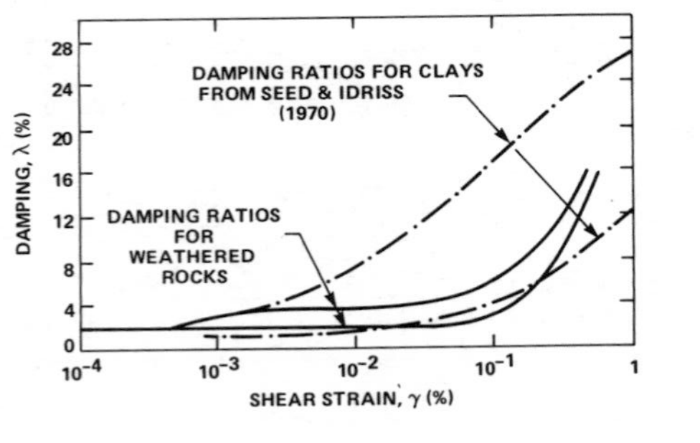

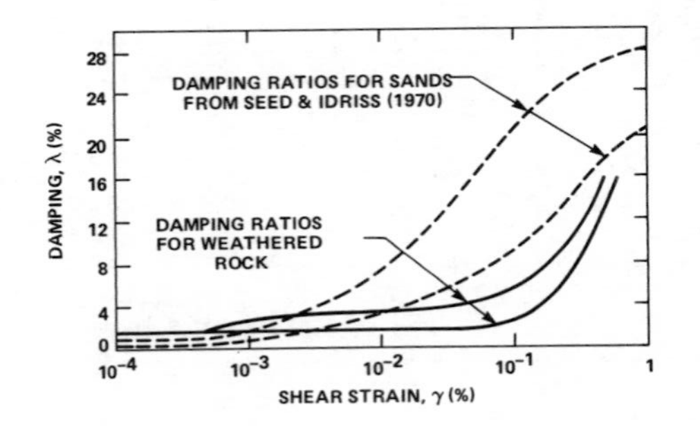

FOUNDATION ENGINEERING AT THE HARTSVILLE NUCLEAR PLANT

Ronald G. Domer,[1] F. ASCE, H. K. McLean,[2] D. R. Millward,[3] and W. M. Seay[4]

A. INTRODUCTION

TVA has carried out, for many years, a site investigation program for its generating facilities. Depending upon the type of facility, certain site characteristics become more or less important. One of the more important characteristics of interest for any major generating facility is the geology and its relationship to the structures (including embankments, ponds, etc.) being considered. For the purposes of this paper this relationship is defined by the term foundation engineering, and it is the objective of this paper to trace this evolutionary relationship from the site study to the construction of TVA's Hartsville Nuclear Plant.

TVA began map reconnaissance studies in the Middle Tennessee area in 1970 and identified several sites, including the Hartsville site, deemed favorable for onsite study. When core drilling equipment moved on the Hartsville site in September 1971 to begin preliminary drilling, a process was begun which is expected to culminate in 1990 when the last of the four 1300-MW boiling water reactor units is placed in service. This huge facility covers an area of 1940 acres including the reactor island complex, turbine/condensers, closed system cooling towers for cooling condenser circulating water, a deep water intake facility for makeup water, discharge facilities and essential service water (ESW) storage, circulation

---

[1] Assistant Manager of Engineering Design, TVA

[2] Geologist, Civil Engineering and Design Branch, TVA

[3] Civil Engineer, Civil Engineering Branch, TVA

[4] Geologist, Civil Engineering and Design Branch, TVA

and cooling systems. Direct access to the site is by barge, including a rail-to-barge unloading/loading facility, and by highway. A layout of the project is shown in figure 1.

The ESW system, which is made up of four elliptical-shaped ponds of 4.6 acres each with associated pumping and spray cooling facilities, is a critical element of the plant's nuclear safety system. The ESW system is the source of cooling water for plant systems required to shut the plant down and maintain it in a safe configuration. Thus, the ponds are required to be low-leakage reservoirs capable of withstanding a severe earthquake (site Safe Shutdown Earthquake [SSE] of 0.20 g). Structures for the reactor island are of a standard design based upon an SSE of 0.30 g.

Relatively shallow overburden and the layout of the reactor island complex dictated from the beginning that these structures would be founded directly in the underlying rock. TVA's longstanding and successful experience in building massive structures on limestone formations supported the prudence of this decision.

B. EXPLORATION

1. SITE STUDIES

- Regional Physiography, Geology, and Structure

The Hartsville Nuclear Plant is located on the northeast edge of the Nashville Basin, an elliptical topographic depression whose axis trends approximately N 30° E (figure 2). This basin is an erosional feature developed by prolonged solution weathering of limestones along the crest of a broad structural dome.

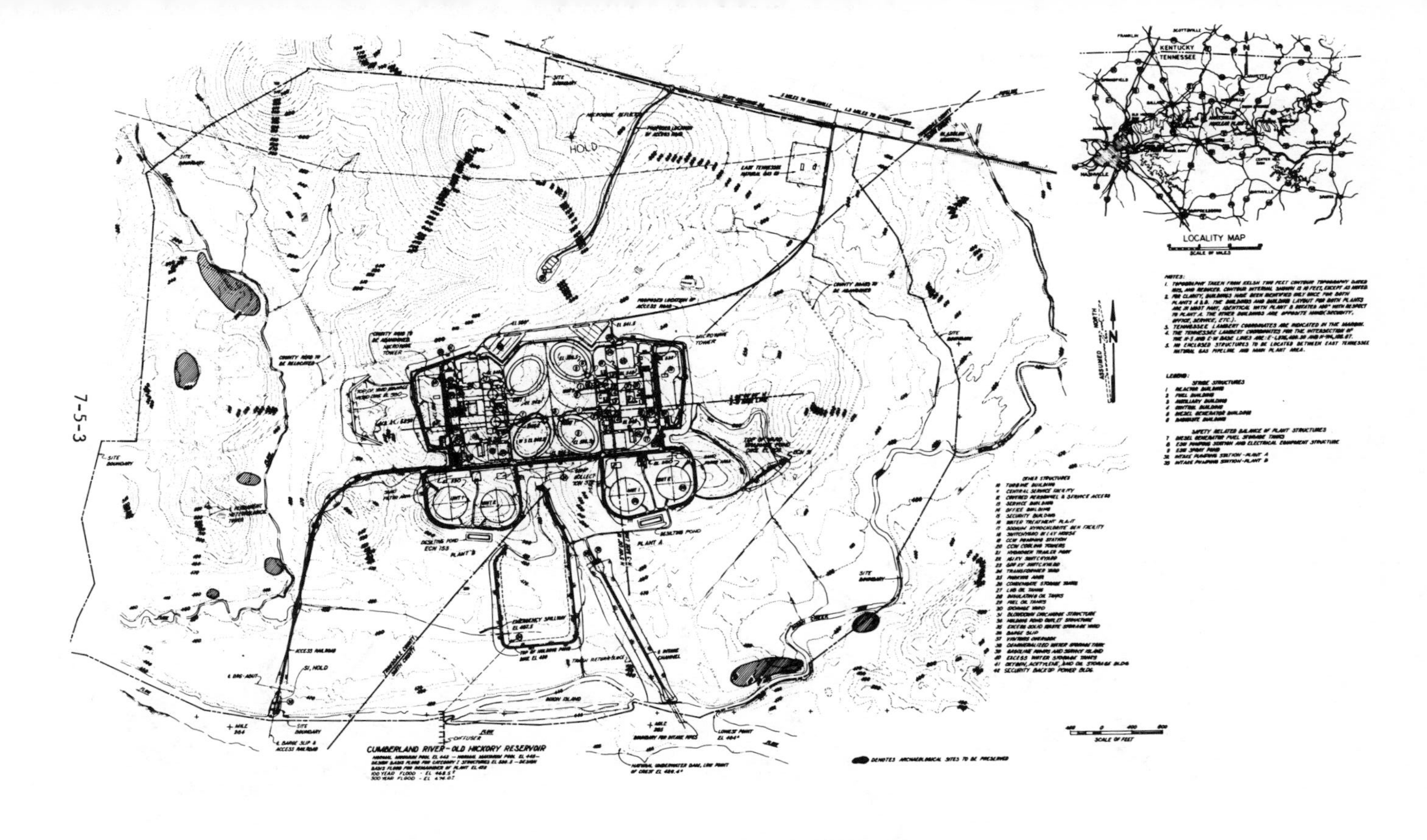

FIG. 1 -- LOCATION OF STRUCTURES

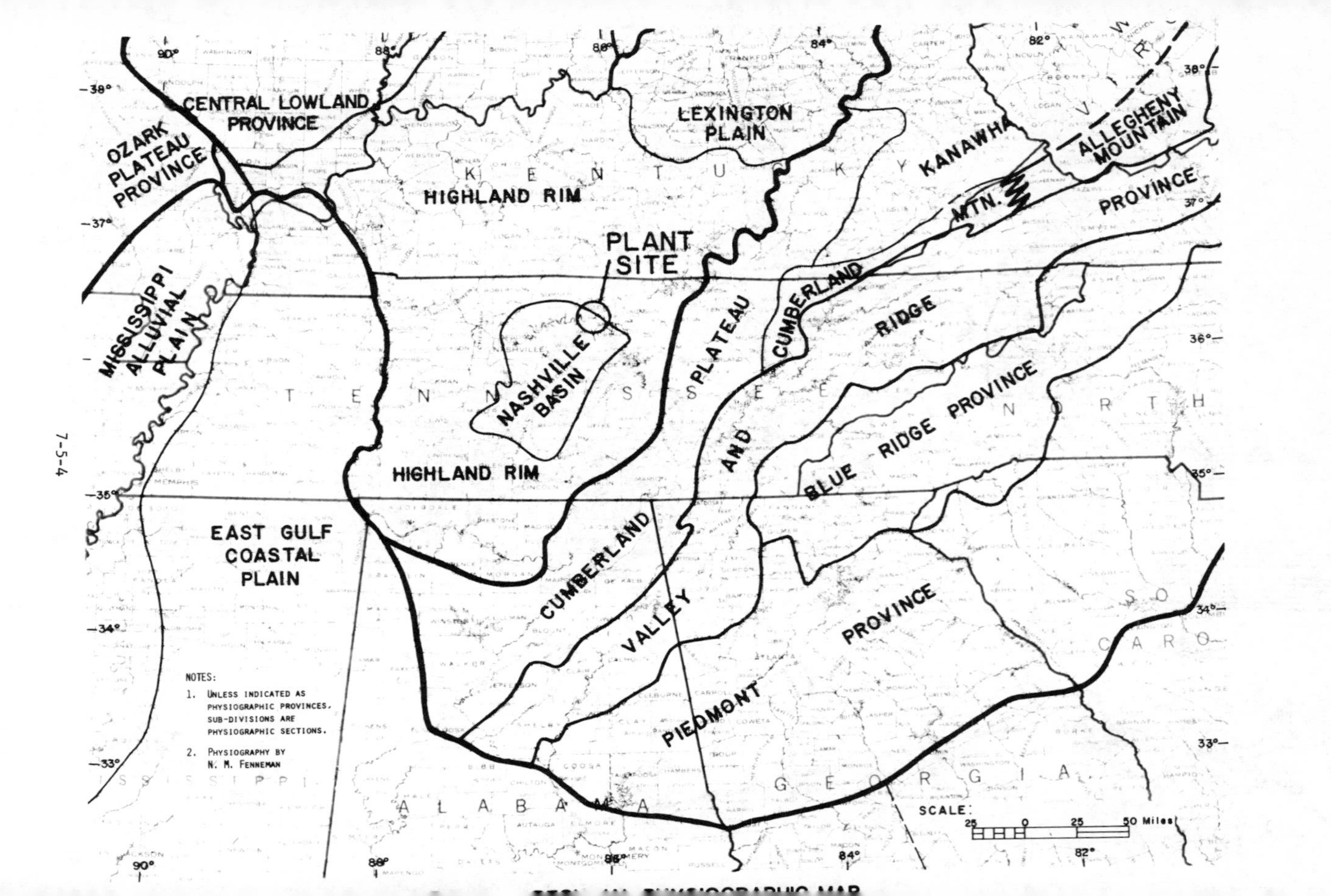
CENTRAL LOWLAND PROVINCE
OZARK PLATEAU PROVINCE
LEXINGTON PLAIN
KANAWHA
ALLEGHENY MOUNTAIN
PROVINCE
MTN.
HIGHLAND RIM
PLANT SITE
MISSISSIPPI ALLUVIAL PLAIN
NASHVILLE BASIN
PLATEAU
CUMBERLAND
RIDGE
AND
BLUE RIDGE PROVINCE
HIGHLAND RIM
EAST GULF COASTAL PLAIN
CUMBERLAND
VALLEY
PROVINCE
PIEDMONT
K E N T U C K Y
T E N N E S S E E
N O R T H
S O
C A R O
G E O R G I A
A L A B A M A
NOTES:
1. UNLESS INDICATED AS PHYSIOGRAPHIC PROVINCES, SUB-DIVISIONS ARE PHYSIOGRAPHIC SECTIONS.
2. PHYSIOGRAPHY BY N. M. FENNEMAN
SCALE:
25 0 25 50 Miles
90°
88°
86°
84°
82°
38°
37°
36°
35°
34°
33°

Surface rocks of the Nashville Basin region include both unconsolidated and consolidated sediments. The unconsolidated strata are flood plain alluvium and terrace deposits of relatively recent geologic age (less than 1 m.y.). The consolidated sedimentary rocks are mainly limestones and cherty limestones, dolomite, and shale, which range in age from Middle Ordovician (450± m.y.) to Mississippian (300± m.y.) (figure 3).

The inclination or "dip" of the strata within the Nashville Basin is generally so slight that it is likely to be undeterminable in any one outcrop. Steeper dips can be observed locally, but such occurrences are the exception. Faults are comparatively rare and generally have vertical displacements of less than 50 feet.

- Regional Seismology

Although light earthquakes occur in the Appalachians to the east, and a few light shocks have been reported in the Nashville Basin region, there are no indications of any geologically recent earthquakes.

Two earthquakes were considered in defining the SSE at Hartsville. The Modified Mercalli (MM) intensity VII-VIII earthquake that occurred at Anna, Ohio, in 1937 is assumed to occur at the site. The MM intensity XI event that represents the postulated occurrence of the largest earthquake in the New Madrid (1811-1812) series is assumed to occur 110 miles from the site.

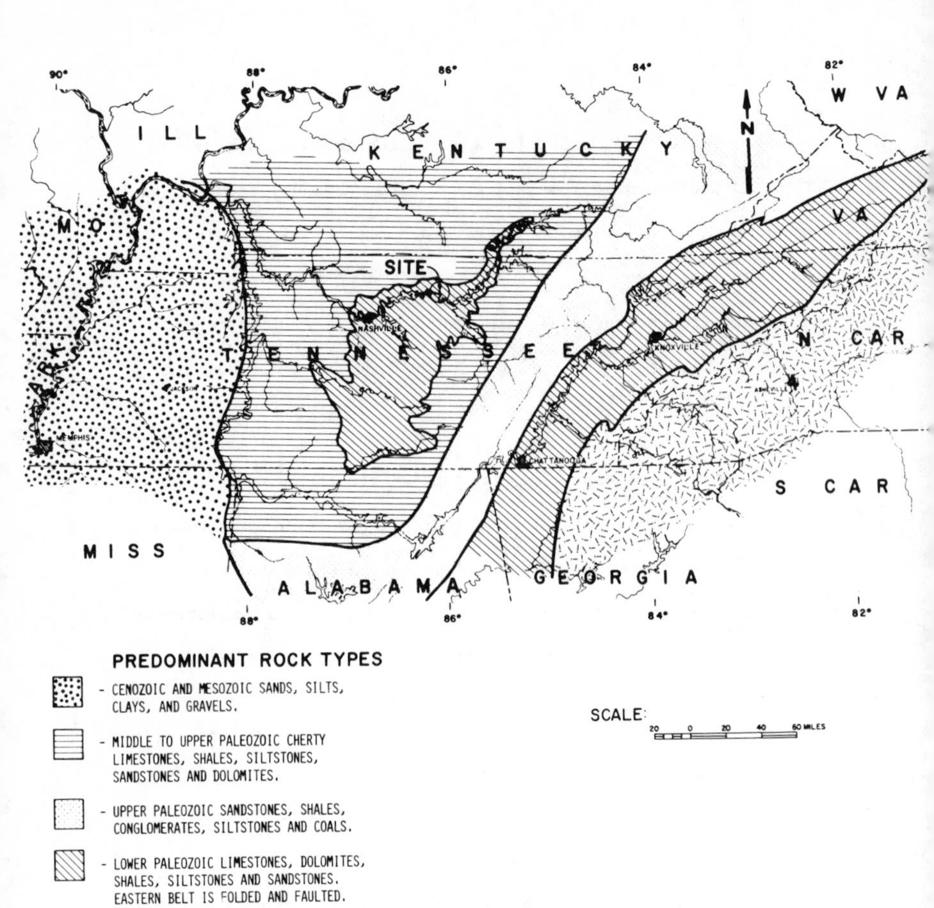

FIG. 3 -- GENERALIZED REGIONAL GEOLOGIC MAP

- Site Physiography

The topography of the Hartsville site is controlled mainly by differential weathering characteristics of the almost flat-lying strata. Knobs are held up by more resistant beds. Lowlands between knobs appear to have been formed by weathering along near-vertical joints that have breached the higher, more resistant beds. Dissolution and erosion concentrated along these jointed areas and eventually created the areas of lower relief.

- Site Structure

The attitude of the strata beneath the Hartsville site is remarkably flat lying and uniform. The site is located along the axis of an extremely shallow syncline which dips southward at about 1 degree. Faults or tight folds were not encountered during exploration and have not been observed during construction.

- Site Lithology

Of the numerous sedimentary formations of Paleozoic Age (230 m.y.± to 600 m.y.±) exposed at the surface in the Hartsville site area and encountered by drilling, only two of Ordovician Age (450 m.y.±) - the Hermitage Formation and Carters Limestone - are involved in the foundations of the major structures (figure 4).

The Hermitage Formation is a medium- to dark-gray, thin-bedded limestone that is laminated with shale stringers. The relatively low carbonate content of the rock accounts

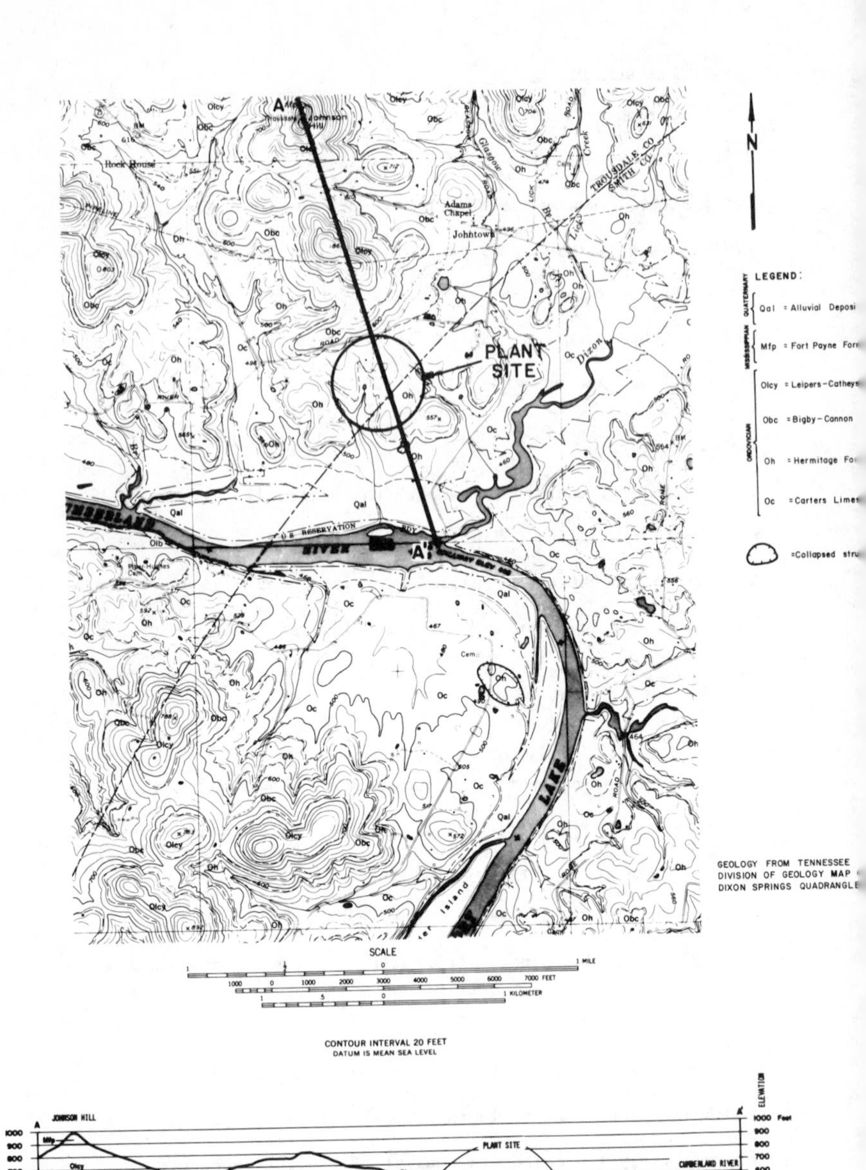

FIG. 4 -- GEOLOGIC MAP OF PLANT AREA

for the gradational weathering without the formation of dissolution voids or cavities. The formation weathers from sound rock to residual soil by way of a transitional zone of "saprolitic" material which is silty, has had the carbonate leached out, has not lost noticeable volume, and may exhibit relict bedding. Unlike purer limestones, the Hermitage does not weather to produce a distinct contact between sound rock and residual soil.

The maximum thickness of Hermitage penetrated during site exploration drilling was 69 feet. The average thickness in the plant A area was 34 feet and in the plant B area was 19 feet. Testing showed the Hermitage to have an average unconfined uniaxial compression strength of 11,200 lb/in^2 and an average unit weight of 170 lb/ft^3.

Stratigraphically beneath the Hermitage Formation is the Carters Limestone, a light-gray to tan, thin- to medium-bedded limestone that contains scattered shale laminae. It has a relatively high carbonate content and is therefore susceptible to the development of cavities. Joints and partings become enlarged by dissolution, and a distinct contact between sound rock and residual soil is developed. An 0.8-foot-thick metabentonite (indurated volcanic ash) occurs within the Carters about 25 feet below its contact with the overlying Hermitage. This layer is continuous throughout the entire site, except where it has been removed by erosion.

In contrast to geologically younger western bentonites, the metabentonite at Hartsville has lost its swelling properties.

The total stratigraphic thickness of the Carters Limestone at the site is 135 feet, but only about the upper half was exposed in plant excavations. Testing showed the Carters to have an average unconfined uniaxial compressive strength of 11,300 lb/in^2 and an average unit weight of 167 lb/ft^3.

Exploration drilling at the site revealed a thickness of 0 to 91 feet of overburden consisting of residual silts and clays. The average thickness was 12 feet beneath plant A and 11 feet beneath plant B. Generally the top of rock surface closely, though more subtly, followed the overlying ground surface.

- Foundation Exploration

Exploration drilling at the Hartsville site was started in September 1971 and continued intermittently through March 1974. During this period 236 Nx wireline core holes totaling 41,723.6 linear feet were drilled and logged; and 213 percussion holes totaling 24,303.0 linear feet were drilled and logged by a variety of downhole geophysical probes including sonic and caliper. Drilling began initally on 400-foot spacing. As engineering studies progressed and subsurface information became available, the spacing was first reduced to 200 feet and finally to holes 50 feet on centers under the major structures (figures 5 and 6).

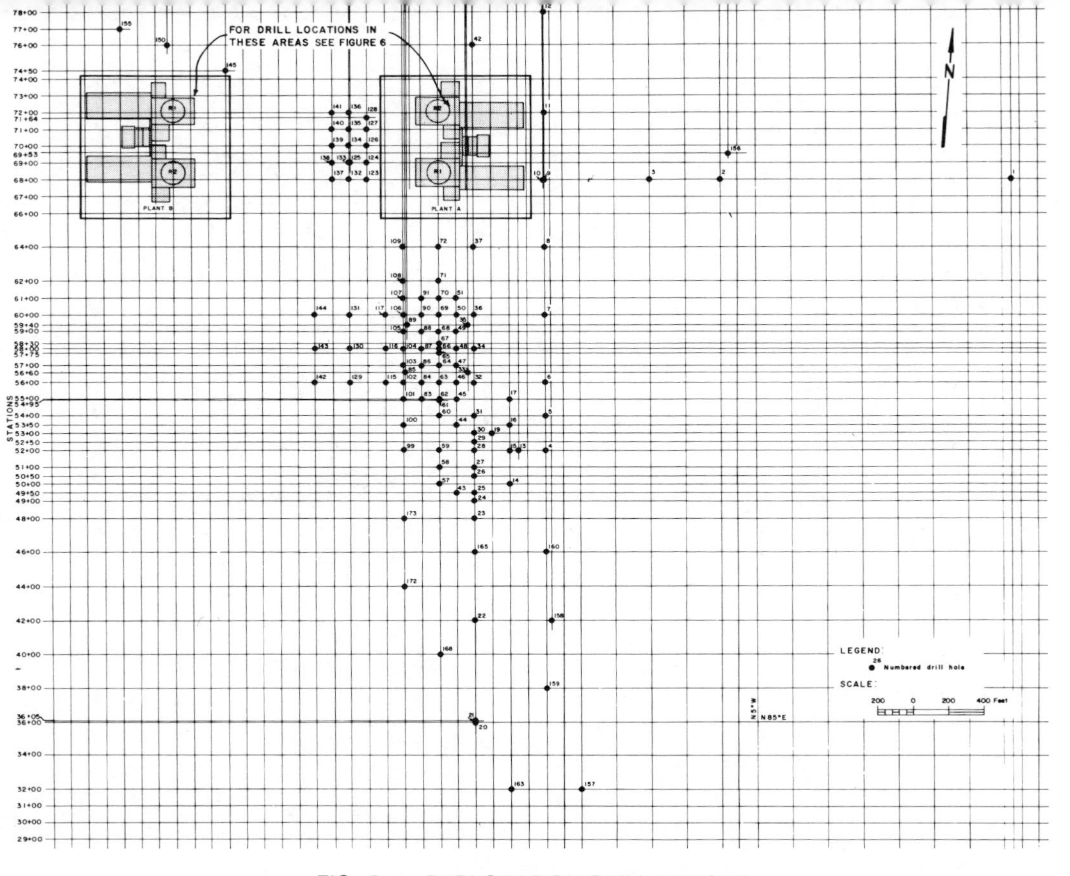

FIG. 5 -- EXPLORATION DRILL LAYOUT

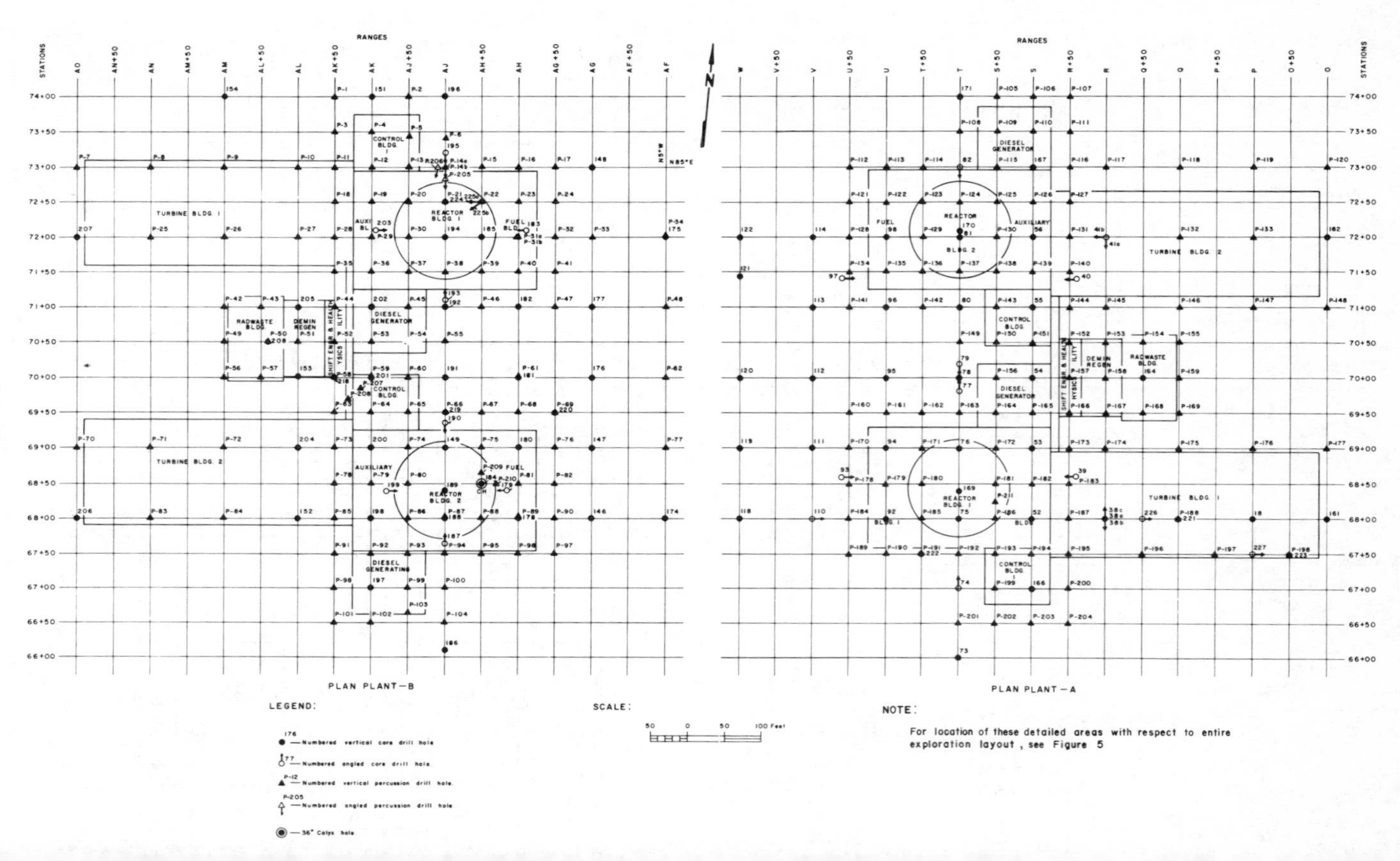
RANGES
STATIONS
TURBINE BLDG. 1
TURBINE BLDG. 2
REACTOR BLDG. 1
REACTOR BLDG. 2
CONTROL BLDG.
DIESEL GENERATOR
RADWASTE BLDG.
DEMIN REGEN
AUXILIARY
FUEL
PLAN PLANT – B
PLAN PLANT – A
LEGEND:
— Numbered vertical core drill hole
— Numbered angled core drill hole.
— Numbered vertical percussion drill hole.
— Numbered angled percussion drill hole.
— 36" Calyx hole
SCALE:
50 0 50 100 Feet
NOTE:
For location of these detailed areas with respect to entire exploration layout, see Figure 5

The depth of drilling ranged from 44 to 1700 feet. Core holes were generally drilled to 100 feet below the metabentonite, and percussion holes were drilled 100 feet below the top of rock.

A 36-inch-diameter calyx core hole was drilled beneath the location of reactor building B-2, at the same location where an exploration hole had encountered a cavity. The core permitted visual inspection of the geologic condition that created the cavity and also facilitated collection of a large, fresh, undisturbed sample of the metabentonite for laboratory testing.

In-situ dynamic testing was performed in the rock beneath the plants. Uphole sonic and caliper logs were run on two selected holes beneath each reactor location, and then crosshole sonic tests were run to evaluate the rock between the two holes. From these tests, compressional velocities, shear velocities, densities, and dynamic moduli were determined for various elevations of rock strata.

- Soil Investigation

Initial soil investigations were conducted in an area between and to the south of the two main plants during late April 1974. The purpose of this investigation was to sample soils that would comprise the spray pond foundations and to study the typical soils of the Hartsville site.

A total of 645 linear feet of drilling and sampling was done in 52 split-spoon borings and 13 undisturbed sample borings. Borings were placed in an irregular grid at 200 to 300 feet on centers. In addition, piezometers were installed in 24 borings to allow long-term water level observations.

This exploration showed that the overburden is composed of a thin blanket of residual clays of Ordovician Age which developed from the underlying limestone. Soil types are relatively uniform with lean clays (CL) and silt (ML) predominating. Some fat clays (CH) and pockets of cherty gravel were also observed. Overburden depths varied from 5 to 17 feet and averaged 10.1 feet. No uniform pattern of soil layers could be established.

Laboratory classifications revealed the absence of sands or sensitive clays. Engineering-oriented testing showed generally stable foundation conditions characterized by medium dry densities and void ratios, medium to high shear strengths, and low compressibility.

- Earthfill

Soil to be used as safety-related earthfill (spray pond fill and backfill around Category I structures) was initially to be obtained from grade or structural excavations. Field explorations in the largest designated cut area, the plant A and plant B main plant areas and cooling tower areas, were completed in the fall of 1974.

Borrow investigations in these areas consisted of a total of 63 auger borings on 200-foot centers. Borings extended to bedrock in areas of structural excavations and to the proposed final grade elevation in the cooling tower areas. These investigations were not combined with geologic investigations.

Soil conditions were similar to those found in the in-situ soil studies. It was estimated that 900,000 yd^3 of soils acceptable for use as safety-related earthfill were available from these excavations.

Since 1974, three additional onsite earthfill borrow investigations have been conducted, and an offsite borrow investigation is currently in progress. All the additional investigations were dictated by shortages of one or more of the various earthfill types (earthfill is typed by the compaction level required to achieve a specified minimum shear strength). Although the actual quantities of earthfill required for construction have steadily risen, each earthfill shortage was precipitated primarily by overestimation of recoverable soils of earthfill quality and in some cases early construction use of approved soils for other purposes.

TVA's approach to field sampling has been based on the use of auger borings and split-spoon borings in ratios of 3:1 to 4:1 in 200 feet to 400 on center grids. A total of 285 borings onsite and 60 borings offsite provided the total TVA field borrow investigation.

Laboratory tests for classifying and indexing borrow soils were based on Atterberg limits, grain size, and standard compaction tests. Design properties and compaction control were based on specific gravity, moisture-density relationships, moisture-penetration, triaxial shear tests, resonant column tests, consolidation, permeability, and clay dispersion tests.

- Bedrock Foundation Conditions

Except for that generally encountered at the top of rock, drilling indicated that weathering is controlled by three dominant interrelated factors: topography, the Hermitage-Carters contact, and vertical joints.

Because the strata at the Hartsville site are nearly horizontal and the top-of-rock profiles closely follow those of the surface, areas of low topographic expression are accompanied by a subsequent thinning of the Hermitage Formation (see figure 7). In such areas the relatively insoluble Hermitage is not an effective protective cover, and the underlying Carters is more susceptible to weathering by solution. Vertical weathered joints were encountered in the Carters Limestone beneath areas of low topographic expression. In areas where the Hermitage-Carters contact crops out, percolating waters moving laterally along the Hermitage-Carters contact result in weathering upward into the basal Hermitage and also downward into the upper Carters. This weathering characteristic diminished laterally inward toward topographic highs, where almost no weathering was found at the contact.

The weathering conditions identified during exploration can generally be represented by five zones that are diagrammatically shown on figure 7.

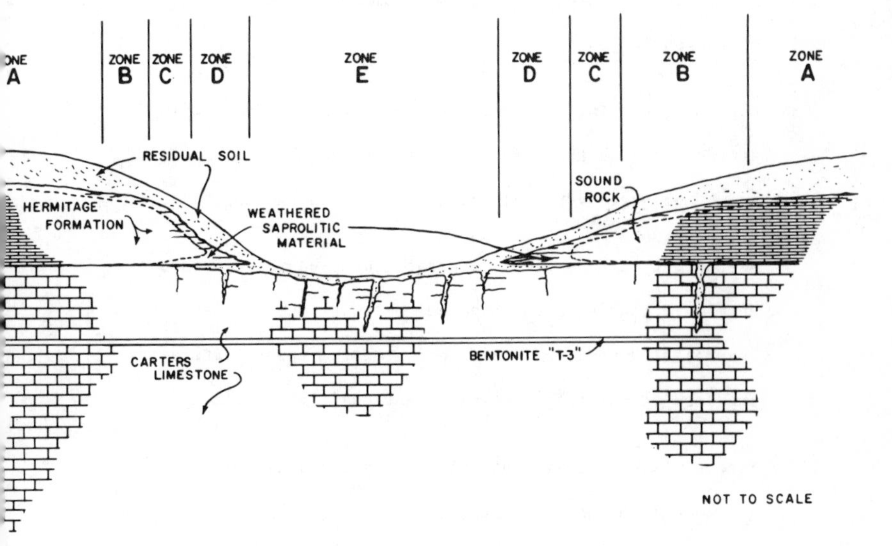

7 -- DIAGRAMMATIC GEOLOGIC SECTION SHOWING WEATHERING CHARACTERISTICS

Zone A weathering is limited to the "saprolitic" residuum encountered at the top of a thick section of Hermitage bedrock. Sound, unweathered rock continues for the remainder of the boring.

Zone B weathering is characterized by residual weathering at the top of a thick sequence of Hermitage, as in Zone A, and by enlarged vertical joint cavities in the upper Carters.

Weathering in Zone C is characterized by basically sound Hermitage bounded both above and below by the weathered "saprolitic" residuum. Weathering of the upper Carters in the form of enlarged vertical joint cavities is likely.

Zone D weathering is characterized by residual soil both above and beneath a zone of weathered, "saprolitic" Hermitage. Considerable weathering of the upper Carters is likely.

Zone E is characterized by areas where the Hermitage has been removed by weathering. Having lost all protection formerly provided by the Hermitage, the Carters is weathered considerably. Enlarged vertical joints are extensively developed in the upper Carters.

Sufficient exploration was performed to determine that there was not a horizontal layer of badly decomposed rock or sheet cavities beneath Zones A and B. However, enlarged joint cavities in the upper Carters, such as those encountered during calyx drilling beneath reactor building B-2 and seen in outcrop around the site, were suspected to exist in Zone B.

Weathering below the metabentonite was found to be extremely limited throughout the area investigated, because the metabentonite tends to act as an impermeable membrane that protects the underlying strata and reduces the percolation of dissolving ground water. Only in areas to the south of the site, where the metabentonite crops out, was significant weathering of the lower Carters seen.

2. PRECONSTRUCTION

- Preliminary Safety Analysis Report (PSAR)

TVA's preliminary safety analysis of the Hartsville Nuclear Plant foundation features based upon the exploratory program and knowledge gained from construction of other large facilities in the area concluded that the geology and foundation conditions of the site were adequate for the safe construction and operation of the facility. Features considered in this analysis were the support and settlement of all safety-related structures (rock supported, compacted earth supported, and compacted granular fill rockfill supported); the ESW spray ponds (compacted earthfill); and the compacted rockfill embankment which stabilizes ESW spray pond areas.

- Rock-Supported Structures

Site studies indicated that the five identified zones of bedrock weathering were adequately defined by the exploration and that the treatment of bedrock problems could be handled on an individual basis for each structural excavation. The excavation process was intended to be exploratory in nature. It was recognized that solution features would be encountered during excavation and it was planned to evaluate these in context and to use these evaluations to develop further exploration, excavation, or treatment as the circumstances dictated. It should be noted that the foundation slabs of the rock-supported Category I reactor island structures vary from 6 to 16 feet in depth, and no problem was anticipated in spanning cavities of the size previously discovered (maximum width 2 feet).

Beneath rock-supported structures, voids in the exposed rock will be cleaned of residual material, treated with dental concrete, and inspected prior to placing of fill concrete or backfill.

- Soil-Supported Structures

  One set of diesel generator fuel storage tanks will be founded on Category I compacted earthfill. Prior to placement of the earthfill, the in-situ overburden will be removed to the top of sound bedrock, and the bedrock will be treated according to its observed condition. Earthfill supporting these structures would be compacted to 95 percent of the maximum standard (proctor) dry density at optimum moisture content ±2 percent.

- Granular Fill-Supported Structures

  The diesel generator buildings, control buildings, and some of the diesel generator fuel storage tanks will be founded on compacted well-graded crushed stone fill extending to bedrock. The maximum particle size for this fill would be 1-1/4 inches. Compaction will be such that the average relative density of the fill is 85 percent. Postconstruction settlements of this fill are neglible.

- Spray Ponds

The spray ponds are located in an area which was a natural drainage feature. The two northern ponds, on the upstream end of the area, and a portion of the southeastern pond will be excavated into or founded on bedrock. The southwestern pond will be constructed above in-situ soils. A bedrock supported rockfill embankment with a 2-stage filter drain will provide downstream support for the southern spray ponds.

Each pond including a 5-foot-thick liner will be constructed of CL and CH clays compacted to 100 percent of their standard maximum dry density at optimum moisture content ±2 percent.

The two northern spray ponds which are cut into the Hermitage Formation will have a substantial thickness of Hermitage beneath the clay blanket. Voids and joints at the excavated surface of the Hermitage will be treated with dental concrete, so the clay blanket is not underlain by localized open zones.

The southeastern spray pond would be supported on approximately equal areas of in-situ soils (underlain by the Hermitage Formation) and the Hermitage Formation. The thickness of Hermitage beneath the pond is adequate to support the spray pond loads.

The southwestern spray pond will be supported entirely on in-situ overburden underlain by weathering Zones C, D, and E. To investigate this area TVA planned to remove the in-situ overburden where the rockfill embankment would be located and study the exposed rock. If any unusual conditions are encountered, further excavation or exploratory drilling will be performed to define the extent of the conditions.

Licensing Concerns

The Nuclear Regulatory Commission's (NRC) review of the Hartsville PSAR stated five safety concerns involving foundation engineering. These concerns remained unresolved until just prior to the scheduled issuance of the NRC's Safety Evaluation Report in the summer of 1976.

- General Solutioning. The principal NRC safety issue was the occurrence of solution cavities in the Carters Limestone. It was the NRC's contention that the random and unpredictable nature of solution cavities made it impossible to be sure that all cavities in the Carters Limestone were confined to narrow zones adjacent to vertical joints. NRC contended that extensive solutioning of the Carters Limestone was not statistically disproved by the site investigation.

- Underseepage. NRC viewed the degree of foundation treatment proposed by TVA beneath the ESW spray ponds as insufficient. The concern was that uncontrolled underseepage from a spray pond into the solutioned limestone could cause internal erosion of the spray ponds.

- Clay Liners. Due to fluctuations of spray pond fill depth related to the topology of the top of rock, the clay liners of the two southern spray ponds were postulated to be susceptible to cracks caused by differential settlements.

- Rockfill. The rockfill embankment which stabilizes the southern spray ponds and the supporting earthfill was determined to be stable during the design seismic event only when it was unsaturated. NRC requested specific material quality assurances that the first several lifts of rockfill would be free draining.

- General Fills. NRC requested specific quality control of construction procedures to assure that both granular fill and earthfill would be uniformly compacted to the required densities.

Licensing Commitments

Each of the five NRC licensing concerns listed in the previous section generated one or more TVA commitments for additional investigations or more stringent foundation specifications. Listed below are the subheadings denoting NRC licensing concerns and the commitments agreed upon to resolve them.

- General Solutioning. A program of drilling and grouting which the NRC proposed was implemented beneath all rock-supported Category I structures to locate and fill foundation voids. This was in lieu of the original proposal. The general procedures used in this program are fully explained in the section on Construction Activities and Findings.

- Underseepage.

  1. A program of angled drilling and grouting around the southern perimeter of the two southern spray ponds was implemented to locate and fill cavities and open joints. This program is more fully explained later in the paper.
  2. TVA will fill the spray ponds and observe them for signs of leakage for at least 1 year prior to fuel loading.
  3. TVA will observe daily the effluent, when it exists, from the toe drain of the rockfill for signs of soil piping into the rockfill.

- Clay Liners.

  1. To ensure ductility in the clay liner of the southwestern spray pond and the southern portion of the southeastern spray pond, TVA has committed to use only CL and CH type clays with liquid limits greater than or equal to 30. This was in lieu of using earthfill with a minimum plasticity index of fifteen for all four spray ponds.
  2. Settlement monuments will be constructed along the southern face of the southern spray ponds. A settlement monitoring program will be implemented upon completion of the spray pond construction to provide early detection of settlement or subsidence.
  3. Several lines of piezometers will be installed along the critical sections (related to slope stability) of the southern spray ponds to monitor pore water pressures. This data will be used to estimate percent consolidation of the fill and preoperational stope stability.

- Backfill. To ensure a free-draining rockfill, TVA committed to limit the material passing the No. 200 sieve by weight to 6 percent.

- General Fills. Test fills for Category I earthfill and granular fills were performed for every type compaction equipment to be used in actual construction. Minimum requirements for loose lift thicknesses and roller passes were established to ensure uniform and adequate compaction.

C. IMPLEMENTATION PLANS

TVA has developed and used construction specifications for grouting, placement of fill materials, and foundation preparation for many years. These construction specifications are prepared by TVA's design organization but must be reviewed and concurred in by TVA's construction organization. They are, thus, an excellent vehicle for communication between the design and construction organizations. During the lengthy licensing process many commitments were made and agreements reached between NRC and TVA in the area of foundation engineering. Other commitments and restrictions were identified due to environmental and internal procedures.

Such restrictions run the gamut from archaelogical clearances to compaction requirements to blasting restrictions. With all these requirements contained in various documents, it became apparent that a special Project Construction Specification (PCS) on foundation engineering issues was required for consolidation and to be certain that nothing "fell in a crack." This PCS as it was originally issued in January 1977 addressed the following areas:

1. Design and construction responsibilities.
2. Assignment of a foundation inspection team and definitions of responsibilities.
3. Borrow materials including stockpiling.
4. Rockfill.
5. Filter materials.
6. Placement of granular, earth, and rockfill.
7. Test fills.
8. Blasting and rock overbreak.
9. Foundation treatment including grouting and "dental" work.
10. Protection requirements for buried piping.
11. Quality control.
12. Location of environmentally sensitive areas, including archaeological areas.
13. Location of approved material sources.
14. Records, including foundation mapping.

The respective project managers for design and construction are responsible for carrying out the PCS. A geologist assigned to the construction site, technically responsible to the design office, monitors and assists in day-to-day operations of grouting and dental treatment as well as mapping the foundation rock. The construction engineering organization at the site carries out the requirements of the PCS. The Foundation Inspection Team (FIT) is headed by the Head Civil Project Engineer in the design office and composed of a geologist, soil and rock mechanics specialists, the lead design engineer for the feature whose foundation, etc., is being inspected, and a civil engineer who has participated in the licensing of the plant.

The FIT, located in Knoxville, makes frequent visits to the site and is required to approve foundations prior to placement of concrete or fill materials and to make judgments on matters such as removing or leaving in place questionable materials, dental treatment, need for additional exploration, and other matters. Although construction engineering is not a member of the FIT, they are an important part of all inspections by the FIT and their input to decisionmaking is considerable. Thus, the four corners of the foundation engineering program are:

- The PCS
- The Construction Engineering Organization
- The Project Geologist
- The FIT

D. CONSTRUCTION ACTIVITIES AND FINDINGS

Earth excavation at Hartsville Nuclear Plant began in May, 1976; rock excavation began the following month. Most of the rock excavations involved the Hermitage Formation; the Carters Limestone was excavated in the deeper cuts in plant A and in plant B, as well as in some of the outlying structures.

The Hermitage Formation presented two problems to the blasting operation: both the thin beds and the tight high angled joints provided natural planes of weakness. In the first case, large masses of rock sometime shifted horizontally as much as a foot. If these occurred under structures, they had to be removed and replaced with fill concrete. In the second case, since joints rarely coincided with design walls, rock was also overexcavated in some walls and backfilled with concrete.

Category I structures founded in rock were grouted, using the following general grouting procedure:

1. Percussion drilled, 2-3/4-inch holes were placed at a 45-degree angle on 25-foot spacings. Since the holes were oriented at 90 degrees to each other, the holes were spaced on 25-foot centers at or just below the Hermitage-Carters contact.
2. Holes were drilled with percussion drills. Inspectors observed the drilling operations to record locations of cavities in each hole.
3. Packers were set just below the top of the hole, and the holes were grouted at 5 lb/in^2.
4. Holes were grouted at an initial water:cement ratio of 2:1 by volume. If a hole accepted grout, the grout mix was thickened accordingly. All holes were grouted to refusal.
5. Holes that accepted more than three bags of grout were "split" by adding two holes on either side at half the original spacing.

- Major Structures

The condenser cooling water (CCW) trenches under the turbine buildings in plant A were excavated early to provide actual geologic sections through the plantsite. These two trenches extended a short distance into the Carters Limestone. What was encountered was basically what was expected from exploration. The density and size of clay-filled joints in the Carters decreased as the thickness of the Hermitage Formation increased. The greatest development of cavities occurred in turbine building A1 (figure 8). The joints were as wide as 6 feet

FIG. 8 -- THE EAST CCW TRENCH UNDER PLANT A, SHOWING SOLUTION FEATURES IN THE CARTERS LIMESTONE AND THE HERMITAGE FORMATION EXPOSED IN THE TRENCH WELLS

in this area but narrowed quickly with depth. These cavities, as were all cavities beneath the turbine buildings and all Category I structures, were cleaned out to a minimum depth equal to twice the width of the joint and backfilled with concrete.

Cavities usually begin at or just below the Hermitage-Carters contact and rarely extend below the metabentonite. Cavities did not extend into the Hermitage Formation. Over wider joints, however, an arch-shaped stoped or collapsed zone developed in the Hermitage Formation. The height of these zones is approximately equal to the width of the cavity at the contact. A stoped zone consists of beds of weathered limestone and saprolite which have partially slumped into the cavity (figure 9).

Toward the north in the CCW trenches, the number and size of clay-filled joints gradually decreased until only solid rock was encountered under turbine building A2. Also, the west trench shows less severe weathering than the east trench in plant A.

In places the CCW trenches in plant B were excavated below the meta-bentonite. Here it was observed that clay-filled joints did not generally penetrate the metabentonite. In fact, very little weathering or joint development was encountered below this layer.

When the excavation of reactor complex A1 was complete, mapping of the final rock surface revealed tight joints in the Hermitage Formation that were oriented in a north-south direction. One joint, however, was oriented toward the northeast and was weathered with

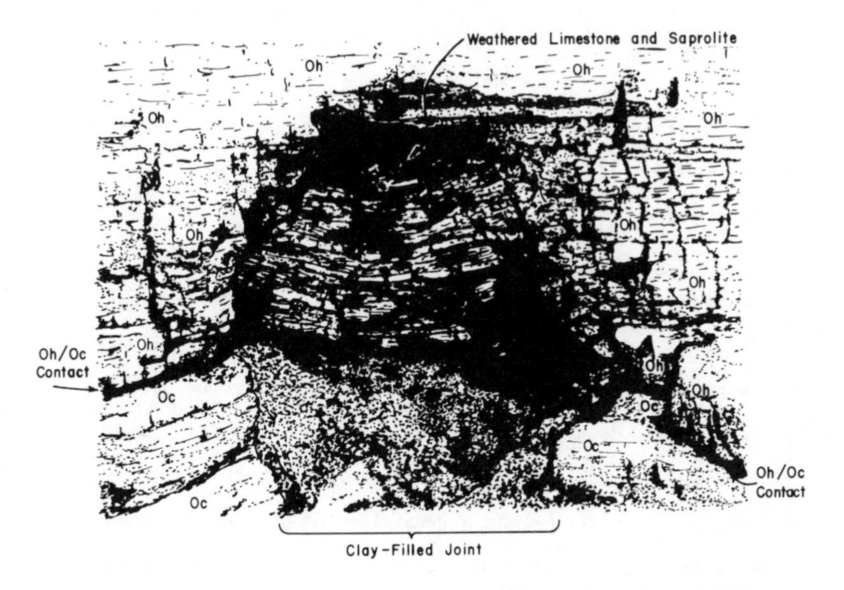

FIG. 9 -- STOPED ZONE IN THE HERMITAGE FORMATION

minor solution development. Exploring this feature by core drilling revealed cavities in the Carters Limestone 12 feet below.

During the drill and grout program required by the NRC, 160 percussion holes were drilled throughout the A1 complex. This program involved split holes in some areas that were less than 1 foot apart. Holes consistently encountered cavities along the northeast-trending weathered joint. To better define the magnitude of this system, a massive drill program was carried out. Two hundred and eight vertical 4-inch percussion holes were drilled on 2-foot centers and arranged in eleven lines crossing the feature. In addition, an exploratory pit, approximately 5 by 25 feet, was excavated beneath the reactor building to 3-1/2 feet below the Hermitage-Carters contact. Both of these exploratory programs revealed that solution was confined mainly to narrow features bounded by sound rock. Since the reactor building base slab is up to 16 feet thick and there is 12 feet of Hermitage between the slab and the cavities in the Carters, the exploratory trench was backfilled with concrete and construction was allowed to continue. Since the program for the other units was similar, they are not discussed further here.

In most areas of the project, except the intake pumping station, solution features rarely extended below the metabentonite. The intake pumping station, however, is the exception. Here, the original top of rock surface was eroded to just above the metabentonite bed. With so little protection, major solution features, including an open cave, extended well below the metabentonite. These features had narrowed to less than

3 feet at the bottom of the excavation, but the features exposed in the walls were backfilled with concrete before construction was continued.

- ESW Spray Ponds

Early in construction, core holes were drilled under each of the ESW pumping stations to determine the condition of the foundation rock and along the southern perimeter of the two southern spray ponds to determine the depth and quality of the Hermitage Formation. Also, earth was excavated from the rockfill embankment to provide an area for rockfill testfills and to provide an early visual examination of the Hermitage-Carters contact and the extent of weathering in the Carters Limestone in this area.

The core drilling under the northern ESW pumping stations A2 and B1 indicated no serious weathering. These structures were excavated according to design without any special treatment other than the 25-foot-on-center grout program. The cores from ESW pumping station A1 showed serious top of rock weathering in the Hermitage Formation. To ensure a sound foundation, this weathered rock was removed and the excavation was extended an additional 9 feet to sound rock. The grout program, done after excavation, indicated no solution development in the Carters Limestone. Fill concrete was then placed and construction started.

Cores from ESW pumping station B2 indicated well developed cavities in the Carters Limestone as well as stoped zones in the Hermitage Formation. The excavation was deepened by 19 feet to get below this weathering. After the excavation was completed, the cavities and stoped zones could be seen in the walls; but the joints had narrowed down to less than a foot wide in the bottom of or on the excavation. The drilling and grouting in this structure indicated no further weathering with depth. After the fill concrete was placed, holes were drilled into each cavity found in the walls and backfilled with grout.

The rock excavations for the ESW ponds A1, A2, and B1 have been completed, but placement of earthfill has not begun. Prior to placing fill, vertical rock faces greater than 1 foot in height will either be cut off on a 45-degree slope or capped by a concrete fillet with a slope of 1:1 to facilitate compaction of earthfill.

Four-inch percussion holes along the southern perimeter of spray ponds A1 and B2 were drilled on 50-foot centers and grouted as required by the NRC. Two rows of holes, oriented toward the north and west, were drilled along spray pond A1 and three rows of holes, oriented toward the northeast and northwest, were drilled along spray pond B2. The holes were drilled perpendicular to the major orientations of clay-filled joints as indicated in the early earth excavation of the rockfill embankment. Each hole was visually logged during drilling to locate cavities. All holes were cased through the overburden, and the packers were set just below the bottom of the casing.

A total of 126 holes were drilled through the metabentonite and pressure grouted in the two spray ponds. Cavities were encountered randomly throughout the area. Also, grout takes were usually low to moderate, but larger takes were scattered throughout the area.

TVA was not committed to split grout takes but was committed to trace out any cavity encountered during the drilling; but this proved to be difficult, if not impossible, due to the wide spacing of the grout holes and the random nature of joints in the Carters Limestone. A TVA concern that these cavities could cause localized pond failure over the life of the plant by (a) gradual removal of the clay that filled the solution features by piping and erosion from the higher future ground water conditions caused by the ponds or (b) differential settlements in the earthfill created by differences in the support capabilities of the soft clay in the cavities as opposed to the adjacent rock caused a reevaluation of the treatment program for the B2 spray pond foundation rock.

An exploratory trench was excavated through the overburden in the B2 spray pond along the minor axis. This trench revealed a very uneven surface in the Carters Limestone cut by a complex system of clay-filled joints. With the observations from the trench and with TVA's commitments to the NRC in mind, a decision was made to strip the overburden from the areas that contain less than 5 feet of Hermitage Formation (mainly along the south side of spray pond B2). This program would allow tracing out each clay-filled joint by mapping and treating each cavity with dental concrete prior to placement of earthfill. Areas that contained more than 5 feet of Hermitage (e.g., spray pond A1) would not require special treatment because TVA felt that such thicknesses of Hermitage would adequately span any cavity encountered in the Carters Limestone in the spray pond drilling.

From mapping the spray ponds and rockfill embankment, the density of clay-filled joints in the Carters Limestone was again shown to be a result of the thickness of the protective Hermitage Formation and the relative length of time the Carters Limestone has been exposed. The number of joints is greatest in the south and decreases northward on either side of spray pond B2 or along the two natural topographic lows in this area.

- Cooling Towers

A preliminary study was made of the cooling towers in 1976. Eleven 4-inch percussion holes were drilled in the plant A towers and seventeen were drilled in the plant B towers. Each hole was logged geophysically using uphole sonics, gamma-gamma density, mechanical caliper, and natural gamma ray. In 1978, a detailed exploration program was carried out. In this program, a 4-inch hole was drilled in the center of each support or caisson location (approximately 210 holes per tower). Each hole was logged geophysically as before. In some areas, large or numerous clay-filled cavities prevented penetration of the original percussion holes. Therefore, additional offset percussion and core holes were drilled.

The geologic data from the exploratory holes basically agreed with the actual geology seen during the caisson drilling. There were some discrepancies, however: (1) Elevations of key horizons in some holes were off as much as 10 feet for the top of rock and 2 feet for the metabentonite and (2) boundaries between weathering zones were difficult to distinguish. In the latter case, a zone of weathered rock in the exploration holes may actually include badly decomposed rock (BDR) in the caisson; and BDR may include clay-filled cavities.

In the cooling towers, the Hermitage Formation is relatively thin; along the west side of cooling tower B1, there is no Hermitage cover. Therefore, there is extensive weathering in the form of clay-filled joint-controlled cavities in the Carters Limestone and stoped zones in the Hermitage. But as in other areas of the project, there are no broad areas of horizontal weathering which may cause general subsidence. All caissons were drilled below any solution features to sound rock as indicated by geophysical logs and by actual inspection of the bearing surface.

- Granular Testfills

Granular fill testfills were begun October 31, 1978, for the purpose of determining the most effective construction procedures for compacting 1032 gradation crushed limestone.

The testfills were constructed on heavily compacted in-situ soils. Each testfill consisted of two base lifts and two test lifts.

One large vibratory roller (Raygo 400) and two small vibratory rollers (Bomag BW35 and Bomag BW60S) were used in the test.

For all compactors, except the Bomag BW60S, four passes provided the required compaction (an average relative density of 85 percent with a minimum of 80 percent). Fewer roller passes result in inconsistent results. Five roller passes with the Bomag BW60S provided the required compaction. The optimum loose lift thickness was determined to be 10 inches for the Raygo and 6 inches for both Bomags.

- Earthfill Testfills

Earthfill testfills were conducted in the late spring of 1978 to define construction procedures which would guarantee uniform compaction to at least the minimum earthfill requirements.

Soil used in the testfills included the full range of soils approved for use as Category I earthfill. These consisted predominately of lean clays with some fat clays.

The testfills were constructed on well-compacted, inplace soils. Each testfill consisted of two base lifts and two test lifts. Two large tamping rollers (Rex Pactor 3-50 and Ferguson 120) were used for the tests.

A summary of the testfill results showed that both compactors achieved the required compaction (100 percent standard maximum dry density) on all soil types tested. Since materials from the site showed pronounced variability of soil classes, a conservative minimum number of roller passes was recommended. Ten passes with the Rex Pactor 3-50 on an 8- or 9-inch loose lift would consistently provide the required percent compaction for all soil type. The Ferguson 120 required twelve roller passes on a 7-inch loose lift to achieve the desired compaction for all soil classes.

C. PRESENT STATUS OF CONSTRUCTION

Present Status of Excavations

Earth and rock excavations have been completed in all areas of the Hartsville project. Construction in plant A, in the four ESW pumping stations, and in the intake pumping station is well under way. (Construction in plant B was halted in summer 1979 because TVA felt that the demand for electricity will increase at a slower rate than had originally been anticipated.)

The 25-foot-on-center grout program remains to be done in the control building (except unit B1) and in the diesel generating buildings for each unit.

The cooling towers at Hartsville are in the early stages of construction. Approximately 80 percent of the caissons in tower A1 have been completed; approximately two-thirds of the prefab veil supports have been erected. Drilling in cooling towers A2 and B1 has begun but no caissons have been completed. Drilling in tower B2 has not been started.

Present Status of Spray Ponds

Earth and rock excavations have been completed in spray ponds A1, A2, and B1 and in the rockfill embankment; earth excavation has been completed in spray pond B2. Foundation treatment is complete for all spray ponds and the rockfill embankment.

Placement of earthfill has begun in the area between the four ponds and along the southern edge of spray pond B2. No earthfill has been placed in spray ponds A1, A2, or B1. No rockfill has been placed in the rock-fill embankment.

F. CONCLUSIONS AND OBSERVATIONS

General

Construction excavations have confirmed the findings of the exploration program in determining that the Hartsville site is a satisfactory location for the two 2-unit plants and their appurtenant structures. Although some "surprises" were seen, conditions anticipated from the exploration program were generally substantiated; however, some of the NRC concerns, with resulting commitments and expenses, were found to be warranted; others, such as the foundation grouting program, were not.

Specific

- Exploration

1. In retrospect, once the site geologic structure beneath the plant locations had been determined, more of the subsequent exploration borings should have been drilled inclined from the vertical. This would have increased the chances of intersecting vertical joint cavities.

2. The identification of the metabentonite horizon as an aquiclude was substantiated.

3. Geophysical logging of percussion drilled holes was found to be a fast inexpensive, and effective method of foundation exploration. This process has become standard in TVA, having replaced most core drilling

4. In general, field exploration for borrow soils provided overly optimistic quantity estimates. In retrospect, perhaps backhoe trenches would have been a more prudent sampling technique than augering.

- Licensing

1. Postgrouting excavations have revealed that little was gained toward increasing foundation integrity by implementing the NRC-imposed grouting program. Cavities that were the object of the grouting program were found to be of insignificant widths considering the ability of the overlying structures to span them. The mud filling in the cavities prevented effective grouting. Hardened grout was seen lying atop poorly consolidated cavity and filling. The additional exploration information obtained by monitoring the grout drill holes was of more benefit than the actual low-pressure grouting.

2. Although necessitating some changes in construction planning and scheduling, the decision to excavate the CCW trenches and the spray pond rockfill embankment foundation early in construction proved to be justified. This permitted firsthand, visual assessment of the sizes of the cavities and the conditions that created them.

Acknowledgments

The authors wish to acknowledge the efforts of TVA's design and construction organization and particularly Robert O. Barnett and Dr. James H. Coulson who participated in the project from its inception.

Ground Failure Displacement and Earthquake Damage to Buildings

T. Leslie Youd[1], M. ASCE

INTRODUCTION

Ground failure is a major cause of damage during most large earthquakes. For example, about 60 percent of the damage generated by the 1964 Alaska earthquake ($300 million, 1964 value) was directly caused by ground failure (Youd, 1978). During that event, massive landslide movements in the city of Anchorage destroyed many buildings and other works. Smaller ground displacements inflicted costly damage to structures that they intersected. Similar damage has been generated by other large earthquakes around the world. This report examines earthquake damage to buildings caused by differential ground displacement. Particular attention is given to the threshold of displacement initiating damage.

The author has conducted several investigations of damage caused by large earthquakes. In addition, he has reviewed written reports about many other earthquakes. Specific case histories of damage to buildings plus general observations on the behavior of structures are used here to examine the relation between damage to buildings and amount of ground displacement. Because earthquake damage to structures caused by ground dispacement is independent of the source of displacement, damage from both secondary ground failure and tectonic surface faulting are considered.

Many buildings have been intersected by ground failure during recent earthquakes. Nevertheless, there are some foundation types, primarily mat and deep foundations, that

---

[1]Research Civil Engineer, U.S. Geological Survey, Menlo Park, California, 94025

have not yet been thoroughly tested by ground displacement. Also, to the author's knowledge, none of the buildings described below was specifically designed to resist ground failure. Hence, the results and conclusions given are provisional and should be revised as new experience is gained. The results provide information for estimating possible damage to buildings in areas susceptible to ground rupture and may also provide guidance for design to resist ground displacement.

BUILDINGS ON FOOTINGS

Several factors influence amount of damage caused by differential ground displacement to buildings on spread or continuous footings. These factors include amount of displacement, mode of displacement (extension, compression, shear, vertical), type and strength of foundation, and type and strength of building. Primary attention is given to amount and mode of displacement and to foundation type.

Extensional ground displacement.--Extensional displacements commonly occur at heads of landslides, incipient landslides, and lateral spreads, as tension cracks behind steep slopes, and in zones of extension associated with tectonic faulting. Extensional cracks are also common in areas of flat ground beneath which liquefaction has occurred. In many instances, extensional cracks are known to open and close repeatedly in response to earthquake shaking. Closures often involve dynamic impact between ground units and structural elements on either side of the crack. These dynamic impacts may lead to progressive damage.

Extensional displacements greater than about 2 ft (0.6 m) have been generally destructive to buildings founded on footings. For example, the steel-frame, corrugated-metal industrial building shown in Fig. 1 was pulled apart by 3 ft (0.9 m) of extensional displacement during the 1964 Niigata, Japan earthquake (M = 7.5) (Kawasumi, 1968, p. 187). The extension was caused by horizontal ground displacement associated with liquefaction of a subsurface soil layer.

Fig. 1.--Steel-frame industrial building pulled apart by about 3 ft (0.9 m) of ground extension. Extension was caused by liquefaction during 1964 Niigata earthquake (after Kawasumi, 1968, p. 187).

The building was apparently founded on perimeter or spread footings, and the floor was a concrete slab. Although the building did not collapse, the steel frame in the right end of the building was twisted and distorted beyond repair, and the foundation and floor slab were ruptured and split apart. The utility of the building was lost. During the Niigata earthquake, many low-rise industrial, commercial, and residential buildings were damaged beyond repair by ground displacements greater than 2 ft (0.6 m) (Kawasumi, 1968, p.183-255).

Several buildings at the San Fernando Valley Juvenile Hall facility were astride ground ruptures generated by a lateral spread that developed during the 1971 San Fernando, California, earthquake (M = 6.4). Fig. 2 shows the interior of a single-story, reinforced concrete-block building that was founded on perimeter and spread footings with a concrete floor slab. The building was split apart by about 2 ft (0.6 m) of extensional displacement that ruptured the perimeter footing and the floor slab and distorted the walls and roof. The structure did not collapse but was deemed

Fig. 2.--Reinforced-block building pulled apart by about 2 ft (0.6 m) ground extension. Extension was generated by lateral spreading during 1971 San Fernando earthquake.

unrepairable, partly because additional ground displacements were considered likely during future large earthquakes.

Low-rise buildings commonly have withstood 4 in to 2 ft (0.1 m to 0.6 m) of extensional displacement with repairable damage. For example, the steel-frame, corrugated-metal industrial building shown in Fig. 3 (Kawasumi, 1968, p. 249) was intersected by the fissure in the foreground and spread apart by about 1.5 ft (0.45 m). Liquefaction of subsurface soils caused the fissure. Extension beneath the building distorted the steel frame in the right end of the building. The exterior view indicates that the damage to the floor, steel frame, and siding in the right end of the structure might be repairable at less than replacement cost.

Hundreds of tension cracks developed in areas behind

Fig. 3.--Extensional fissure in foreground split foundation and pulled building apart by about 1 ft (0.3 m). Extension was caused by lateral spreading toward a nearby channel (after Kawasumi, 1968, p. 249).

Fig. 4.--Ground fissure, 2 in (5 cm) wide, passed beneath block building causing no significant damage. Fissure was cased by extension of ground toward nearby New River during 1979 Imperial Valley earthquake (after Youd and Wieczorek, in press).

the large landslides that developed in Anchorage, Alaska, during the 1964 earthquake. Hansen (1966, p. A60) reported that tension cracks opened behind the head of the Turnagain Heights landslide causing severe structural damage to houses in the Turnagain Heights subdivision, disrupting all underground utilities, and seriously damaging streets and curbs. These fissures were as wide as about 2 ft (0.6 m). Houses intersected by these cracks were generally repairable, but at appreciable expense.

Many small buildings and large flexible structures have withstood as much as 4 in (0.1 m) extensional displacement with minor or no damage. For example a 2-in (0.05-m) wide fissure passed beneath the small building shown in Fig. 4 causing no damage except hairline cracks. The masonry block building is built on a perimeter foundation and has a concrete slab floor. The fissure was caused by incipient slumping or lateral spreading of the ground toward the nearby New River during the 1979 Imperial Valley, California, earthquake (M = 6.6) (Youd and Wieczorek, in press).

Long, brittle buildings on poorly reinforced foundations have sustained significant damage from extensional displacements of less than 4 in (0.1 m). A case in point is a building at the Escuela Normal in Caucete, Argentina, that was damaged during the 1977 western Argentina earthquake (M = 7.5) (Youd and Keefer, in press). During the earthquake, extension occurred beneath a 220-ft (66.6-m) long wing as a consequence of liquefaction. The wing is a reinforced concrete frame with block infilled walls below window level, capped by a heavy, reinforced-concrete roof (Fig. 5). The foundations are lightly reinforced continuous-wall footings. Extensional displacement cracked and spread apart the foundations and block walls above. Cumulative opening across the cracks was 2.8 in (7.0 cm) in 230 ft (70 m) of building length. Most of the displacement, however, was in the right end of the building as shown in Fig. 5C. The rigid roof section remained intact and attached to the left end of the building. Displacement between the foundation

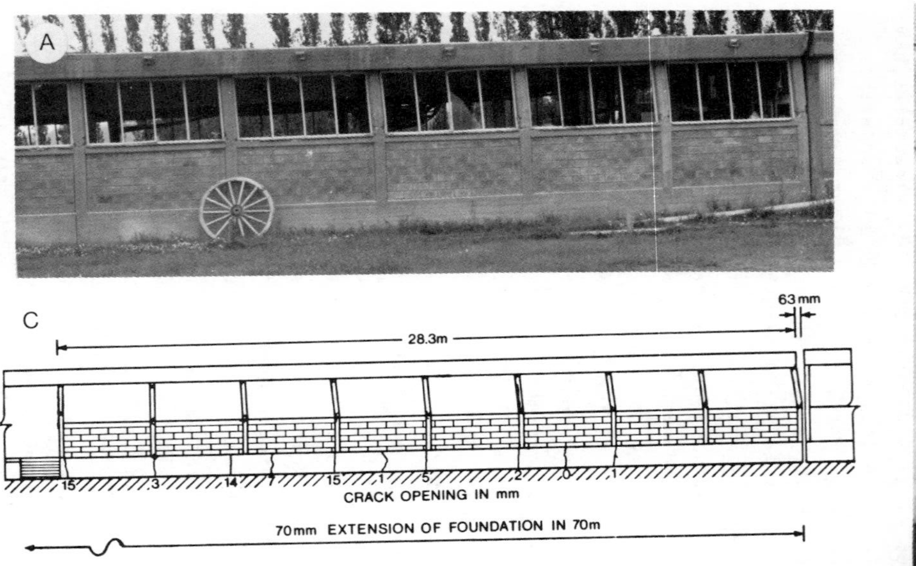

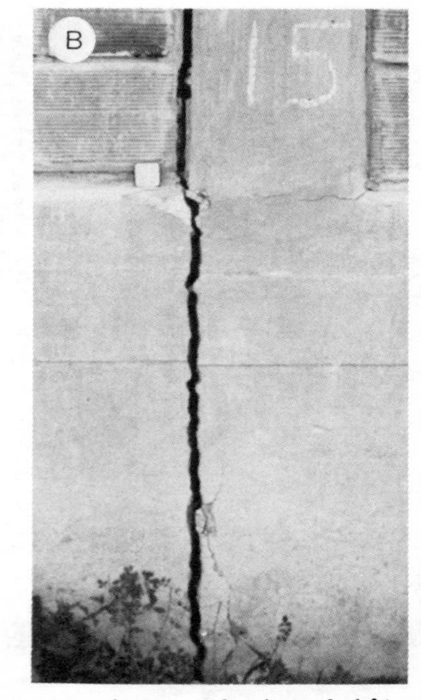

Fig. 5.--Building pulled apart by 2.8 in (7.0 cm) differential extension producing drift in columns through window level. Building was damaged during 1977 western Argentina earthquake. A. Front view of building showing tilted columns. B. Close up of a foundation crack caused by ground extension. C. Sketch showing displacements and damage to building (after Youd and Keefer, in press).

and the roof in the right end of the building was accomodated by rupture and tilting of the columns through the window level. The drift in the column on the right end of the building was 2.5 in (6.3 cm), roughly equivalent to the extension of the foundation. The amount of column drift decreases with each column leftward until the eighth, column which is undamaged. Rather costly repairs would be required to rehabilitate this building.

Compressional ground displacements.--Compressional displacements commonly occur at the toes of landslides and lateral spreads and along thrust faults. Buildings on continuous footings are generally more resistant to damage from compressional displacement than from extensional displacement. Compressional displacements, however, often have associated vertical components that intensify the damage.

Compressional behavior of a single-story building on perimeter and wall footings and a concrete floor slab is illustrated in Fig. 6. This building was subjected to about 40 in (1 m) of compressional thrust faulting during the 1971 San Fernando, California, earthquake (M = 6.5). The compressional deformation was absorbed by slippage between the foundation and floor slabs and the ground. This slippage dozed up the sod in front of the building as shown in Fig. 6A and diagrammed in Fig. 6B. The floor slabs were fractured at a few locations primarily to accomodate differential vertical displacements. No attempt was made to repair the building (Youd and others, 1978).

Several industrial and commercial buildings were intersected by thrust faulting during the 1971 San Fernando, California, earthquake. In most of these instances, compressional displacements of greater than 2 ft (0.6 m) caused displacement or buckling of footings and floor slabs and structural damage that was uneconomical to repair. In a few instances, particularly where the faulting affected only part of a building, the damage was repairable.

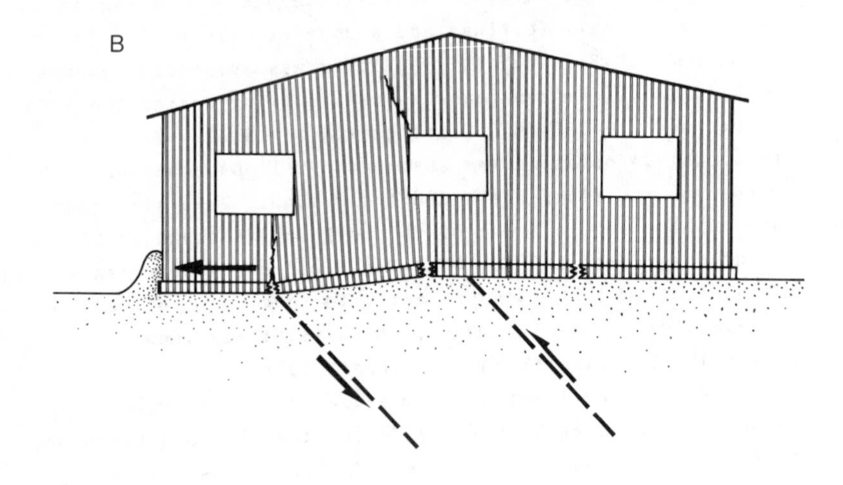

Fig. 6.--Foothill Nursing Home damaged by about 40 in (1.0 m) compression across a reverse dip-slip fault during 1971 San Fernando earthquake. A. Plowed-up sod in front of building caused by slippage between ground and foundation. B. Sketch of ground failure and building displacements (after Youd and others, 1978, p. 1118).

During the San Fernando earthquake, buildings on well-reinforced footings were generally able to withstand as much as 1 ft (0.3 m) of compression of underlying ground with little damage. Weaker foundations buckled under the increased compressive stress. For example, the concrete floor slab of the house shown in Fig. 7 buckled in the garage as a consequence of about 1 ft (0.3 m) of compresssion across a fault (Youd and others, 1978). Several other houses astride the surface rupture during the San Fernando event sustained minor to readily repairable damage from compressional displacements as large as several inches.

Shear deformation.--Shear deformations commonly occur at the lateral margins of landslides and lateral spreads and across strike-slip faults. Deformations may be concentrated across a single fissure or narrow zone, or they may be spread across a wide zone. Generally, the more concentrated the deformation, the more intense the damage.

Shear deformations with total displacement greater than about 2 ft (0.6 m) have been severely damaging to most buildings that completely span the shear zone. For example, the west wing of the administration building at the San Fernando Valley Juvenile Hall was astride a shear zone that developed at the margin of a lateral spread. The lateral spread was caused by liquefaction during the 1971 San Fernando earthquake. Beneath the building, about 30 in (0.76 m) of shear displacement occurred across a 30-ft (9-m) wide zone. The single-story, reinforced-concrete frame building collapsed (Fig. 8). Additional information on the construction and damage to this building is given by Thompson (1973).

Another wing of the Juvenile Hall facility, the east wing of the living and reception units, was distorted by about 20 in (0.5 m) of shear displacement across a 20-ft (6-m) wide zone and an additional 1 ft (0.3 m) across a 200-ft (60-m) wide zone to the west. The structure was a reinforced concrete frame with masonry-block infilled walls on spread footings (Fig. 9). Thompson (1973, p. 150) describes

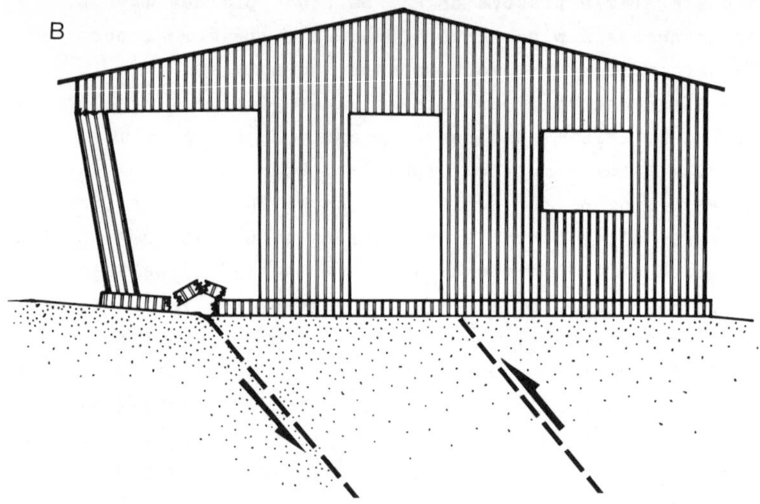

Fig. 7.--House damaged by about 1 ft (0.3 m) compression across a reverse dip-slip fault during 1971 San Fernando earthquake. A. Front view of distorted house. B. Sketch showing ground displacements and damage to structure (after Youd and others, 1978, p. 1117).

Fig. 8.--Administration building at San Fernando Valley Juvenile Hall collapsed by 1971 San Fernando earthquake. Building is astride a 30-ft (9.0-m) wide zone in which 3 ft (0.9 m) of horizontal shear displacement was generated by lateral spreading (photograph by J.F. Meehan, Office of Architecture and Construction, State of California).

Fig. 9.--Horizontal shear deformation beneath wing of San Fernando Valley Juvenile Hall. Building damage was greatest where shear deformation was greatest.

damage to this building as follows: "Damage to the west part of the structure was relatively light with moderate cracks in walls and fooor slabs. The east portion of the building had several areas of extremely heavy damage to floors, walls and the roof structure. Damage at the separation joint on the east end was very severe." Thus, damage was relatively minor across the wide zone of low shear distortion but was relatively intense where 2 ft (0.6 m) occurred across a narrow zone. The structure was torn down and replaced after measures were taken to stabilize the foundations (Youd, 1978). Several other buildings in the Juvenile Hall complex sustained shear distortions of a few inches across a wide zone. These buildings showed only minor damage from the shear distortion.

Shear deformation also occurred across the Sylmar Converter Station during the 1971 San Fernando earthquake. This facility is located about 0.4 mi (0.6 km) southwest of the Juvenile Hall and was subjected to some of the same lateral spreading movements as the Juvenile Hall (Youd, 1973). The major building on the site is a 541-ft (165-m)-long single-story structure in two units, a 212-ft (65-m)-long control facility on the south and a 329-ft (100-m)-long valve hall on the north. Both units are on wall and spread footings. The control facility has a full basement and is connected to the valve hall through a common wall. The valve hall has no basement except for a control corridor down the axis of the building. The two units are physically separated above ground but are joined by a common wall below grade. Shear displacement of probably not much more than 1 ft (0.3 m) in the length of the structure displaced the north end of the valve hall about 1 ft (0.3 m) westward. There was no significant damage to the valve hall, but there was considerable damage to the common wall (Schoustra and Yann, 1973).

Vertical ground displacements.--Differential vertical ground displacements are very common within and at the margins of landslides and lateral spreads, in areas with sediments undergoing compaction, and across normal and thrust faults. Loss of bearing strength beneath footings is another important cause of differential settlement. As with shear deformation, building damage is dependent on the width of the zone across which displacement occurs, the damage generally being more intese across narrower zones. Damage is also dependent on whether the foundation fractures or only tilts. Tilting without fracture is common for strong continuous footings on relatively soft ground, such as sites that lose of bearing strength owing to liquefaction. In firmer ground, only very strong foundations can be expected to undergo more than a few inches of differential vertical ground displacement without rupture. Differential vertical displacement with and without foundation rupture are considered separately below.

Vertical displacement with foundation rupture.--Few buildings founded on footings have retained their utility when undercut by displacments greater than 1 ft (0.3 m) across a single scarp or narrow zone. Displacements greater than 2 ft (0.6 m) across a single scarp generally have been destructive to all but the strongest buildings and foundations. An extreme example of damage due to vertical displacememt is the Government Hill School that was destroyed during the 1964 Alaska earthquake (Fig. 10). This building was split into several sections by vertical displacements as great as 12 ft (3.7 m) (Hansen, 1966). Many building failures occurred on the 4th Ave. landslide. That slide generated scarps and grabens throughout a 14-block commercial area of Achorage containing numerous 1- to 3-story, primarily wood-frame buildings. On the north side of 4th Ave. between B and D streets, Hansen (1966, p. A43) reported that the damage was total, indicating that those buildings were not economically repairable. Scarps along that section ranged from 2.5 ft to 11 ft (.76 m to 3.4 m). Many additional buildings on other parts of the slide were astride 1-ft to 3-ft (0.3-m to 0.9-m)-high scarps. Few of these buildings were repaired and reused after the earthquake.

Fig. 10.--Government Hill School, Anchorage, destroyed by landslide during 1964 Alaska earthquake (photograph by M.G. Bonilla, U.S. Geological Survey).

Fig. 11.--House damaged beyond repair by 6 in (15 cm) differential displacement across a scarp generated by liquefaction. Damage occurred during 1977 western Argentina earthquake (After Youd and Keefer, in press).

Differential settlement occurred beneath many buildings during the 1964 Niigata, Japan, earthquake, either as a consequence of massive ground displacement or as a consequence of loss of bearing strength. Most buildings that sustained more than 1 ft (0.3 m) differential settlement and a ruptured foundation were unrepairable or required major repairs (Kawasumi, 1968, p. 183-309). In a few instances, foundation rupture and building damage were confined to sections near expansion joints, leaving other parts of the building relatively undamaged and making repairs more economically feasible. For example, the 420-ft (128-m)-long, 4-story, reinforced-concrete Nuttari High School (foundation type not reported) sank unevenly by 16 in (40 cm). Damage was confined, however, to one section near an expansion joint (Kawasumi, 1968, p. 265).

If the foundation fractures, structural damage generally has occurred, even for displacements of only a few inches. The seriousness of the damage is dependent on the brittleness of the structure. For example, the house in Fig. 11 is astride a 7.8-in (20-cm) scarp in the ground. That scarp fractured and offset the foundation and floor slab of the house by 6 in (15 cm) and ruptured several masonry walls of the house. The house was economically unrepairable. This house is located in Caucete, Argentina, and the scarp was caused by liquefaction generated during the 1977 western Argentina earthquake (Youd and Keefer, in press).

Another example of damage caused by a few inches of vertical displacement is the Van Gough Elementary School, Mission Hills, California. This school was astride two or three fissures with scarps as high as 2 in (5 cm). These scarps formed the head of a lateral spread generated by the 1971 San Fernando earthquake (Youd, 1971; 1973). Differential vertical displacement fractured footings and floor slabs in the building and caused minor damage to the walls and roof. The damage was repairable at a cost of about $145,000 (1972 value) (Meehan, 1973, p. 675).

Vertical displacement without foundation rupture.--In Niigata, many buildings on footings tilted up to 6 degrees and settled up to 5 ft (1.5 m) with little damage to the superstructure. A few buildings tipped several tens of degrees without major damage to the building frame. In these instances, foundations did not rupture but tilted and settled as a unit. The most spectacular of these failures were the Kawagisi-cho appartment buildings (Fig. 12). These buildings were reinforced concrete structures three and four stories high. The buildings settled slowly and tilted as much as 60 degrees immediately after the earthquake. More than 200 buildings settled and tipped during the Niigata earthquake with little damage to the superstructures (Kawasumi, 1968, p.355-374). In most of these instances, tilting and settlement were due to loss of bearing strength caused by liquefaction of soils immediately below the foundation. Apparently, these softened soils allowed foundations to settle and tip without developing fracture-producing stress concentrations, as commonly occurs across scarps in more brittle soils. Most of the buildings were releveled, repaired, and reused.

BUILDINGS ON MAT FOUNDATIONS

Few large buildings on mat foundations have been subjected to ground failure displacements during recent earthquakes, and the author has no well-documented case histories to report for these buildings. Mats are generally more strongly reinforced and better tied together than are continuous footing foundations and, hence, should be more resistant to fracture than continuous footings. Mats that do not rupture should perform equally as well as continuous footings cited above that did not rupture. Structures other than buildings, such as tanks, have been founded on mats that have been subjected to ground failure displacement. For example, the tank and foundation shown in Fig. 13 were undercut by several feet of extensional and some vertical displacement during the 1964 Niigata, Japan, earthquake.

Fig. 12.--Four-story, reinforced-concrete apartment bearing strength during 1964 Niigata earthquake. The building and foundation remained intact despite tilting (after Kawasumi, 1968).

Fig. 13.--Tank and mat foundation in zone of ground failure comprising several feet of extensional displacement. Other than severance of service connections, tank and foundation were undamaged. Ground failure was generated by 1964 Niigata earthquake (after Kawasumi, 1968, p. 350).

The tank and foundation were undamaged although service conections to the tank were severed (Kawasumi, 1968, p. 350).

BUILDINGS ON DEEP FOUNDATIONS

Several factors influence amount of damage buildings on deep foundations sustain in areas of ground failure, including type of failure, depth of foundation, strength of foundation, and strength of structure above foundation level. Not many deep foundations beneath buildings have been intersected by ground failure movements during recent earthquakes. Hence, only limited information is available on the behavior of such foundations.

Extensional, compressional, and shear displacements.--There are few case histories of buildings on deep foundations that have been subjected to horizontal ground displacements during earthquakes. Thus, there are insufficient data seperately to consider damage due to extension, compression, and shear. To supplement the meager information available, some pertinent information from behavior of bridges on deep foundations is included.

During the Niigata earthquake, large extensional movements occurred beneath a single-story, steel-frame, arched-roof, industrial building on concrete piles (Kawasumi, 1968, p. 250). The extension was due to lateral spread of the ground toward a nearby canal. In the central section of the structure, the column bases and piles on one side of the building were displaced horizontally about 6.2 ft (2 m) in a length of 108 ft (33 m). The extension caused by this displacement split the foundation and the floor slab, wrenched the frame, and buckled and fractured chord members. Severe, if not irreparable, damage was inflicted on the structure. Information is not given on the length of the piles. It is clear, however, that the piles shifted horizontally with the ground.

Large lateral displacements have occurred beneath many bridges as a consequence of lateral spreading of flood-plain deposits toward stream channels. For example, during the

1964 Alaska earthquake, 266 bridges over rivers sustained compressional damage from shifting flood-plain deposits (Youd, 1978). During the 1964 Niigata earthquake several bridge piers were shifted, fractured, and tilted owing to lateral movement of sediments toward river channels (Kawasumi, 1968, p. 431-461). In each of these instances, pile foundations were unable to resist the lateral press of the spreading soil and were displaced inward toward the river channel. These displacements buckled decks, thrust stringers past abutments, and otherwise damaged the bridges. Although liquefaction was the cause of ground failure in most of these instances, there are no known cases where liquefied soil flowed through a pile group rather than displacing it.

Strong bridge foundations have been able to withstand lateral movements of less than a foot with little damage to the superstructure. For example, foundations of several bridges over the New and Alamo Rivers penetrated sediments that slumped several inches toward the river during the 1979 Imperial Valley, California, earthquake. Some bridges, particularly older timber structures sustained abutment and deck damage due to the ground movements. Conversely, several well-built, reinforced-concrete bridges, resisted several inches of soil displacement without damage. Inspection of piles and piers beneath these bridges revealed that soil had compressed against piles on up-slope sides and pulled away as much as several inches on the down-slope sides (Youd and Wieczorek, in press).

Several high-rise buildings on deep foundations were in ground failure zones generated by the 1906 San Francisco, California, and the 1964 Niigata, Japan, earthquakes (Youd and Hoose, 1978, p. 125-131; Kawasumi, p. 311-317). Little quantitative information is available on horizontal ground movements beneath these buildings. Open ground cracks and minor street deflections near some of these buildings suggest that several inches of horizontal ground movement may have intersected the foundations of these structures. A few of the buildings settled a little, but none of the reported damage indicated differential horizontal displacement within the structures.

Vertical ground displacement.--Deep foundations extending down to competent layers have been effective in preventing damage as a consequence of differential vertical ground movements. For example, the the ground and street adjacent to the Aetna Building in San Francisco settled nonuniformily as much as a few feet during the 1906 earthquake, but the steel-frame building on piles was virtually undamaged (Youd and Hoose, 1978, p. 129). Takada (in Kawasumi, 1958, p. 314-315) lists several buildings in sections of Niigata that were heavily damaged as a consequence of liquefaction during the 1964 Niigata earthquake. Of the buildings on piles, only the Toei Hotel was damaged by differential settlement and that damage was minor. In areas near these buildings,

Fig. 14.-- Three-story reinforced-concrete building settled differentially by 6.5 ft (2.0 m) and tilted 8 degrees owing to liquefaction during 1964 Niigata earthquake. Building is founded on 20-ft (6-m) long timber piles (after Kawasumi, 1968, p. 188).

many structures on shallow foundations were severely affected by tipping and settling.

Pile foundations have not prevented differential settlement in instances where the piles bore in layers that later liquefied. For example, the 3-story building shown in Fig. 14 settled and tipped even though it was founded on 20-ft (6-m) piles. Liquefaction apparently occurred around or below the piles, allowing the foundation to penetrate downward and the building to settle and tip. The building tipped during the 1964 earthquake (Kawasumi, 1968, p. 188). Some longer and wider buildings founded on piles suffered severe differential settlement during the Niigata earthquake when supporting soils lost strength and piles penetrated into the weakened sediments.

CONCLUSIONS

On the basis of past experience and specific case histories of building performance in areas of ground failure generated by earthquakes, the following general conclusions are drawn with respect to thresholds of damage for various foundation types and modes of differential ground displacement. Because of the limited data on which these conclusions are based and because most of the structures examined were not specifically designed to resist ground failure displacements, these conclusions should be used only in the context in which they are given, that is, as general information for estimating possible damage to buildings in areas susceptible to ground failure and for guidance to engineers designing buildings to resist ground displacement.

Buildings on Footings

Extensional ground displacement.--Flexible buildings on well-reinforced foundations can generally withstand as much as 4 in (0.1 m) of extension without significant damage, and perhaps as much as 2 ft (0.6 m) of displacement with repairable damage. Only unusually strong buildings can withstand more than 2 ft (0.6 m) of displacement without damage beyond repair. Brittle structures on poorly reinforced foundations may sustain damage requiring repairs at displacements as small as 2 in (5 cm). Few such structures can withstand displacement greater than 1 ft (0.3 m) without serious damage.

Compressional ground displacement.--Foundations are generally stronger under compression than extension. However, differential vertical movements that commonly accompany compression, such as vertical displacements associated with thrust faulting, may severely reduce the capability of a foundation to resist compression. Buildings on strong foundations may withstand as much as 1 ft (0.3 m) of compression without damage, and commonly as much as 2 ft (0.6 m) with repairable damage. Few structures can withstand more than 2 ft (0.6 m) of compression without severe damage. Most poorly reinforced foundations can withstand as much as 6 in (0.15 m) compression with minor damage but may be damaged beyond repair by 1 ft (0.3 m) or more of compression.

Shear deformation.--Foundation damage due to shear deformation is a function of the width of the shear zone, the amount of shear displacement, and whether the foundation ruptures or merely rotates. Strong foundations and flexible buildings may sustain little damage from shear displacements as large as 1 ft (0.3 m) across a concentrated zone. Displacements greater than 2 ft. (0.6 m) across a concentrated zone generally cause severe damage.

Differential vertical ground displacement.--Differential vertical displacements are usually very damaging if the foundation ruptures. If the foundation does not rupture but merely settles and tilts as a unit, relatively large displacements can be accomodated with repairable damage. Footings usually fracture when intersected by ground failure movements producing scarps in firm ground. Continuous footings may not fracture when displacement results from loss of bearing strength due to softening of soils beneath the footings, such as by liquefaction.

Footings that fracture.--Flexible buildings on strong foundations may withstand up to 4 in (0.1 m) of vertical displacemnt with little building damage. Two inches (0.05 m) of vertical displacement can fracture weak foundations and cause damage to the overlying structure. More than 2 ft (0.6 m) of differential vertical displacement with foundation fracture is severely damaging to almost all buildings. This conclusion generally agrees with the consensus opinion drawn by a panel of experts (Swiger, 1978) that it is doubtful if even small (1 ft (0.3 m) or less) vertical components of displacement can be accomodated without severe damage to low buildings.

Footings that do not fracture.--For footings that settle and tip without fracture, building damage depends on the strength of the superstructure. During the 1964 Niigata, Japan, earthquake, many buildings settled and tipped as much as 6 degrees without damage to the superstructure. Some buildings tipped by as much as 60 degrees with little damage to the superstructure. Thus, for strong structures, settlements of several feet and tilting of several degrees without foundation rupture can generally be accomodated with repairable damage.

Buildings on Mat Foundations

To the author's knowledge, few large buildings on mat foundations have been subjected to ground failure displacement during past earthquakes. Buildings on mat foundations that settle and tip without fracture are subjected to the same stresses as those on footings that settle and tip without rupture and hence should perform equally well. If the mat ruptures, damage would be similar to that of buildings on footings that rupture. However, because mats are generally more strongly reinforced and better tied together than footings, they should be less susceptible to fracture, and hence should generally perform better. This conclusion agrees with the consensus opinion of a panel of experts (Swiger, 1978, p. 1466) that mat foundations can be designed to withstand small (as much as 1 ft (0.3 m)) shear displacements. The panel, however, doubted that even small (less than 1 ft (0.3 m)) differential vertical displacements could be accomodated without severe damage. This study agrees that if the foundation ruptures, displacements as great as 1 ft (0.3 m) generally cause serious damage. However, if the foundation does not rupture, tilting and vertical displacements as large as several feet in well-built buildings may be accomodated with little damage.

Buildings on Deep Foundations

Differential horizontal ground displacements.--The meager information available suggests that strong deep-foundations can withstand as much as 1 ft (0.3 m) horizontal ground displacement with little damage. Displacements greater than 1 ft (0.3 m) are likely to rupture ties between foundation elements and damage the superstructure.

Differential vertical ground displacements.--Foundations that bear in deep, competent materials usually safeguard buildings against damage from differential settlements of as much as a few feet in surficial, less competent layers. The foundation must be strong enough to resist soil drag forces in the upper layers, however. If the deep foundation bears in a layer that later weakens, such as by liquefaction, damage due to differential settlement is very likely to occur.

REFERENCES

Hansen, W.R., 1966, "Effects of the Earthquake of March 27, 1964 at Anchorage, Alaska," U.S. Geological Survey Professional Paper 542A, 68 p.

Kawasumi, Hiroshi, Ed., 1968, General Report on the Niigata earthquake of 1964, Tokyo Electrical Engineering College Press, Japan, 550 p.

Meehan, J.F., 1973, "Public School Buildings," San Fernando Earthquake of February 9, 1971, U.S. Department of Commerce, Vol. 1, Pt. B, p. 667-684.

Schoustra, J.J., and Yann, J.K., 1973, "Performance of Foundations at the Sylmar Converter Station," San Fernando Earthquake of February 9, 1971, U.S. Department of Commerce, Vol. 1, Pt. B, p. 821-824.

Swiger, W.F., 1978, "Specialty Session on Design for Fault Displacement," Earthquake Engineering and Soil Dynamics, ASCE, Vol. 3, p. 1464-1468.

Thompson, J.H., 1973, "San Fernando Valley Juvenile Hall," San Fernando Earthquake of February 9, 1971, U.S. Department of Commerce, Vol. 1, Pt. A, p. 147-162.

Youd, T.L., 1971, "Landsliding in the Vicinity of the Van Norman Lakes," The San Fernando, California, Earthquake of February 9, 1971, U.S. Geological Survey Professional Paper 733, p. 105-109.

Youd, T.L., 1973, "Ground Movements in Van Norman Lake Vicinity during San Fernando Earthquake," San Fernando Earthquake of February 9, 1971, U.S. Department of Commerce, Vol. 3, p. 197-206.

Youd, T.L., 1978, "Major Cause of Earthquake Damage is Ground Failure," Civil Engineering, April, p. 47-51.

Youd, T.L., and Hoose, S.N., 1978, "Historic Ground Failures in Northern California Triggered by Earthquakes," U.S. Geological Survey Professional Paper 993, 177p.

Youd, T.L., and Keefer, D.K, in press, "Liquefaction: Site and Regional Studies", The Western Argentina Earthquake of November 23, 1977, U.S. Geological Survey Professional Paper.

Youd, T.L., and Wieczorek, G.F., in press, "Liquefaction and Secondary Ground Failure," The Imperial Valley Earthquake of October 15, 1979, U.S. Geological Survey Professional Paper.

Youd, T.L., Yerkes, R.F., and Clark, M.M., 1978, "San Fernando Faulting Damage and Its Effect on Land Use," Earthquake Engineering and Soil Dynamics, ASCE, Vol. 2, p. 1111-1125.

SOIL DENSITY VARIATIONS
WASHINGTON PUBLIC POWER SUPPLY SYSTEM
WNP-1 & WNP-4

I. INTRODUCTION

This paper describes the variations encountered during the soil density testing of backfill at the Washington Public Power Supply System Nuclear Projects No. 1 and 4 (WNP-1 and WNP-4) located at the Department of Energy Hanford Reservation near Richland, Washington. The backfill density testing criteria was developed by Shannon & Wilson, Inc., of Seattle, Washington. The Preliminary Safety Analysis report and subsequent contract specifications define that maximum soil density shall be determined in accordance with ASTM D 2049-69. As the compaction testing progressed, it became apparent that compliance could not be effected with certain requirements of the standard. It became necessary to modify the requirements to conform to the characteristics of the available equipment.

II. RELATIVE COMPACTION VS RELATIVE DENSITY

Relative density is one of the parameters governing liquefaction potential of saturated sands on Hanford projects. Soil compaction is checked in terms of relative density of the in-place materials in accordance with ASTM D 2049-69.

The "relative compaction" and "relative density" methods are often confused. As shown in Figure 1, relative density covers a much more narrow range of possible numerical values for dry density than does relative compaction. For instance, a one pound per cubic foot change in dry density, which could be testing error, may change the relative density value as much as 5 percent.

This same change in dry density would change the relative compaction value by only 1 percent. Therefore, small changes in actual dry density would reflect a larger percentage change in relative density than in percent compaction.

In addition to these inherent errors, the time lag caused by laboratory testing while large quantities of fill were being placed made it difficult to maintain good field control.

III. RELATIVE COMPACTION-RELATIVE DENSITY CORRELATION

The relative compaction method for compaction control was consequently adopted for the WNP-1 and WNP-4 projects by Washington Public Power Supply System as a more expedient method and to avoid the quality control problems associated with large variations in values. However, it became necessary to correlate relative compaction with relative density in order to determine what relative compaction values corresponded to the desired 75 percent and 85 percent relative densities.

For each set of maximum and minimum densities determined for a sample, the densities corresponding to 75 percent and 85 percent relative density were determined. These values are plotted on Figure 2 and show that there is a nearly linear relationship between maximum density and the day density required to achieve each relative density value.

On those figures relative compaction lines were drawn and the following correlations were determined:

1) 85% Relative Density = 97% Relative Compaction
2) 75% Relative Density = 95% Relative Compaction

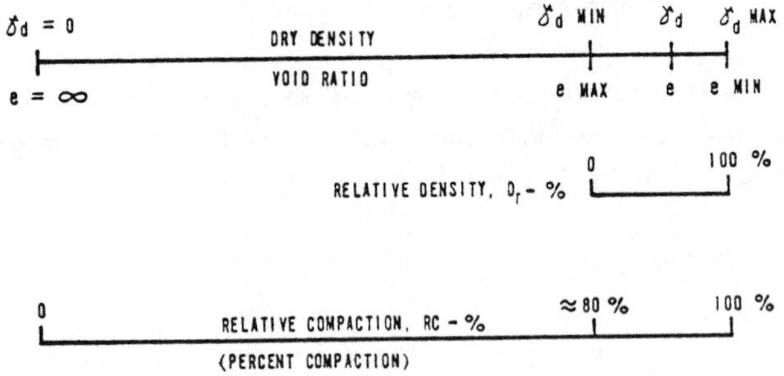

$\gamma_d = 0$ DRY DENSITY γ_d MIN γ_d γ_d MAX

VOID RATIO

$e = \infty$ e MAX e e MIN

RELATIVE DENSITY, D_r - % 0 100 %

0 RELATIVE COMPACTION, RC - % ≈80 % 100 %

(PERCENT COMPACTION)

RELATIVE DENSITY

$$D_r = \frac{e\ MAX - e}{e\ MAX - e\ MIN} \times 100$$

$$D_r = \frac{\gamma_d\ MAX}{\gamma_d} \times \frac{\gamma_d - \gamma_d\ MIN}{\gamma_d\ MAX - \gamma_d\ MIN} \times 100$$

RELATIVE COMPACTION

$$RC = \frac{\gamma_d}{\gamma_d\ MAX} \times 100$$

WHERE:

D_r = RELATIVE DENSITY %
e = ACTUAL VOID RATIO
e MIN = VOID RATIO, DENSEST STATE
e MAX = VOID RATIO, LOOSEST STATE

RC = RELATIVE COMPACTION IN %
γ_d = IN-PLACE DRY DENSITY, PCF
γ_d MAX = MAX DRY DENSITY, PCF
γ_d MIN = MIN DRY DENSITY, PCF

RELATIVE DENSITY
AND
RELATIVE COMPACTION

Figure 1

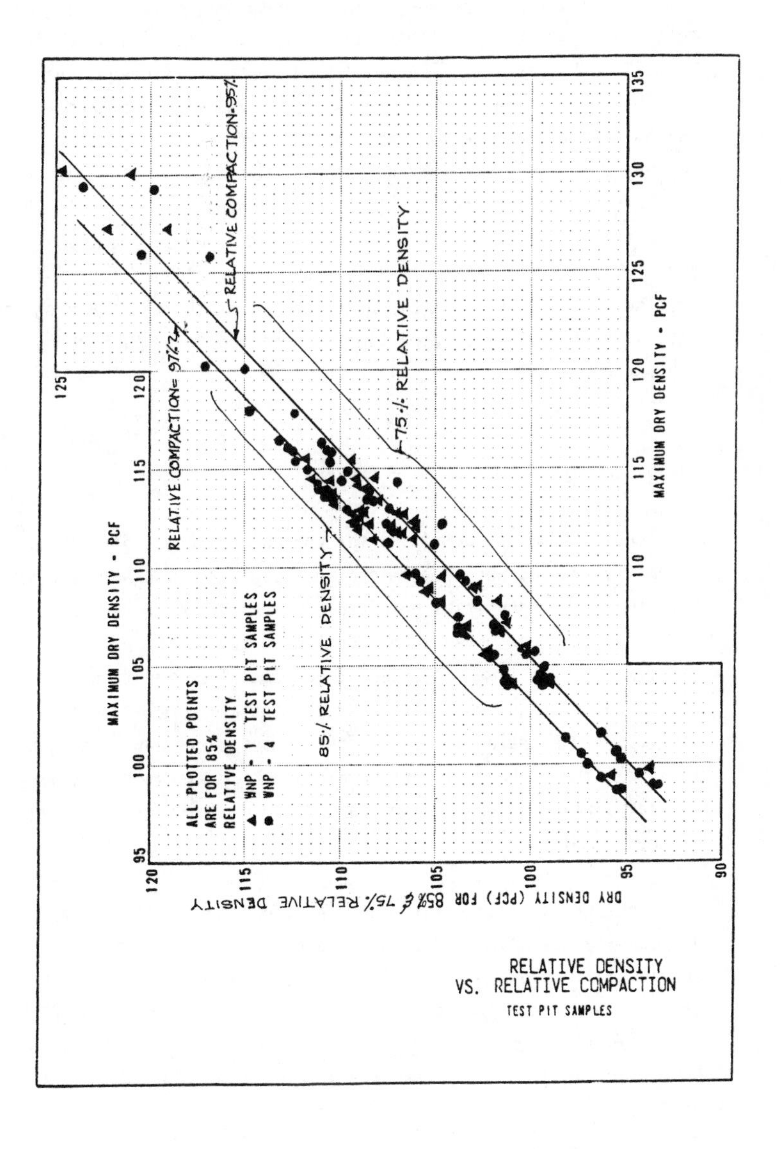

MAXIMUM DRY DENSITY - PCF
ALL PLOTTED POINTS ARE FOR 85% RELATIVE DENSITY
▲ WNP - 1 TEST PIT SAMPLES
● WNP - 4 TEST PIT SAMPLES
RELATIVE COMPACTION= 97%
RELATIVE COMPACTION-95%
75% RELATIVE DENSITY
85% RELATIVE DENSITY
DRY DENSITY (PCF) FOR 85% & 75% RELATIVE DENSITY
RELATIVE DENSITY
VS. RELATIVE COMPACTION
TEST PIT SAMPLES

Figure 2

The soil report presented in the PSAR, gives the results of tests performed on Type A material at relative densities ranging from 71 to 80 percent. The results of these tests demonstrate that backfill compacted to a relative density of 71 percent (94.2% Relative Compaction) will not liquify when subject to loading conditions more severe (about 100 cycles duration) than the Safe Shutdown Earthquake (SSE). For compaction control, 75 percent relative density (95 percent relative compaction) is acceptable for this type of soil. However, realizing a normal testing variation of two (2) pounds per cubic foot, the acceptability limit for relative compaction was defined to be 97 percent (85 percent relative density) in the PSAR and in the construction specifications.

IV. GRAIN SIZE VS MAXIMUM DENSITY CORRELATION

The maximum dry densities for the establishment of relative compaction values were determined in accordance with ASTM D-2049-69 because maximum densities obtained by vibratory methods are greater than those obtained by the impact methods of ASTM D-1557. To eliminate the time lag, compaction control correlations were developed relating grain size characteristics to maximum soil density. Thus, relative compaction in most cases could be obtained in a relatively short time by obtaining the maximum density from the correlation curves for the particular grain size distribution of that sample.

The method of correlating grain size distribution to maximum dry density was developed by D. M. Burmister. This method is applicable to the WNP-1 and WNP-4 soils. In its application, the grain size distribution of a granular soil sample is represented by a mean slope constructed within the distribution. The value C_r (effective grain size range) was calculated using the equation on Figure 3.

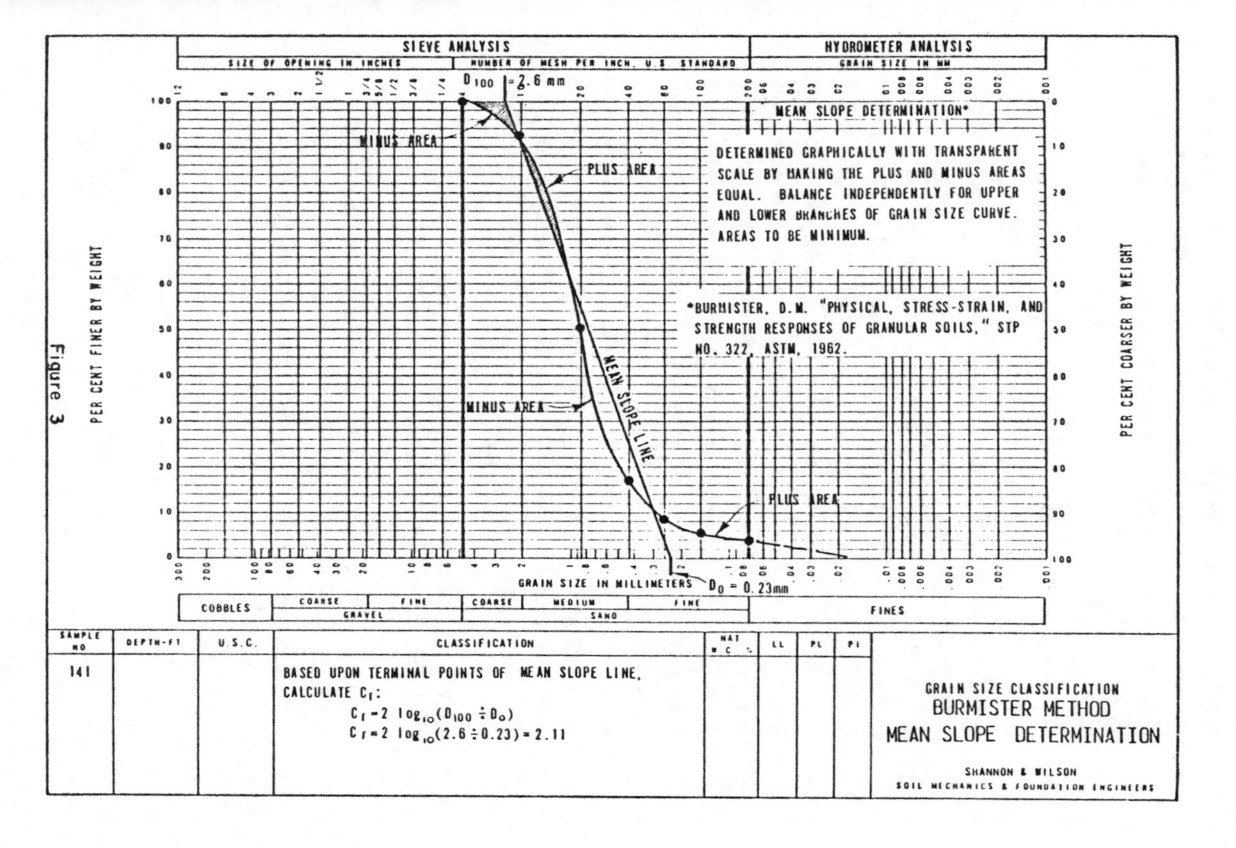
SIEVE ANALYSIS
SIZE OF OPENING IN INCHES
NUMBER OF MESH PER INCH, U. S. STANDARD
HYDROMETER ANALYSIS
GRAIN SIZE IN MM
D100 = 2.6 mm
MEAN SLOPE DETERMINATION*
DETERMINED GRAPHICALLY WITH TRANSPARENT SCALE BY MAKING THE PLUS AND MINUS AREAS EQUAL. BALANCE INDEPENDENTLY FOR UPPER AND LOWER BRANCHES OF GRAIN SIZE CURVE. AREAS TO BE MINIMUM.
*BURMISTER, D.M. "PHYSICAL, STRESS-STRAIN, AND STRENGTH RESPONSES OF GRANULAR SOILS," STP NO. 322, ASTM, 1962.
MINUS AREA
PLUS AREA
MEAN SLOPE LINE
MINUS AREA
PLUS AREA
PER CENT FINER BY WEIGHT
PER CENT COARSER BY WEIGHT
GRAIN SIZE IN MILLIMETERS
D0 = 0.23mm
COBBLES
COARSE
FINE
GRAVEL
COARSE
MEDIUM
FINE
SAND
FINES
SAMPLE NO
DEPTH-FT
U.S.C.
CLASSIFICATION
NAT W.C. %
LL
PL
PI
141
BASED UPON TERMINAL POINTS OF MEAN SLOPE LINE, CALCULATE Cr:
Cr = 2 log10(D100 ÷ D0)
Cr = 2 log10(2.6 ÷ 0.23) = 2.11
GRAIN SIZE CLASSIFICATION
BURMISTER METHOD
MEAN SLOPE DETERMINATION
SHANNON & WILSON
SOIL MECHANICS & FOUNDATION ENGINEERS

Figure 3

Charts were developed plotting the C_r values and the corresponding maximum density values for each of the soil samples used. The charts were further categorized with respect to those soils having less than or more than 10 percent passing the No. 60 sieve.

In Figure 4 for example, a near-maximum curve was constructed. Nearly all the plotted points fall on or below this curve which thereby represents the practical upper limit for maximum dry densities and, in turn, the relative compaction (RC) was taken as 100 percent. The lowest curve on this was constructed to show density values needed to achieve RC = 95 percent (75% relative density). The middle curve was established as RC = 100 percent of the correlation. The line represents a conservative average for the data shown and was constructed such that the lower control curve would be 97 percent of the middle curve. Thus, in-place density values achieving a relative compaction of 97 percent based upon the middle curve (assumed correlation maximum) would not be less than 95 percent based upon the upper curve. For example, soil with C_r value of 1.5 is shown to have a correlation maximum value of 115.8 PCF. A field density test value of 112.4 PCF would then produce a relative compaction (RC) value of 97.1 percent (112.4 ÷ 115.8).

V. PRACTICAL APPLICATION OF THE CORRELATION CURVES TO THE WNP-1 AND WNP-4 PROJECT

In practice, the in-place field density tests were taken and an additional sample of soil was taken to perform grain size distribution analysis. The grain size distribution was checked to verify that the curve fell within the grain size limits for types A and B soils shown on Figure 5.

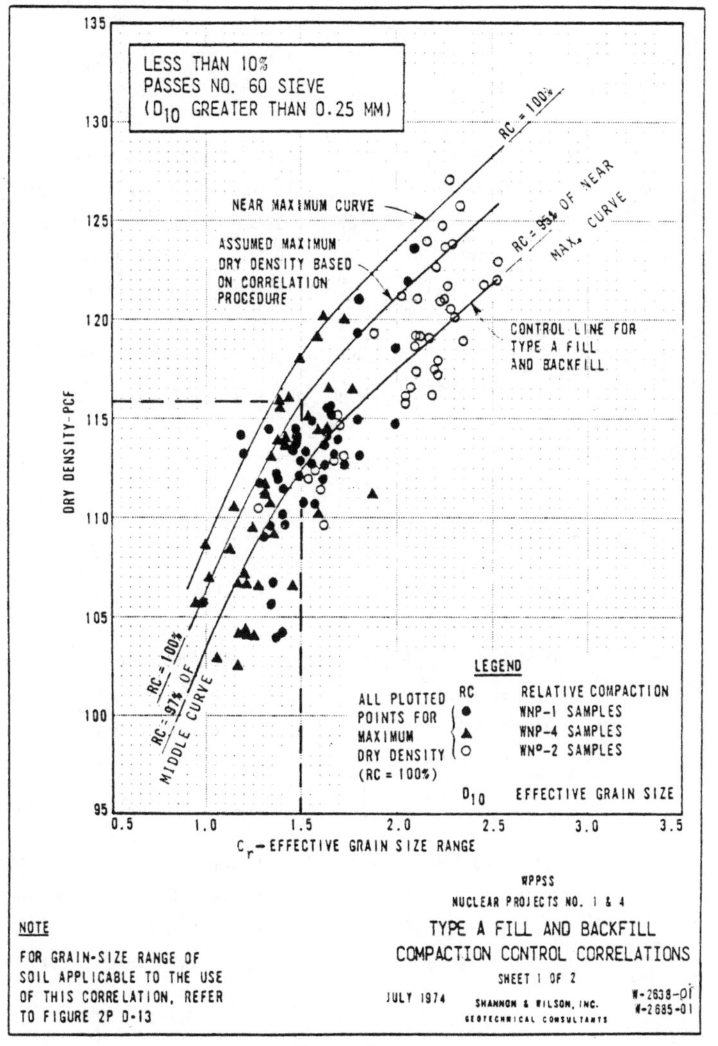
LESS THAN 10%
PASSES NO. 60 SIEVE
(D_{10} GREATER THAN 0.25 MM)
RC = 100%
NEAR MAXIMUM CURVE
RC = 95% OF NEAR MAX. CURVE
ASSUMED MAXIMUM DRY DENSITY BASED ON CORRELATION PROCEDURE
CONTROL LINE FOR TYPE A FILL AND BACKFILL
RC = 100%
RC = 97% OF MIDDLE CURVE
DRY DENSITY-PCF
135
130
125
120
115
110
105
100
95
0.5
1.0
1.5
2.0
2.5
3.0
3.5
C_r – EFFECTIVE GRAIN SIZE RANGE
LEGEND
RC RELATIVE COMPACTION
ALL PLOTTED POINTS FOR MAXIMUM DRY DENSITY (RC = 100%)
WNP-1 SAMPLES
WNP-4 SAMPLES
WNº-2 SAMPLES
D_{10} EFFECTIVE GRAIN SIZE
NOTE
FOR GRAIN-SIZE RANGE OF SOIL APPLICABLE TO THE USE OF THIS CORRELATION, REFER TO FIGURE 2P D-13
WPPSS
NUCLEAR PROJECTS NO. 1 & 4
TYPE A FILL AND BACKFILL
COMPACTION CONTROL CORRELATIONS
SHEET 1 OF 2
JULY 1974
SHANNON & WILSON, INC.
GEOTECHNICAL CONSULTANTS
W-2638-01
W-2685-01

Figure 4

If the grain size curve did not meet the limits shown, then the maximum density of the soil was determined by performing both the wet and dry methods of ASTM D-2049-69.

If the grain size curve met the limits shown, the C_r value for the grain size curve was determined in accordance with the Burmister method. The maximum density of the soil was then obtained and relative compaction RC was calculated.

If the calculated relative compaction value RC was less than the specified compaction requirements of Type A (97%) or Type B (95%) soil, the maximum density was determined in accordance with the wet or dry methods of ASTM D 2049-69 (as modified) depending upon whether the C_r value was less than or greater than 1.65.

1)	1.65 or less	Dry method only
2)	Over 1.65	Both wet and dry methods

The relative compaction was then recalculated.

VI. SOIL DENSITY VARIATIONS ENCOUNTERED ON THE PROJECT

A. Background Information

ASTM D 2049 specifies that an electromagnetic vibratory table, capable of a vertical amplitude variable between .002 and .025 inches (under a 250 pound load) be used. The maximum density was to be determined with the vibrator control at its maximum amplitude setting.

During the initial phase of backfill density testing, it was noted that the maximum density values obtained by the test lab were significantly lower than those obtained during the PSAR development work. The vibratory table used

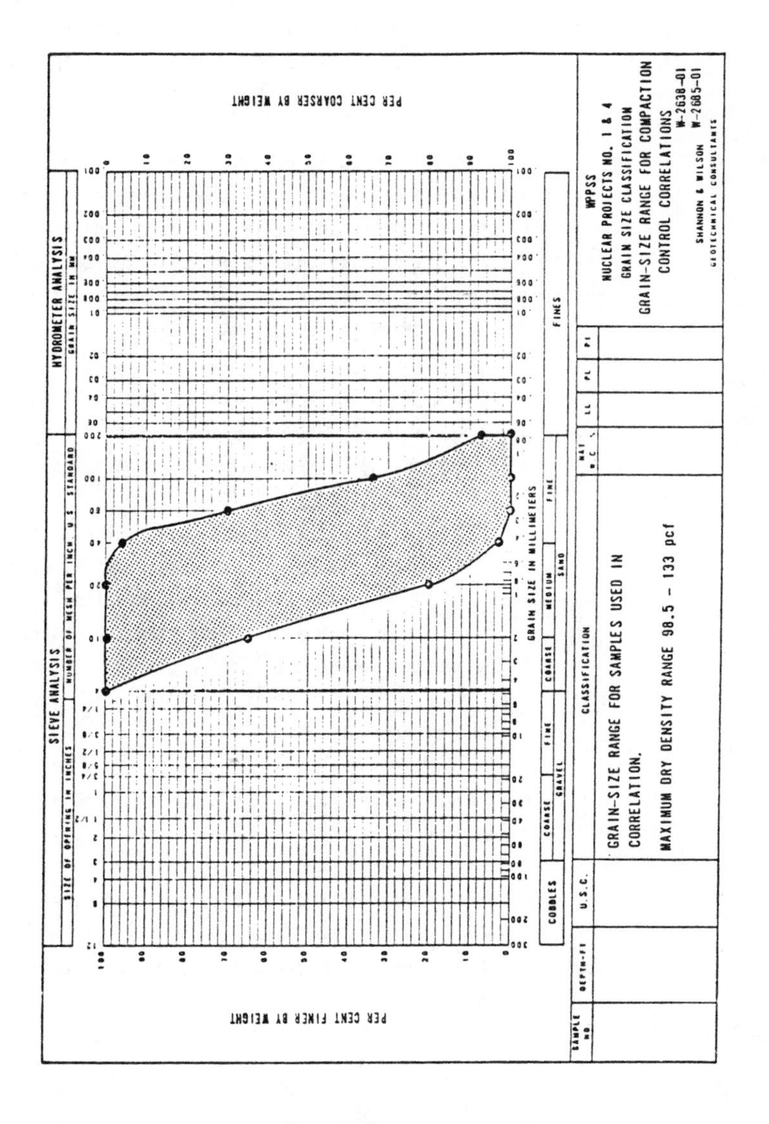

SIEVE ANALYSIS
HYDROMETER ANALYSIS
SIZE OF OPENING IN INCHES
NUMBER OF MESH PER INCH, U.S. STANDARD
GRAIN SIZE IN MM
PER CENT COARSER BY WEIGHT
PER CENT FINER BY WEIGHT
GRAIN SIZE IN MILLIMETERS
COBBLES
COARSE
FINE
GRAVEL
COARSE
MEDIUM
FINE
SAND
FINES
SAMPLE NO.
DEPTH-FT
U.S.C.
CLASSIFICATION
NAT W.C.
LL
PL
PI
GRAIN-SIZE RANGE FOR SAMPLES USED IN CORRELATION.
MAXIMUM DRY DENSITY RANGE 98.5 - 133 pcf
WPPSS
NUCLEAR PROJECTS NO. 1 & 4
GRAIN SIZE CLASSIFICATION
GRAIN-SIZE RANGE FOR COMPACTION
CONTROL CORRELATIONS
W-2638-01
W-2685-01
SHANNON & WILSON
GEOTECHNICAL CONSULTANTS

Figure 5

was a Syntron VP-86. A comparison testing program was undertaken running identical soil samples on the Syntron VP-86 table as well as on the vibratory table which was used during the PSAR preparation work. The results of this density comparison testing showed a difference of approximately eight (8) pounds per cubic foot between the two tables.

The Syntron VP-86 contained an additional steel plate attached to the normal table top. Removal of this plate and oter attempts to improve the table performance decreased the difference in soil density results between the two tables to four (4) pounds per cubic foot but the maximum vertical amplitude could not exceed .002 inches.

Because the vibrating table was not performing as required, a new table, a Syntron VP-181-A1 advertised by the supplier as meeting the ASTM D 2049 requirements was purchased.

This new table was installed and comparison tests were run between it and the Shannon & Wilson table (VP-86-B1) and between the two test lab tables using identical soil samples. These tests (Figure 6) showed that the Shannon & Wilson table and the new test lab table (VP-181-A1) gave maximum density results within plus or minus one (1) pound per cubic foot of each other.

Even though the new table gave comparable results, the test lab was directed to certify that the table was in compliance with the ASTM D 2049 requirements. Tests, using velocity transducers and an oscilloscope, showed the maximum vertical amplitude to be .005 inches. Because of this low amplitude, the supplier and the manufacturer were contacted for assistance.

The supplier obtained measurements using a strobe light and microscope, but could not exceed the .005 inch maximum amplitude verifying that the new table VP-181-A1 could not meet the amplitude requirements of ASTM D-0249.

The ASTM D 2049 Committee was contacted for clarification regarding the specific requirements of the ASTM D 2049 standards that were in question. They stated that ASTM D 2049 was currently under extensive review and revision, and that the amplitude and the vibrator control setting requirements of the Standard do not follow the present practice of the industry. Soil testing laboratories are following the Corps of Engineers Manual EM 1110-2-1906, which requires "amplitudes to a maximum of at least 0.015 inches under a 250 pound load," and a "rheostat setting at which maximum density is produced and then use this setting for subsequent maximum density testing."

These discussions confirmed that the present ASTM Standard could not be complied with and should not be specified without modification.

The backfilling operation was allowed to continue at the site because of the close correlation between the new table VP-181-A1 soil density test results and the Shannon & Wilson table soil density test results. Field compaction was being done with six passes over an eight-inch lift, since previous experience with similar material and procedures had proved adequate. Also, backfill densities were determined from the compaction control correlation curves which were known to be conservative.

CORRELATION TESTS TABLE (VP-181) vs SW Table (VP-86-B1)
TABLE (VP-181) vs Table (VP-86)

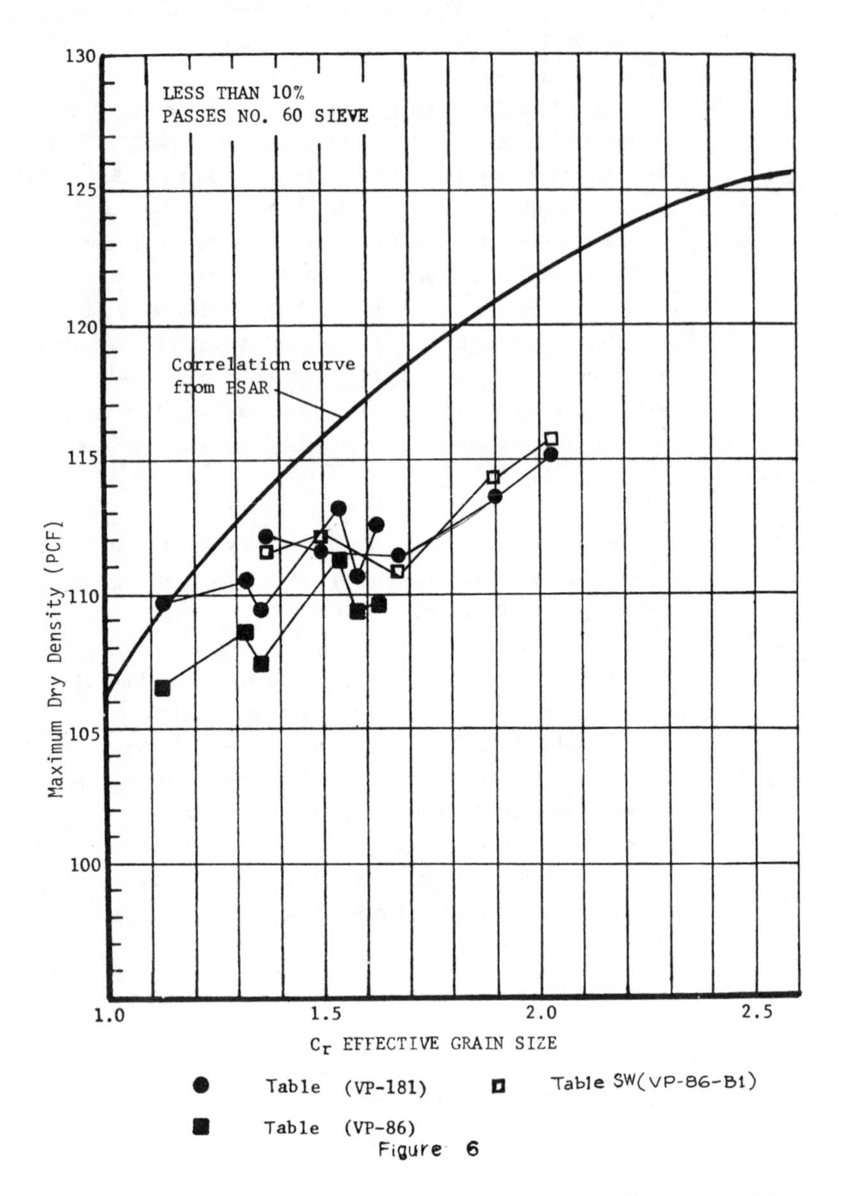

Figure 6

B. Technical Evaluation

During the resolution of these density testing variances, it was recognized that cohesionless soils with various gradations reach maximum density at different amplitudes. The maximum density of a soil is determined by investigating the amplitude range within which the maximum density occurs. Subsequent density tests must be performed at an assigned amplitude within this range.

In order to determine this optimum amplitude for the various soils and to verify that the in-place backfill was acceptable, various testing programs and in-place density verification programs were undertaken.

C. Testing Programs

Soil samples were tested by Shannon & Wilson in their own laboratory and at the University of California facilities. Other soil samples were tested by Woodward-Clyde Consultants and others were tested at the WNP-1/4 test laboratories under the direction of United Engineers & Constructors, Inc.

The samples were tested on various tables:

Woodward-Clyde 100 VP-D
WNP-1/4 Test Lab VP-181-A.1
University of California MTS
University of California VP-200

Figure 7 shows the amplitude versus maximum dry density relationships of soil samples of various gradations tested on those tables.

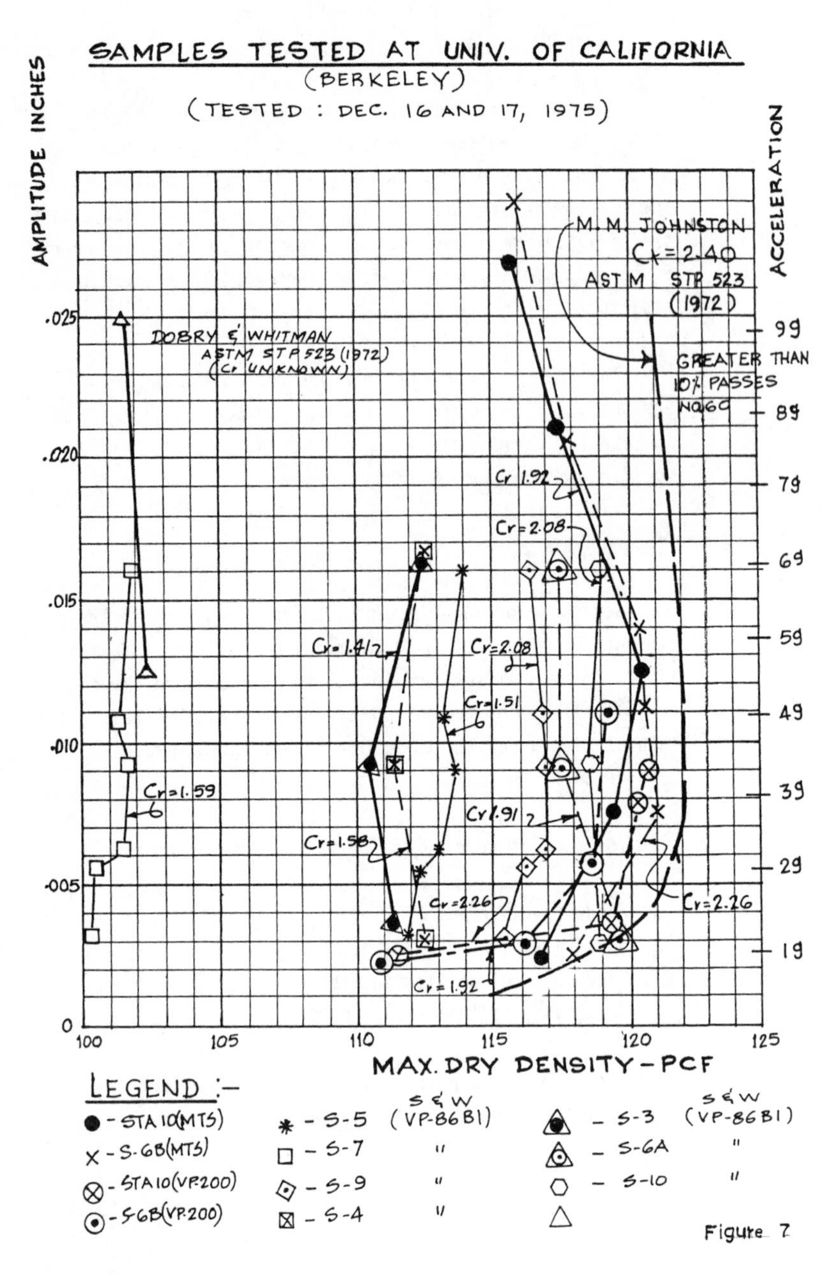

SAMPLES TESTED AT UNIV. OF CALIFORNIA
(BERKELEY)
(TESTED : DEC. 16 AND 17, 1975)
AMPLITUDE INCHES
ACCELERATION
M. M. JOHNSTON
Cr=2.40
ASTM STP 523
(1972)
DOBRY & WHITMAN
ASTM STP 523 (1972)
(Cr UNKNOWN)
GREATER THAN
10% PASSES
NO.60
Cr 1.92
Cr=2.08
Cr=1.41
Cr=2.08
Cr=1.51
Cr=1.59
Cr 1.91
Cr=1.58
Cr=2.26
Cr=2.26
Cr=1.92
.025
.020
.015
.010
.005
0
9g
8g
7g
6g
5g
4g
3g
2g
1g
100
105
110
115
120
125
MAX. DRY DENSITY - PCF
LEGEND :-
- STA 10(MTS)
- S-6B(MTS)
- STA 10(VP-200)
- S-6B(VP-200)
- S-5
- S-7
- S-9
- S-4
S & W (VP-86B1)
"
"
"
- S-3
- S-6A
- S-10
S & W (VP-86B1)
"
"

Figure 7

With the exception of Shannon & Wilson Laboratory, the other vibratory equipment used for testing was capable of varying amplitudes from zero to over .025 inches. However, this equipment did not meet the requirements of ASTM D 2049 as the vibration is created by mechanical rather than electromagnetic means.

Figure 8 shows the C_r versus maximum dry density correlation comparison with the PSAR correlation curve.

The results of this testing showed that maximum densities generally occurred at amplitudes in the range of .005 to .010 inches and that the compaction control correlation curves given in the PSAR are conservative.

D. In-Place Density Verification Programs

To verify the acceptablility of the in-place backfill under the General Services Building (G.S.B.), fourteen (14) soil density retests were performed in the completed backfill area. The samples were taken from various locations and elevations in the existing backfill. The results of five of these retests are shown in Figure 9. The testing techniques and procedure used for the soil density tests were conducted in accordance with the density testing criteria developed by Shannon & Wilson. The density retests were performed using the new test lab table (VP-181-A1) and the WPPSS mechanical vibratory table (100 VP-D). The average relative compaction of these retests were above the 97% relative compaction acceptability limit defined by the PSAR.

For the WNP-1 Spray Pond verification testing program, test location plans were provided for. The average percent relative compaction values for the new test table (VP-181)

MAXIMUM DENSITY CORRELATION RESULTS
From Various Tables

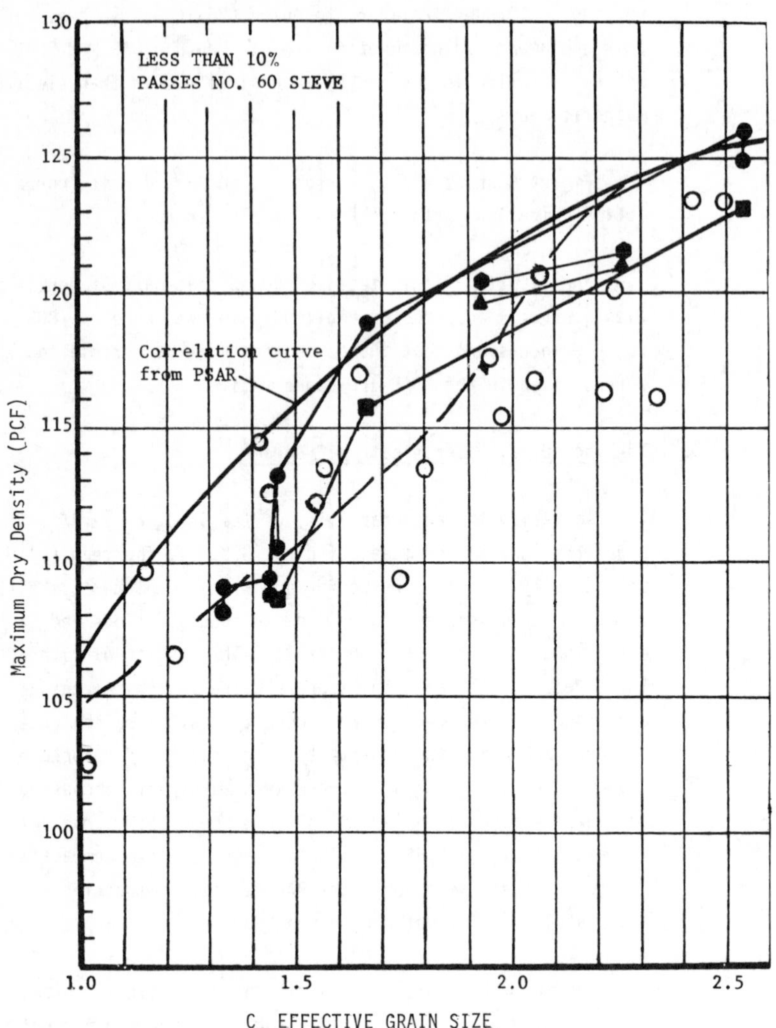

● W/C Table (100VP-D)
■ WNP-1/4 Testing Table (VP-181)
⬢ U/C Table (MTS)
▲ U/C Table (VP-200)
○ WPPSS Table (100 VP-D)

FIGURE 8

BACKFILL RETESTING DATA

Performed on TEST LAB tab.(VP-181) - 1

on WPPSS table (100 VP-D) - 2

	1 TABLE	2 TABLE	CURVE	1 TABLE	2 TABLE	CURVE	1 TABLE	2 TABLE	CURVE
TEST NUMBER	10			11			12		
IN-SITU DENSITY	113.2			112.2			109.1		
C_r VALUE	1.43			1.33			1.30		
MAXIMUM DENSITY	112.5	112.7	114.6	108.3	108.6	112.7	107.5	107.5	112.5
% REL COMP	101	100	99	104	103	100	101	101	97

	1 TABLE	2 TABLE	CURVE	1 TABLE	2 TABLE	CURVE	1 TABLE	2 TABLE	CURVE
TEST NUMBER	13			14					
IN-SITU DENSITY	114.5			111.1					
C_r VALUE	1.50			1.59					
MAXIMUM DENSITY	111.5	111.6	116.0	110.6	110.9	116.9			
% REL COMP	103	103	99	100	100	95			

	1 TABLE	2 TABLE	CURVE	1 TABLE	2 TABLE	CURVE	1 TABLE	2 TABLE	CURVE
TEST NUMBER									
IN-SITU DENSITY									
C_r VALUE									
MAXIMUM DENSITY									
% REL COMP									

Note: all density results in #/F^3

Average % of relative compaction from WNP table: 100.71%

Average % of relative compaction from WPPSS table: 100.46%

Average % of relative compaction from compaction control curves: 97.5%

Figure 9

E. R. Rybarski

and the WPPSS table were 99.62 percent and 97.79 percent, respectively, which is above the 97 percent PSAR requirement. The average relative compaction using the compaction control curves was 96.41 percent, but acceptable because the compaction control curves are conservative.

The test results of the WNP-4 Spray Pond verification testing program, indicate that the average relative compaction for PTL and WPPSS tables are 99.7% and 98.2%, respectively. The average relative compaction using the compaction control curves was 97.5%.

VII. CONCLUSION

The conclusion reached as a result of the testing program, the in-place density verification program, and the construction of a test fill (which shows the acceptability of the backfill placing procedure), is that the in-place backfill was acceptable for the WNP-1 General Services Building area and for the WNP-1 and WNP-4 Spray Pond areas.

After the testing programs, the conclusion was reached that Type A soils reach maximum density generally at amplitudes in the range of 0.005 to 0.10 inches, regardless of the C_r value. The variation of soil density values in this range is within the testing tolerances expected.

For all subsequent soil density testing, in addition to compaction control correlation curves (which have proved to be conservative), the WPPSS mechanical table (Model 100 VP-D) was used with an amplitude setting of 0.0075 inches. During any backfill operations, control tests were performed on three soil samples with various C_r values and at incremental amplitude settings of 0.005 to 0.015 inches to verify the proper amplitude range and setting required to achieve maximum density. These control tests were performed once every two (2) weeks.

In addition to the above, soil density tests were likewise performed for the Type B soils which are placed five (5) feet above any expected groundwater level at the plant site.

For all subsequent testing, exceptions were taken to the ASTM D 2049 to permit the use of the vibrating table 100 VP-D, at the amplitude which gives the maximum density for the soil type tested.

The correlation method did not eliminate laboratory testing for maximum density on all samples. However, the proposed compaction control correlation procedure greatly reduced the amount of time-consuming maximum density testing in the laboratory.

VIII. REFERENCES

1) ASTM Special Technical Publication 523 (Evaluation of Relative Density) 1972.

2) Special Technical Publication No. 322, ASTM, 1962 Burmister, D. M., "Physical Stress-Strain and Strength Responses of Granular Soils."

3) Preliminary Safety Analysis Report, WNP-1, WNP-4, Docket Nos. 50-460, 50-513.

4) Report on Variations to the PSAR Soil Density Testing Criteria, United Engineers and Constructors, Inc., 1976.

FULL-SCALE HORIZONTAL LOAD TESTS
ON DRILLED PIERS

By Syed B. Ahmed[1], M. ASCE, and Ruben O. Hernandez[2], A.M. ASCE

INTRODUCTION

Full-scale horizontal load tests were conducted on drilled piers at the Tennessee Valley Authority's Yellow Creek Nuclear Plant site in northern Mississippi. The objective was to define the pier response to horizontal loads in order to arrive at an adequate design for drilled pier foundations of the turbine buildings. The tests were performed during March 1977.

Six reinforced concrete drilled piers, 3 and 4 feet in diameter and approximately 38 feet long, were constructed in three rows (Fig. 1). Piers were embedded (a) in all compacted earthfill, (b) in all in situ chert residuum, and (c) partially in compacted earthfill and partially in the in situ chert residuum. All piers were socketed into bedrock. Embedment details are shown in Figs. 2, 3, 4, and 5.

Prior to the pier load tests, soil borings and plate bearing tests were performed at locations shown in Fig. 1. Laboratory tests were conducted on the compacted earthfill and the chert residuum.

Horizontal loads up to 90 tons (20 cycles maximum) were applied by jacking apart the two piers in each row. Lateral deflection and slope were measured near the pier top.

Reese's (1975) computer program COM 622 (3) was used to predict resulting pier response. Pier top deflections measured during the load test were correlated with the predicted deflections. Based upon this study Welch and Reese's (1972) formulations (5) were modified.

A parametric analysis was also done to determine the minimum earthfill thickness required for uniform horizontal response between

---

[1] Civil Engineer, Tennessee Valley Authority, Knoxville, Tennessee.
[2] Head Civil Engineer, Tennessee Valley Authority, Knoxville, Tennessee.

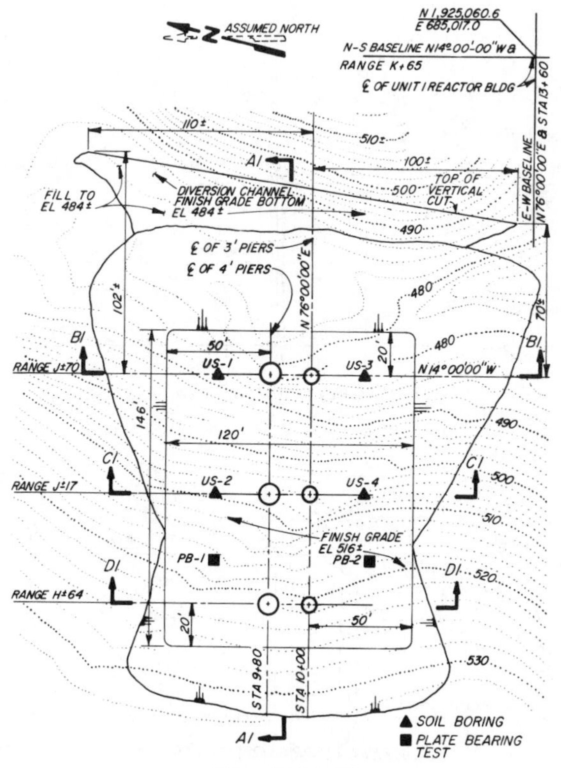

FIG. 1. - TEST SITE

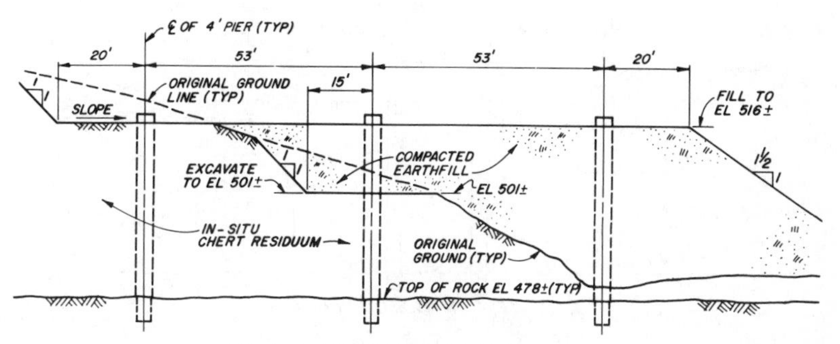

FIG. 2. - TEST PIERS SECTION A1-A1

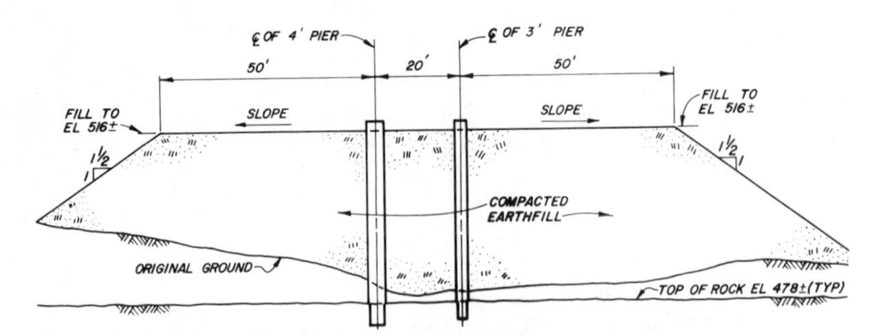

FIG.3.- TEST PIERS SECTION B1-B1

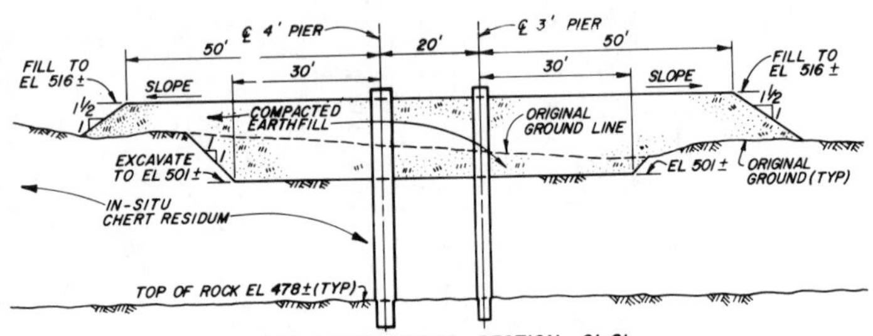

FIG. 4. TEST PIERS SECTION C1-C1

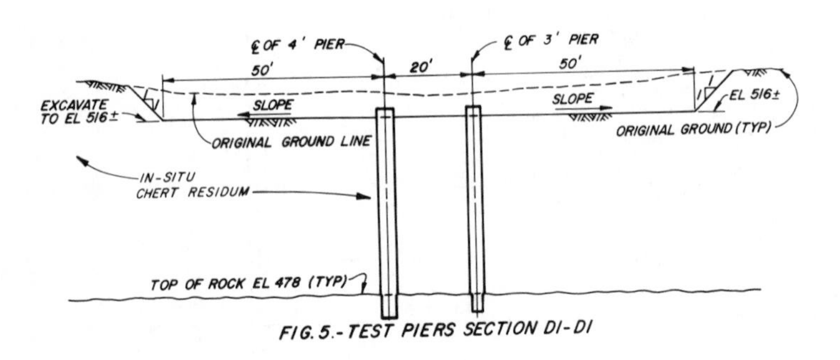

FIG. 5.- TEST PIERS SECTION D1-D1

the piers constructed in all earthfill and those constructed partially in earthfill and partially in the in situ chert.

TEST SITE AND SOIL CONDITIONS

The 120- by 146-foot test site is shown in Fig. 1. The western portion was in cut with a maximum of 38-foot in situ chert residuum over the bedrock. The eastern portion was raised to elevation 516.0 with selected borrow soil compacted to at least 95 percent standard maximum dry density (ASTM D 698). The borrow soil was an alluvial clayey sand (SC). The zone between all cut and all fill areas consisted of partial earthfill and partial in situ chert over the bed-rock. Geologically, the in situ chert residuum belongs to the Upper Fort Payne formation of the Mississippi age. The in situ chert varies from mixtures of sands, gravels, and boulders to fractured blocks and massively bedded thin layers with standard penetration blow counts often exceeding 50.

Underlying the chert residuum is a gray calcarious siltstone bedrock belonging to the Lower Fort Payne formation. Rock sockets for the 3- and 4-foot piers were 2 feet and 3 feet in diameter, respectively. Socket length varied from 4.6 to 5.5 feet.

DRILLING AND CONSTRUCTION

The compacted earthfill was drilled by earth augering with a crane-mounted drill and single flight augers. Penetration rates ranged from 16 to 22 feet per hour.

The in situ chert excavation varied from layers of 2 to 18 inches. A combination of conventional core barrels and drilling buckets and extensive use of a heavy drop bar proved to be the most effective drilling method.

Rock sockets were best done by down-the-hole percussion drilling followed by rotary drilling.

Reinforcing steel fabrication and installation was conducted with no apparent deficiencies or problems. Additional cross bracing was used in the cages during fabrication to alleviate distortion during handling and hoisting. It was removed as the cages were placed in the shaft.

Concrete handling and placement followed the guidelines of ACI (1973), Manual of Concrete Practice (1). All concrete was placed in open dry shafts. Concrete was allowed to free fall through a centralizing hopper with a short trunk. Concrete test cylinders were prepared and tested onsite.

SOIL EXPLORATION AND TESTING

A. Compacted Earthfill

Soil exploration consisted of four undisturbed borings US-1 through US-4, shown in Fig. 1. Undisturbed samples of the in-place compacted earthfill were recovered in 5-inch-diameter shelby tubes.

Representative samples of the compacted earthfill from each boring were tested for index properties and unconsolidated undrained triaxial compression (Q) tests. The Q test samples were trimmed with their axes oriented horizontally. Such trimming was to simulate the direction of loads used in the pier load tests. The Q tests were performed at a confining pressure equal to the overburden pressure.

Table 1 presents a summary of the laboratory test results. The results indicate that the compacted earthfill was basically a (SC) soil, generally containing a trace of gravel. The liquid limits were fairly uniform with an average of 34.6 and extreme values of 27.8 and 38.6 percent. The plasticity index ranged from 8.0 to 20.1 percent with 13.9 percent as an average value. The dry densities were also quite uniform with an average value of 106.9 pcf and extreme values of 101 and 114 pcf.

Fig. 6 indicates the variation of $\frac{\sigma_1 - \sigma_3}{2}_{max}$ (one-half the maximum principal stress difference) with depth. The relationship between the principal stress difference and depth below ground can be expressed as:

$$\frac{\sigma_1 - \sigma_3}{2}_{max} = (2500 + 77.4\ x)\ \text{psf, where x = depth of soil in feet.}$$

This expression was used in developing p-y curves for the piers in compacted earthfill. Strain ε_{50} corresponding to

TABLE 1.——Summary of Laboratory Tests on Compacted Earthfill (SC)

Elevation	Natural Moist. %	Grain Size Analysis				Atterberg Limits		Dry Density pcf	Triaxial Tests	
		Gravel %	Sand %	Silt %	Clay %	Liquid Limit %	Plasticity Index %		$\left(\frac{\sigma_1 - \sigma_3}{2}\right)_{max}$ tsf	ε_{50}
Boring US-1										
515.0–513.0	15.9	0	70	11	19	32.9	12.0	114.0	1.55	0.015
509.0–506.6	18.4	2	54	16	28	38.6	15.9	106.0	1.53	0.012
503.0–501.1	18.6	0	62	14	24	34.6	12.9	105.3	1.79	0.010
497.0–495.1	14.9	2	61	11	26	33.3	11.6	107.8	2.17	0.015
491.0–489.4	18.4	0	65	11	24	35.2	14.7	105.1	2.10	0.010
Boring US-2										
511.4–509.0	18.3	4	55	17	24	38.6	16.2	106.7	1.57	0.008
505.4–503.3	16.7	1	65	9	25	37.6	20.1	108.7	1.43	0.020
502.4–501.3	16.7	6	65	11	18	32.9	11.3	101.0	1.24	0.008
Boring US-3										
512.3–510.0	17.7	0	61	14	25	31.1	10.8	111.0	1.51	0.010
506.3–504.0	16.1	0	52	23	25	34.5	15.4	103.8	1.81	0.010
500.3–498.0	16.7	1	74	7	18	27.8	8.0	107.1	1.71	0.016
494.3–492.2	16.4	1	56	18	25	34.2	13.4	105.9	2.04	0.016
488.3–486.5	16.4	0	62	15	23	33.4	12.9	109.2	2.43	0.015
483.3–481.6	18.3	8	47	18	27	36.6	16.1	107.0	3.06	0.010
Boring US-4										
511.0–508.9	18.4	1	54	18	27	36.4	14.2	105.1	1.46	0.017
505.0–502.8	17.5	2	60	13	25	36.0	16.6	106.8	1.72	0.016
502.0–500.6	16.2	7	57	14	22	34.7	14.9	107.2	1.49	0.015

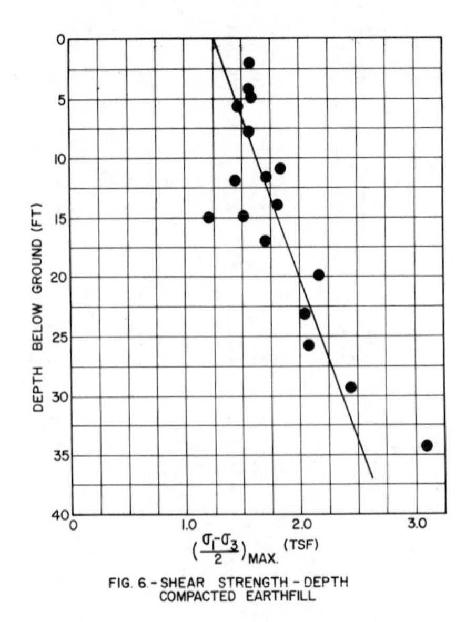

FIG. 6 - SHEAR STRENGTH - DEPTH
COMPACTED EARTHFILL

$\left(\frac{\sigma_1 - \sigma_3}{2}\right)_{max}$ ranged from 0.008 to 0.020 with an average value of 0.013.

B. In Situ Chert Residuum

For predicting the horizontal response of drilled piers embedded in cohesionless material like the in situ chert, the following soil properties are required: (1) density, (2) angle of internal friction, and (3) modulus of elasticity. Several attempts were made to obtain undisturbed samples of the in situ chert from the borings. Trials were also made to conduct Menard pressure meter tests on the walls of the borings made in the chert residuum. Such attempts and trials were not successful due to the nature of the in situ chert (containing large-size gravels and fractured boulders). Therefore, horizontal plate bearing tests were conducted at two locations (Fig. 1) to determine the modulus of elasticity of the in situ chert.

The modulus of elasticity was evaluated from the following equation:

$$S = q\ B \frac{(1 - \mu^2)}{E} I_w \quad (2)$$

where q = intensity of contact pressure; S = settlement of the plate; B = width of the plate; μ = poisson's ratio; I_w = shape factor; and E = modulus of elasticity of soil. Poisson's ratio (μ) and the shape factor (I_w) were assumed to be 0.20 and 0.82, respectively.

Undisturbed block samples of the in situ chert were obtained by hand trimming from the trenches excavated in the main power plant area. Direct shear strength tests were conducted on the samples remolded at the in situ density and natural moisture content. Table 2 summarizes the properties of chert residuum.

TABLE 2.——Properties of Chert Residuum

Moist Density γ	Friction Angle ϕ	Modulus of Elasticity E
117 pcf	40° - 45°	23,000 - 69,000 psi

PIER LOAD TESTS

Three identical pier load tests were conducted. A calibrated hydraulic jack (150-ton capacity) was used to simultaneously load a 3-foot-diameter and a 4-foot-diameter pier in the horizontal direction in each row. A calibrated electronic load cell (150-ton capacity) was used to record the horizontal loads. The resulting slope and deflection were measured near the top of the piers.

A. Loading Apparatus, Instrumentation, and Test Set-up

Figs. 7 and 8 illustrate the loading apparatus, instrumentation, and test set-up.

Pier deflection was measured by dial gages having an accuracy of 0.001 inch. These gages were mounted on a reference beam supported by stakes located 15 feet on either side of the centerline.

The slope of the pier was measured with the device shown in Fig. 8. The essential elements of this device were a supporting frame made of 3-inch equal leg angles, a bar with a sensitive level bubble firmly attached to it, and a micrometer screw to adjust the level. The micrometer screw had an accuracy of 0.001 inch. The support frame was firmly attached to the pier

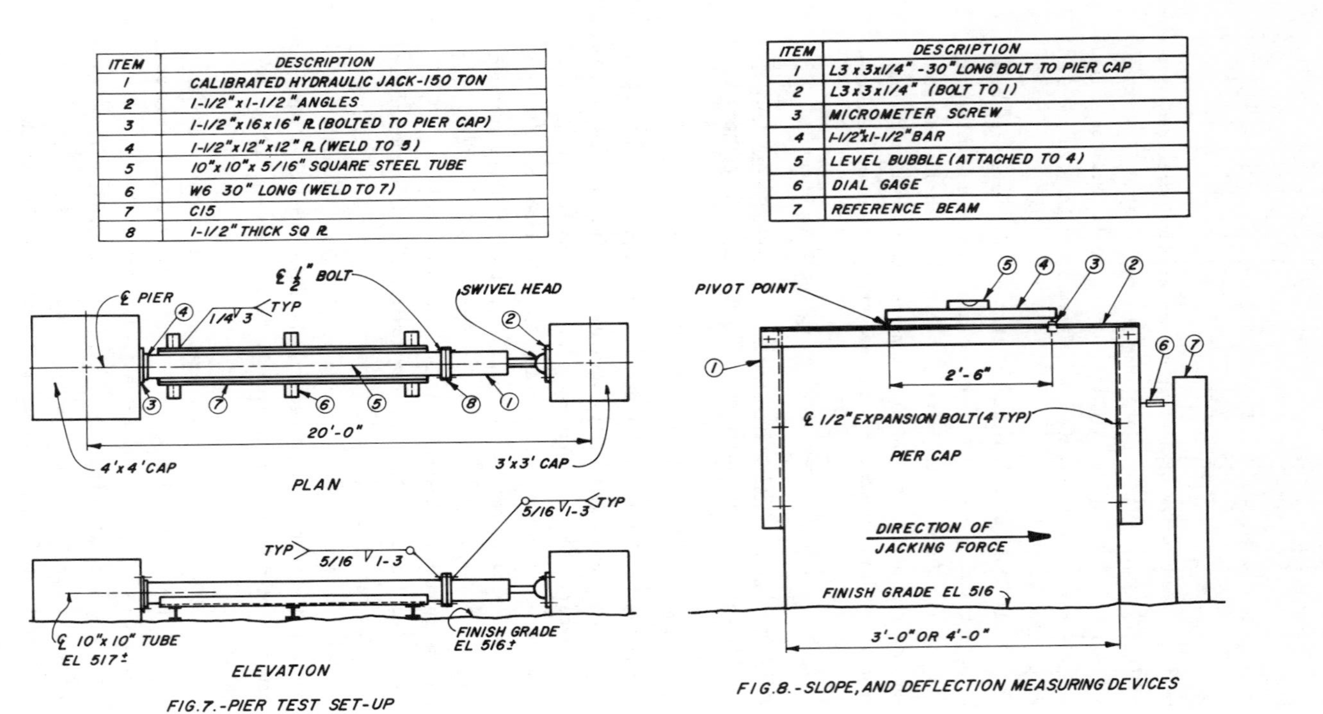

ITEM	DESCRIPTION
1	CALIBRATED HYDRAULIC JACK-150 TON
2	1-1/2"x1-1/2" ANGLES
3	1-1/2"x16x16" PL (BOLTED TO PIER CAP)
4	1-1/2"x12"x12" PL (WELD TO 5)
5	10"x10"x 5/16" SQUARE STEEL TUBE
6	W6 30" LONG (WELD TO 7)
7	C15
8	1-1/2" THICK SQ PL

ITEM	DESCRIPTION
1	L3 x 3x1/4" -30" LONG BOLT TO PIER CAP
2	L3x3x1/4" (BOLT TO 1)
3	MICROMETER SCREW
4	1-1/2"x1-1/2" BAR
5	LEVEL BUBBLE (ATTACHED TO 4)
6	DIAL GAGE
7	REFERENCE BEAM

FIG.7.-PIER TEST SET-UP

FIG.8.-SLOPE, AND DEFLECTION MEASURING DEVICES

cap. The top of the frame was essentially level and parallel to the centerline of the two piers. The bar was supported at one end by a pivot point and the other end by the micrometer screw. The two support points were 30 inches apart.

For slope measurement, the bar was leveled prior to load application. During loading the bar was kept level by adjusting the micrometer screw. The slope (in radians) was calculated by dividing the supported length into the distance the micrometer screw was moved.

B. Loading Procedure

The piers were subjected to loads of 10, 20, 30, 40, 60, 75, and 90 tons in the following sequence.

1. The slope measurement device was leveled and the deflection measurement instrument was set to zero.
2. A 10-ton load was applied. Deflection and slope were measured. The load was removed and the slope and deflection were again measured. Measurements for both piers were taken simultaneously and as soon as possible after the load application or removal.
3. The 10-ton load was reapplied and slopes and deflections were measured. The load was removed and the slopes and deflections were measured again. This procedure was repeated until the deflection did not increase more than 0.005 inch for three successive loadings, or until the load had been applied twenty times.
4. The above procedure was repeated for each load being cycled until the deflection stabilized, as described in step 3, or until the load was applied twenty times.
5. Steps 1 through 4 were repeated for all test piers.

PIER TEST RESULTS

The deflections and slopes measured near the pier top for various loads at the end of the first cycle are presented in Figs. 9 and 10. These figures indicate that, for given horizontal load and soil conditions, deflection and the slope decrease with increasing pier size. The piers embedded in all compacted earthfill exhibited larger deflections and slopes than those embedded in all

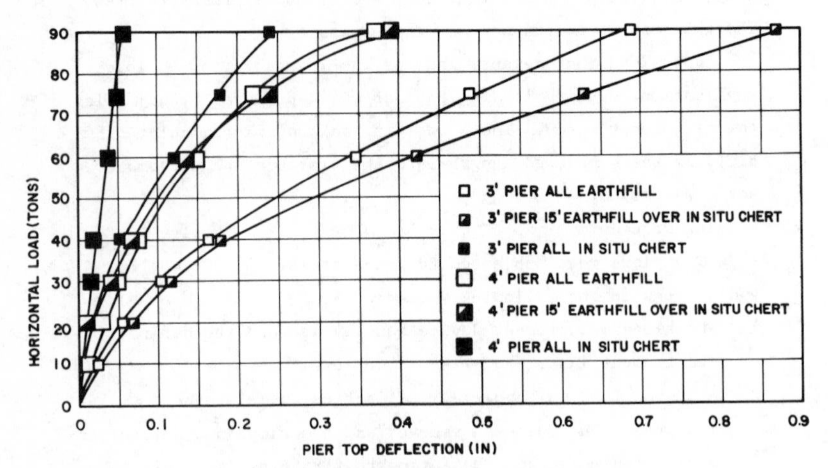

FIG.9 .- LOAD-MEASURED DEFLECTION, ALL TEST PIERS

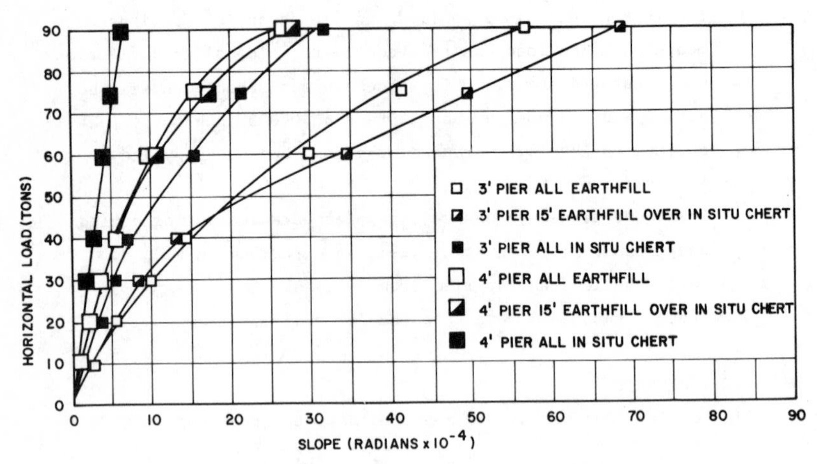

FIG.10.- LOAD- MEASURED SLOPE, ALL TEST PIERS

in situ chert. Measured deflections and slopes for the piers in partial (15 feet) earthfill and partial (23 feet) in situ chert were close to the all-earthfill case.

ANALYSIS OF PIER TEST RESULTS

The pier test results were analyzed for short-term static loads only.

Formulations recommended for computing p-y curves for stiff clay and sand (5) were applied to the compacted earthfill and the in situ chert, respectively.

1. Piers in Compacted Earthfill

   (a) Shear strength of the compacted earthfill was evaluated from (Q) tests on undisturbed samples. Strain, ε_{50}, corresponding to one-half the maximum principal stress difference, was determined from the stress-strain curves.

   (b) The smallest value of the ultimate soil resistance per unit length of pier, p_u, was adopted as given by one of the following equations:

$$p_u = [3 + \frac{\gamma}{c} x + \frac{0.5}{b} x] \, c \, b \quad \text{(1)}$$

$$p_u = 9 \, c \, b \quad \text{(2)}$$

   where γ = average unit weight of soil from ground surface to depth x, c = average shear strength from ground surface to depth x, b = diameter of the pier, and x = depth to p-y curve.

   (c) The deflection y_{50} at half the ultimate soil resistance was computed from the following equation:

$$y_{50} = 2.5 \, \varepsilon_{50} \, b \quad \text{(3)}$$

   (d) p-y Curves were developed from:

$$\frac{p}{p_u} = 0.5 \left(\frac{y}{y_{50}}\right)^{\frac{1}{4}} \quad \text{(4)}$$

   Computer program (3) was used to analyze lateral response of the piers. The input information included the pier diameter, pier length, modulus of elasticity of concrete, moment of inertia of the pier, soil density, p-y curves data, number of finite difference increments, and the horizontal load and slope measured near the pier top as boundary conditions. Computer output included lateral deflection, moment, soil modulus, and soil reaction along the pier length.

Results of analysis indicated that for a given load and similar piers and soil conditions, the deflection computed from formulations 1, 2, 3, and 4 were always higher than the measured deflections. Modifications were considered to improve compatibility among the computed and the measured deflections. Use of formulation 3, which is Skempton's (1951) equation (4), seemed to be reasonable because the ratio of the length to the diameter of the test piers was approximately 10, which is close to the assumption made by Skempton. Therefore, there was no need to modify formulation 3.

Formulations 1 and 2 were candidates for modification because the clays described in reference (5) were different from the compacted earthfill studied in this program, but it was considered much simpler to modify the constants in formulation 4 only. A parametric analysis was made to investigate the effect of varying the coefficient 0.5 and the exponent $\frac{1}{4}$ on the computed deflections. A reasonable agreement between the measured and the computed deflections was achieved by modifying the coefficient to 0.7 and retaining the exponent $\frac{1}{4}$. Formulation 4 was thus revised to $\frac{p}{p_u} = 0.7 \left(\frac{y}{y_{50}}\right)^{\frac{1}{4}}$.

A comparison among the measured deflections, those computed by the referenced formulations (5), and the modified formulation is shown in Figs. 11 and 12.

2. Piers in In Situ Chert Residuum

Techniques for developing p-y curves for predicting the horizontal response of the piers constructed in materials like chert residuum are not available in the literature. However, since the in situ chert is basically a cohesionless granular material, the procedure recommended for predicting the horizontal

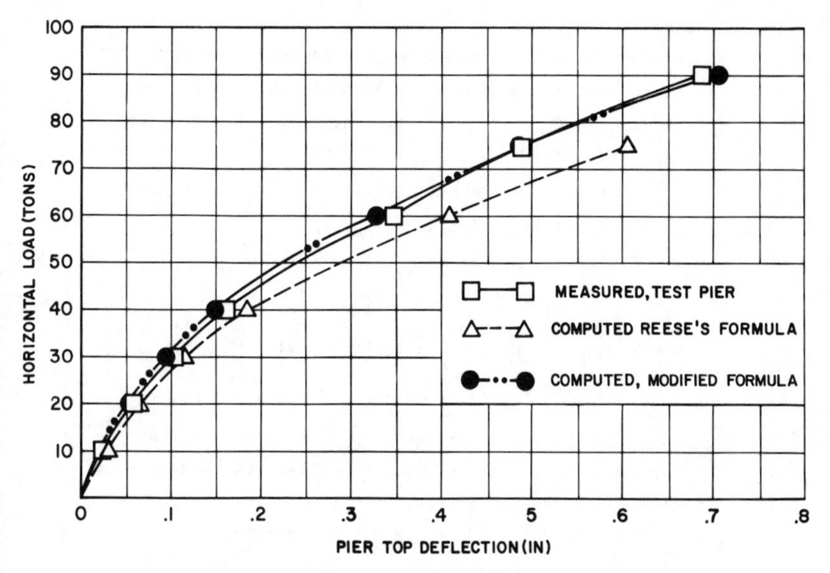

FIG.11. - LOAD-DEFLECTION, 3' PIER ALL EARTHFILL

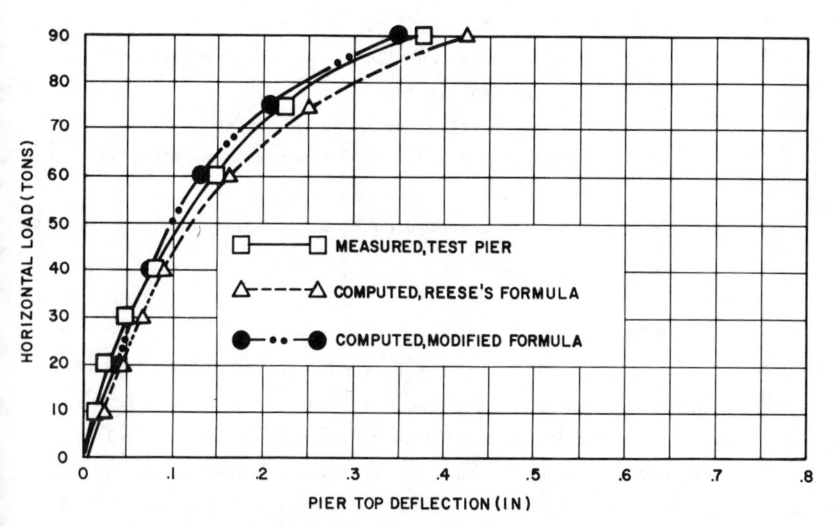

FIG.12. - LOAD-DEFLECTION, 4' PIER ALL EARTHFILL

response of piers in sand (5) was adopted. The p-y curves were developed as follows:

(a) Modulus of elasticity was determined from the horizontal plate bearing tests conducted on in situ chert. Density and angle of internal friction were obtained from the laboratory tests.

(b) The earth pressure coefficients were computed from the following equations:

Coefficient of active earth pressure:

$$K_A = \mathrm{Tan}^2\ (45 - \phi/2) \quad (5)$$

Coefficient of passive earth pressure:

$$K_P = \mathrm{Tan}^2\ (45 + \phi/2) \quad (6)$$

(c) The ultimate soil resistance (p_u) was computed from the following two equations, but a smaller value was used for developing the p-y curves:

$$p_u = \gamma\, x\, [b\, (K_P - K_A) + x\, K_P\, (\tan \alpha \tan \beta) + x\, K_0 \tan \beta\, (\tan \phi - \tan \alpha)] \quad (7)$$

and $$p_u = \gamma\, x\, b\, [K_P^3 + 2\, K_P^2\, K_0 \tan \phi + 2\, K_0 \tan \phi - K_A] \quad (8)$$

where γ = average density from the ground surface to depth x;
x = depth to p-y curves, K_0 = earth pressure coefficient at rest = 0.5
α = angle defining the shape of the wedge in lateral direction = $\phi/2$, and β = angle from bottom of wedge to vertical = $45 + \phi/2$.

(d) p-y Curves data were computed from the following relationship.

$$P = C_1 p_u \tan h \frac{Ey}{1.35 p_u} \quad (9)$$

where $C_1 = 1.0$.

The measured pier top deflections were significantly lower than those calculated from computer program (3) using formulations 5, 6, 7, 8, and 9 recommended in reference (5). Some level of modification was needed in order to improve compatibility between the computed and measured deflections. Since formulations 5 and 6 are standard equations and formulations 7 and 8 are based upon sound assumptions, no attempt at modification was made. However, a parametric study was conducted to modify formulation 9 by varying the constant C_1. Reasonable agreement was achieved between the computed and measured deflections for $C_1 = 2.5$.

Figs. 13 and 14 show the plots of pier top deflections measured during the test and those computed from recommended formulations (5) and the modified formulations for piers embedded in all in situ chert. These figures show a better agreement between the measured deflections and those computed from the modified formulation for the 3-foot-diameter pier than for the 4-foot-diameter pier. The inconsistency in the behavior of the 4-foot-diameter test pier may be caused by variation in the in situ chert at the two test locations. In situ chert was nonhomogeneous, changing from a mixture of sands and gravels to boulders, blocks, and massively bedded residuum within short distances.

3. Piers Partially in Compacted Earthfill and Partially in In Situ Chert Residuum

Pier top deflections were computed using the modified formulations for the compacted earthfill and the in situ chert. Comparison between the measured and computed deflections for 3- and 4-foot-diameter piers are shown in Figs. 15 and 16, respectively.

EARTHFILL THICKNESS FOR A UNIFORM HORIZONTAL RESPONSE

The objective of the study was to achieve uniform horizontal response among piers embedded in all compacted earthfill and those partially in compacted earthfill and partially in the in situ chert at the turbine buildings' construction site. This, in turn, requires determining a minimum thickness of earthfill over the in situ chert. A computer aided parametric study focused on the horizontal response of 3.5- and 4-foot-diameter, 38-foot-long piers in earthfill embedments of 3, 6, 9, 12, and 15 feet over the in situ chert. The maximum deflection and the maximum bending moment were computed for each case. The pier loads and boundary conditions were: axial load = 1500 k; horizontal load at pier top = 100 k and 150 k; and rotational restraint at pier top = 1×10^7 ft-k/radian. Figs. 17 and 18 show that earthfill

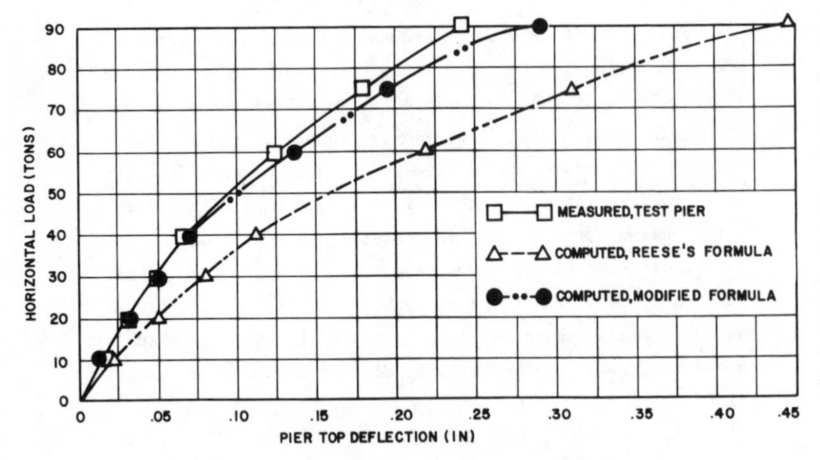

FIG.13. - LOAD-DEFLECTION, 3' PIER ALL IN SITU CHERT

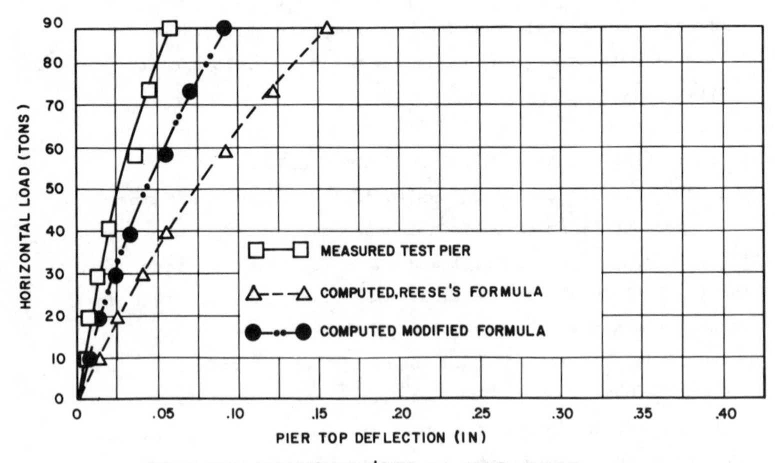

FIG.14. - LOAD-DEFLECTION, 4' PIER ALL IN SITU CHERT

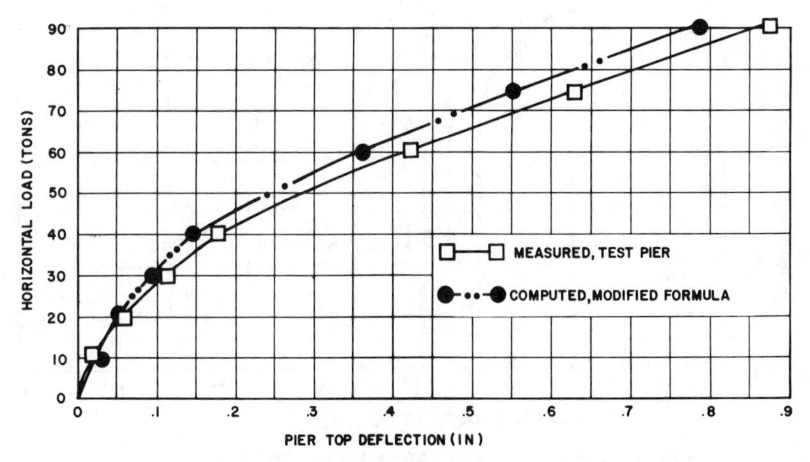

FIG. 15.- LOAD-DEFLECTION, 3' PIER 15' EARTHFILL OVER IN SITU CHERT

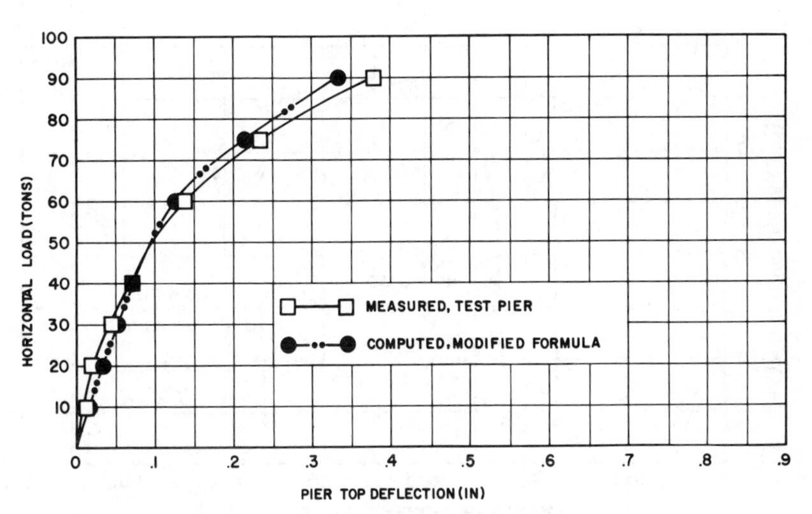

FIG. 16.- LOAD-DEFLECTION, 4' PIER 15' EARTHFILL OVER IN SITU CHERT

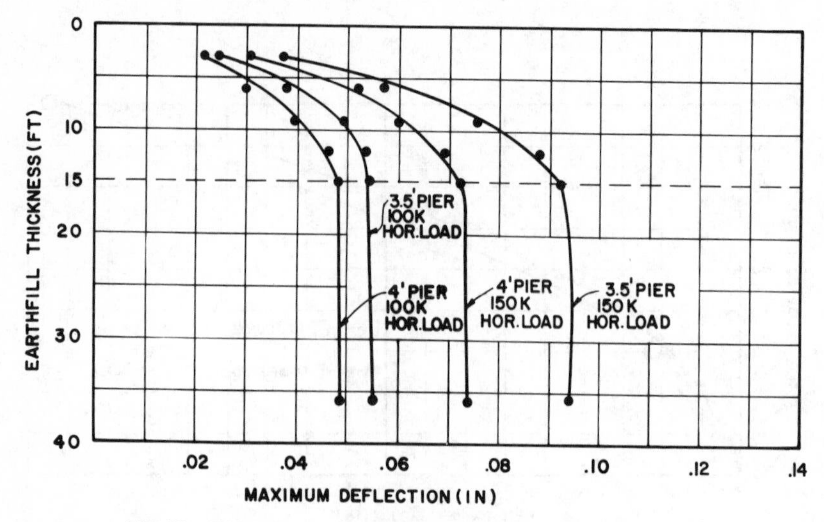

FIG. 17.-EARTHFILL THICKNESS-MAX. DEFLECTION

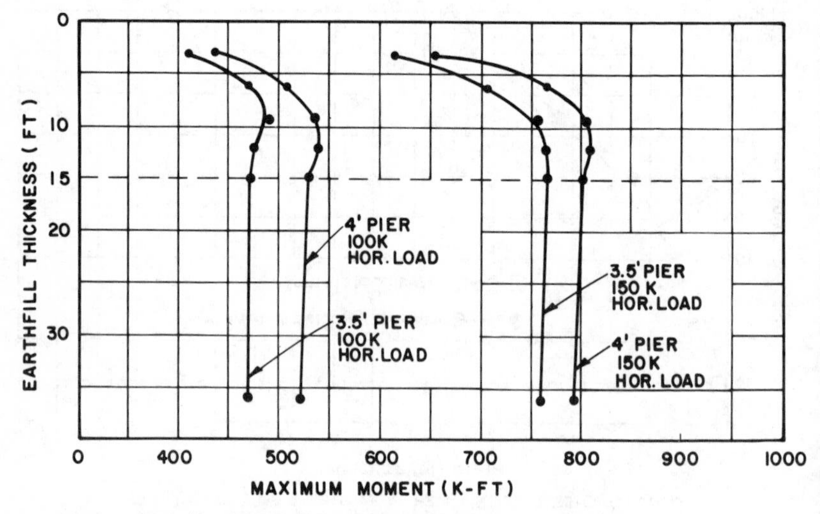

FIG. 18.-EARTHFILL THICKNESS-MAX. MOMENT

in excess of 15 feet, i.e., approximately 4 times the pier diameter, did not have a significant effect on the horizontal response of the piers. Therefore, a minimum earthfill thickness of 15 feet was considered necessary to achieve a uniform horizontal response among the piers constructed in all earthfill and those constructed partially in earthfill and partially in the in situ chert.

CONCLUSIONS

The following conclusions are limited to the test conditions, soil parameters, and pier sizes used in the testing program:

1. The horizontal response of the test piers embedded in compacted earthfill and in situ chert residuum can be obtained successfully by measuring the applied horizontal load, deflection, and slope near the pier top. The test data can be extended for use in the design and analysis of laterally loaded drilled piers.
2. For all test piers, deflections measured near the pier top were lower than those computed from the procedure recommended in reference (5).
3. Reasonable agreement was achieved between the measured and the computed deflections of the test piers by modifying the recommended formulations (5) for p-y curves for stiff clay and sand. The equation for stiff clay was modified to $\frac{p}{p_u} = 0.7 \left(\frac{y}{y_{50}}\right)^{\frac{1}{4}}$ and applied to piers in compacted earthfill. The formulation for sand was modified to $p = 2.5\ p_u \tanh \frac{Ey}{1.35p_u}$ and applied to piers in the in situ chert.
4. A 15-foot-layer of earthfill (approximately 4 times the pier diameter) was the minimum required for a uniform horizontal response among the piers constructed in all earthfill and those constructed partially in earthfill and partially in the in situ chert.

ACKNOWLEDGMENTS

This study was conducted under the direction of Robert O. Barnett, Chief, Civil Engineering Branch, TVA. The testing work was done by TVA's Singleton Materials Engineering Laboratory in Knoxville, Tennessee.

Roy E. Hoekstra and James W. Rosteck assisted in the computer analysis and the test setup. Cooperation and encouragement received from R. J. Hunt, Principal Engineer, TVA, in preparing this presentation is appreciated.

APPENDIX 1 - REFERENCES

(1) ACI Manual of Concrete Practice, 1973, Part 2, American Concrete Institute.

(2) Bowles, J. E., "Foundation Analysis and Design," McGraw-Hill Book Company, Incorporated, New York, 1977.

(3) Reese, L. C., "Analysis of Laterally Loaded Piles," Software Documentation, GESA Report No. D-75-7, UCCC Report No. 75-10, Department of Civil Engineering, The University of Texas at Austin, July 1975.

(4) Skempton, A. W., "The Bearing Capacity of Clays," Proceedings, Building Research Congress, Division 1, London, England, 1951.

(5) Welch, R. C., and Reese, L. C., "Lateral Load Behavior of Drilled Shafts," Research Report 89-10, Project 3-5-65-89, Center for Highway Research, The University of Texas at Austin, May 1972.

SOIL-STRUCTURE INTERACTION AND IDENTIFICATION OF STRUCTURAL MODELS

By J. Enrique Luco[1]

INTRODUCTION

The San Fernando earthquake of 1971, in which a large number of strong motion records in structures was obtained, has prompted an increased interest in the identification of structural models from earthquake records. Because of the possible lack of uniqueness in the identification of structural characteristics such as stiffness and damping matrices, the emphasis has been placed on the identification of modal quantities such as modal frequencies, modal damping ratios and modal participation factors. One of the most interesting results obtained from identification studies is that the calculated modal frequencies appear to vary during the earthquake response and that the values obtained for the modal frequencies during moderate earthquakes are significantly lower than those obtained from low-amplitude forced or ambient vibration tests (Udwadia and Trifunac, 1974). In addition, the calculated modal damping ratios for different structures subjected to different intensities of shaking seem to increase with amplitude of shaking and are considerably higher than the values obtained from vibration tests (Hart and Vasudevan, 1975). On the other hand, some apparent reduction in energy dissipation capacity during the strongest shaking portion of the response of the Millikan Library building has been noted by Iemura and Jennings (1974). These apparent variations of modal characteristics are typically interpreted as resulting from structural degradation.

---

[1]Assoc. Prof., Dept. of Applied Mechanics and Engineering Sciences, Univ. of California, San Diego, La Jolla, California.

Parametric and non-parametric structural identification techniques are generally based on the assumption that the effects of soil-structure interaction are negligible. More precisely, since typically only the translational components of motion at a point on the foundation are recorded, it is necessary, for the purpose of the identification, to assume that the rocking and torsional components of motion at foundation level are negligible. In cases in which this assumption is not valid the modal quantities determined from typical identification techniques reflect not only the characteristics of the superstructure but also the characteristics of the foundation and of the underlying soil. Depending on the particulars of the identification technique employed, the calculated modal frequencies and damping ratios may correspond to the modal quantities for the complete structure-foundation-soil system or to some other values intermediate between those for the complete system and those for the superstructure.

The principal aim of this study is to describe the effects of soil-structure interaction on the modal quantities obtained by typical identification techniques in which the interaction effects are excluded. In particular, the physical meaning of the apparent modal quantities is examined. A second objective of the study is to determine the range of system configurations for which the interaction effects can be neglected. Finally, the possibility of interpreting the observed variations of modal characteristics as resulting from variations in soil properties is explored.

This study is addressed to the analysis of identification techniques based on the use of strong-motion records. The interpretation of the results of forced vibration tests has been presented elsewhere (Luco, Wong and Trifunac, 1980).

BASIC INTERACTION EQUATIONS

To determine the physical meaning of the results obtained by different identification techniques it is convenient to start by considering a planar elastic structure supported on a flat rigid foundation resting on a viscoelastic half-space. The equation of motion for the superstructure, in the frequency domain, can be written in the form

$$-\omega^2[M]\{U_T\} + i\omega[C]\{U\} + [K]\{U\} = \{0\} \tag{1}$$

in which [M], [C] and [K] are the fixed-based mass, damping and stiffness matrices for the superstructure. The displacement vector $\{U_T\}$ corresponds to the total displacement of the mass points, while $\{U\}$ represents the displacement of the mass points relative to a moving frame of reference attached to the rigid foundation. The displacements $\{U_T\}$ and $\{U\}$ are related by

$$\{U_T\} = \{1\}\ (U_g + U_o) + \{h\}\ \phi_o + \{U\} \tag{2}$$

in which U_g corresponds to the free-field motion on the ground surface, U_o to the additional translational motion of the foundation associated with interaction, and ϕ_o to the angle of rocking of the foundation. The vectors $\{1\}$ and $\{h\}$ appearing in Eq. 2 are defined by $\{1\}^T = (1,1,\ldots,1)$ and $\{h\}^T = (h_1,h_2,\ldots,h_n)$ where h_i denotes the height of the ith mass. In writing Eq. 2 it has been assumed that the seismic excitation corresponds to vertically incident SH-waves.

Neglecting the mass of the foundation and the rotatory inertia of the superstructure it is possible to write the equations of motion for the foundation in the form

$$-\omega^2\{1\}[M]\{U_T\} + Gr\ [(1+2i\xi_s)k_H + i\omega c_H]\ U_o = 0 \tag{3}$$

$$-\omega^2\{h\}[M]\{U_T\} + Gr^3\ [(1+2i\xi_s)k_R + i\omega c_R]\ \phi_o = 0 \tag{4}$$

in which G is a shear modulus of reference for the soil, r the equivalent radius of the foundation, ξ_s the hysteretic damping constant for the soil, k_H and k_R the frequency-dependent horizontal and rocking stiffness coefficients for the foundation, and c_H and c_R the frequency-dependent horizontal and rocking radiation damping coefficients for the foundation. In Eqs. 3 and 4, the coupling impedance functions for the foundation have been assumed negligible.

For vibrations in the neighborhood of the fundamental fixed-base natural frequency of the superstructure it is possible to neglect the contributions of the higher modes. In this case, the relative displacement $\{U\}$ can be approximated by

$$\{U\} = \{\psi_1\} U_b \tag{5}$$

where U_b represents the relative displacement at the top of the structure and $\{\psi_1\}$ denotes the fundamental fixed-base mode normalized to unity at the top of the structure.

Substitution from Eqs. 2 and 5 into Eqs. 1, 3 and 4 leads to the following system of equations for U_b, U_o and ϕ_o:

$$[D(\omega)] \begin{Bmatrix} U_b/\beta_1 \\ (\omega_H/\omega_1)U_o \\ (\omega_R/\omega_1)H_1\phi_o \end{Bmatrix} = \begin{Bmatrix} 1 \\ (\omega_1/\omega_H) \\ (\omega_1/\omega_R) \end{Bmatrix} U_g \tag{6}$$

in which the matrix $[D(\omega)]$ is given by

$$[D(\omega)] = \begin{bmatrix} 1-(\omega/\omega_1)^2+2i(\omega/\omega_1)\xi_1 & -(\omega/\omega_1)(\omega/\omega_H) & -(\omega/\omega_1)(\omega/\omega_R) \\ -(\omega/\omega_1)(\omega/\omega_H) & 1-(\omega/\omega_H)^2+2i[\xi_s+(\omega/\omega_H)\xi_H] & -(\omega/\omega_H)(\omega/\omega_R) \\ -(\omega/\omega_1)(\omega/\omega_R) & -(\omega/\omega_R)(\omega/\omega_H) & 1-(\omega/\omega_R)^2+2i[\xi_s+(\omega/\omega_R)\xi_R] \end{bmatrix} \tag{7}$$

In Eqs. 6 and 7, ω_1 and ξ_1 represent, respectively, the fixed-base natural frequency and modal damping ratio of the superstructure, while β_1 and H_1 defined by

$$\beta_1 = \{\psi_1\}^T[M]\{1\}/\{\psi_1\}^T[M]\{\psi_1\} \tag{8}$$

and

$$H_1 = \{\psi_1\}^T[M]\{h\}/\{\psi_1\}^T[M]\{1\} \tag{9}$$

represent the modal participation factor and modal height, respectively. The frequency-dependent quantities $\omega_H(\omega)$ and $\omega_R(\omega)$ are defined by

$$\omega_H^2 = Gr\ k_H(\omega)/M_1 \tag{10}$$

$$\omega_R^2 = Gr^3\ k_R(\omega)/M_1H_1^2 \tag{11}$$

in which M_1 corresponds to the modal mass

$$M_1 = \beta_1^2\ \{\psi_1\}^T[M]\{\psi_1\}\ . \tag{12}$$

Finally, the frequency-dependent quantities $\xi_H(\omega)$ and $\xi_R(\omega)$ are defined by

$$\xi_H = \omega_H c_H(\omega)/2k_H(\omega) \tag{13}$$

$$\xi_R = \omega_R c_R(\omega)/2k_R(\omega)\ . \tag{14}$$

It should be noted that the approximations

$$\{1\}^T[M]\{1\}\sim M_1,\ \{h\}^T[M]\{1\}\sim M_1H_1,\ \{h\}^T[M]\{h\}\sim M_1H_1^2 \tag{15}$$

have been used in the derivation of Eq. 6. For most structures, $\beta_1\sim 1.3\text{-}1.5$, $H_1/H\sim 0.6\text{-}0.7$ and $M_1/M_b\sim 0.6\text{-}0.8$ where M_b is the total mass.

An approximate solution of Eq. 6 obtained by neglecting terms involving products of the small quantities ξ_1, ξ_s, ξ_R and ξ_H is given by

$$\begin{Bmatrix} U_b \\ U_o \\ H\phi_o \end{Bmatrix} = \frac{(\omega/\omega_1)^2}{\Delta_o(\omega)} \begin{Bmatrix} \beta_1 \\ (1-2i[\xi_s-(\omega/\omega_1)\xi_1+(\omega/\omega_H)\xi_H])\ (\omega_1/\omega_H)^2 \\ (1-2i[\xi_s-(\omega/\omega_1)\xi_1+(\omega/\omega_R)\xi_R])\ (\omega_1/\omega_R)^2(H/H_1) \end{Bmatrix} U_g \tag{16}$$

in which H is the height of the structure and

$$\Delta_o(\omega) = 1-(\omega/\tilde{\omega}_1)^2 + 2i\tilde{\xi}_1 \tag{17}$$

where

$$1/\tilde{\omega}_1^2 = 1/\omega_1^2 + 1/\omega_R^2 + 1/\omega_H^2 \tag{18}$$

and

$$\tilde{\xi}_1(\omega) = (\omega/\omega_1)^3\xi_1 + [1-(\omega/\omega_1)^2]\xi_s + (\omega/\omega_H)^3\xi_H + (\omega/\omega_R)^3\xi_R + $$
$$+ [1-(\omega/\tilde{\omega}_1)^2]\ (\xi_1\omega/\omega_1-\xi_s) \tag{19}$$

It is apparent from Eqs. 16 and 17 that the peak response of the system occurs in the vicinity of the frequency $\tilde{\omega}_1$, given by Eq. 18, which corresponds to the fundamental frequency of the complete soil-structure system. The equivalent damping ratio for the complete system corresponds to $\tilde{\xi}_1(\tilde{\omega}_1)$ which can be obtained from Eq. 19 by setting $\omega=\tilde{\omega}_1$. In the particular case of a single-story structure ($\beta_1=1$, $H_1=H$, $M_1=M_b$), the results given by Eq. 16 are equivalent to those obtained by Jennings and Bielak (1973). In the general case of multistory structures, the expressions for U_o and ϕ_o given above differ from those presented by Jennings and Bielak.

Having obtained a solution for U_b, U_o and $H\phi_o$ in terms of the free-field motion U_g it is possible to obtain other quantities of interest such as the total translational response at foundation level $U_{To}=U_g+U_o$, the total translational response at the top of the structure $U_{Tb} = U_{To} + H\phi_o + U_b$, and the translational response at the top relative to the total translation of the base $U_{Tb}-U_{To}$. The resulting expressions are

$$U_{To} = [1-(\omega/\tilde{\omega}_1^*)^2 + 2i\tilde{\xi}_1^*]\ U_g/\Delta_o(\omega) \tag{20}$$

$$U_{Tb} = \{\beta_1(\omega/\tilde{\omega}_1^*)^2 + 1-(\omega/\tilde{\omega}_1^*)^2 + 2i\tilde{\xi}_1^* + (\omega/\omega_R)^2\ (H/H_1-\beta_1) + $$
$$+ 2i(H/H_1)[(\omega/\omega_1)\xi_1-\tilde{\xi}_1^*]\}U_g/\Delta_o(\omega) \tag{21}$$

$$U_{Tb}-U_{To}=\{\beta_1(\omega/\tilde{\omega}_1^*)^2+(\omega/\omega_R)^2(H/H_1-\beta_1)+2i(H/H_1)[(\omega/\omega_1)\xi_1-\tilde{\xi}_1^*]\}U_g/\Delta_o(\omega) \tag{22}$$

in which $\tilde{\omega}_1^*$ and $\tilde{\xi}_1^*(\omega)$ are defined by

$$(\tilde{\omega}_1^*)^{-2} = (\omega_1)^{-2} + (\omega_R)^{-2} \tag{23}$$

and

$$\tilde{\xi}_1^*(\omega)=(\omega/\omega_1)^3\xi_1+[1-(\omega/\omega_1)^2]\xi_s+(\omega/\omega_R)^3\xi_R+[1-(\omega/\tilde{\omega}_1^*)^2](\xi_1\omega/\omega_1-\xi_s) \tag{24}$$

Eq. 20 indicates that the total translational response of the base U_{To} will exhibit a peak in the vicinity of the system frequency $\tilde{\omega}_1$ and a minimum in the vicinity of the frequency $\tilde{\omega}_1^*$ given by Eq. 23.

Finally, it is interesting to note that $U_{To}+H_1\phi_o$ given by

$$U_{To}+ H_1\phi_o = [1-(\omega/\omega_1)^2+2i\xi_1(\omega/\omega_1)]\ U_g/\Delta_o(\omega) \tag{25}$$

exhibits a minimum in the vicinity of the fixed-base natural frequency ω_1.

The basic results given by Eqs. 16, 20, 21, 22, and 25 are used next to interpret the results of structural identification techniques which neglect the effects of soil-structure interaction.

INTERPRETATION OF STRUCTURAL IDENTIFICATION TECHNIQUES

It is convenient to consider first a hypothetical identification technique based on the transfer function between the free-field motion U_g and the total translational response at the top of the structure U_{Tb}. This is a hypothetical situation since the free-field motion is rarely recorded. In this case, and neglecting the effects of soil-structure interaction, the transfer function would be given by

$$\left(\frac{U_{Tb}}{U_g}\right)_{r.s.} = \frac{\beta_1(\omega/\omega_1)^2 + 1-(\omega/\omega_1)^2 + 2i\xi_1(\omega/\omega_1)}{1-(\omega/\omega_1)^2 + 2i(\omega/\omega_1)\xi_1} \tag{26}$$

The actual transfer function given by Eq. 21 is

$$\frac{U_{Tb}}{U_g} = \frac{\beta_1(\omega/\tilde{\omega}_1^*)^2 + 1-(\omega/\tilde{\omega}_1^*)^2 + 2i\tilde{\xi}_1^*}{1-(\omega/\tilde{\omega}_1)^2 + 2i\tilde{\xi}_1} \tag{27}$$

where the last two terms appearing in Eq. 21 have been neglected. Comparison of Eq. 26 and 27 reveals that the modal frequency and modal damping ratio calculated by this identification technique would actually correspond to the system frequency $\tilde{\omega}_1$ and to the equivalent system damping ratio $\tilde{\xi}_1(\tilde{\omega}_1)$. Since $\tilde{\omega}_1^*$ is not very different from $\tilde{\omega}_1$, the calculated participation factor would correspond to the actual participation factor β_1.

As a second case, consider a more common identification technique based on the transfer function between the total translational motion at foundation level U_{To} and the total translational response at the top of the structure U_{Tb} (Udwadia and Trifunac, 1974). The transfer function, if the effects of soil-structure interaction are neglected, is also given by Eq. 26 (when soil-structure interaction is neglected, U_{To} is assumed to be equal to U_g). The actual transfer function as obtained from Eqs. 20 and 21 is

$$\frac{U_{Tb}}{U_{To}} = \frac{\beta_1(\omega/\tilde{\omega}_1^*)^2 + 1-(\omega/\tilde{\omega}_1^*)^2 + 2i\tilde{\xi}_1^*}{1-(\omega/\tilde{\omega}_1^*)^2 + 2i\tilde{\xi}_1^*} \tag{28}$$

where, again, the last two terms in Eq. 21 have been neglected. Comparison of Eqs. 26 and 28 indicates that, in this case, the calculated modal frequency and damping ratio would correspond to the apparent system frequency $\tilde{\omega}_1^*$ and to the apparent system damping ratio $\tilde{\xi}_1^*$ $(\tilde{\omega}_1^*)$. In this case, the calculated participation factor would also correspond to the actual participation factor.

An identification technique based on the transfer function between the total translation of the base U_{To} and the relative translation $U_{Tb} - U_{To}$ at the top, leads to similar results, i.e., the calculated modal quantities correspond to $\tilde{\omega}_1^*$, $\tilde{\xi}_1^*(\tilde{\omega}_1^*)$ and β_1.

To isolate the modal characteristics of the superstructure it would be necessary to consider the transfer function between $U_{To}+H_1\phi_o$ and $U_b = U_{Tb}-(U_{To} + H\phi_o)$. From Eqs. 16 and 25, the transfer function in this case is

$$\frac{U_b}{U_{To} + H_1\phi_o} = \frac{\beta_1(\omega/\omega_1)^2}{1-(\omega/\omega_1)^2 + 2i\xi_1(\omega/\omega_1)} \qquad (29)$$

in which only structural properties are involved. For this technique to be used it is necessary to record U_{Tb}, U_{To}, and the rocking angle ϕ_o (on a common time basis). The typical instrumentation of buildings which does not contain provisions for the recording of the rocking response at foundation level eliminates the possibility of isolating the structural characteristics.

Having established that the present identification techniques lead to results that reflect the characteristics of the complete soil-structure system rather than the characteristics of the superstructure, it is necessary to examine next the conditions under which these differences may be important.

EFFECTS OF SOIL-STRUCTURE INTERACTION ON THE APPARENT SYSTEM FREQUENCIES

The previous discussion indicates the important role that the system frequency $\tilde{\omega}_1$ and the apparent system frequency $\tilde{\omega}_1^*$ play in the calculation of the response of the system. It is of particular interest to determine the deviation of $\tilde{\omega}_1$ and $\tilde{\omega}_1^*$ from the fixed-base frequency of the superstructure ω_1. From Eqs. 18 and 23,

$$\left(\frac{\omega_1}{\tilde{\omega}_1}\right)^2 = 1 + \left(\frac{\omega_1}{\omega_R}\right)^2 + \left(\frac{\omega_1}{\omega_H}\right)^2 \qquad (30)$$

$$\left(\frac{\omega_1}{\tilde{\omega}_1^*}\right)^2 = 1 + \left(\frac{\omega_1}{\omega_R}\right)^2 \qquad (31)$$

which indicate that the deviations of $\tilde{\omega}_1$ and $\tilde{\omega}_1^*$ from ω_1 depend on the ratios (ω_1/ω_R) and (ω_1/ω_H). At this point it should be noted that ω_R corresponds to the frequency of the system if the superstructure is rigid and the foundation is prevented from translating. Similarly, the frequency ω_H corresponds to the resonant frequency of the system if the superstructure is rigid and the foundation is prevented from rocking. The frequency $[\omega_R^{-2} + \omega_H^{-2}]^{-1/2}$ corresponds to the frequency of the system if the superstructure is rigid and the foundation is free to translate and rotate.

The ratios (ω_1/ω_R) and (ω_1/ω_H), calculated by use of Eqs. 10 and 11, are given by

$$(\omega_1/\omega_R)^2 = (\omega_1 r/\beta)^2 \, (M_1/\pi r^2 \rho_s H_1)(H_1/r)^3 \, (\pi/k_R) \tag{32}$$

$$(\omega_1/\omega_H)^2 = (\omega_1 r/\beta)^2 \, (M_1/\pi r^2 \rho_s H_1)(H_1/r) \, (\pi/k_H) \tag{33}$$

where β is a shear wave velocity of reference for the soil and ρ_s is the soil density $(G=\rho_s\beta^2)$. Eqs. 32 and 33 indicate that the deviations of $\tilde{\omega}_1$ and $\tilde{\omega}_1^*$ from ω_1 will depend on the relative stiffness of the structure measured by $(\omega_1 r/\beta)$, on the mass ratio $(M_1/\pi r^2 \rho_s H_1)$ and on the slenderness ratio H_1/r. The slenderness ratio H_1/r is approximately equal to 0.67 H/r and takes values in the range from 0.5 to 3.0. The mass ratio $(M_1/\pi r^2 \rho_s H_1)$ is approximately equal to $(M_b/\pi r^2 \rho_s H)$ and for most structures has a value of the order of 0.2. The relative stiffness parameter $(\omega_1 r/\beta)$ typically varies in the range from 0. to 2. . For a ten story building with a fixed-base frequency of 1 Hz, resting on a foundation with an equivalent radius r = 15 m supported on a soil with a characteristic shear wave velocity β = 400 m/sec, the value of $\omega_1 r/\beta$ is 0.24. For a containment structure in a nuclear power plant (f_1 = 5 Hz, r = 23 m) supported on a soil with a shear wave velocity of 800 m/sec, the corresponding value of $\omega_1 r/\beta$ is 0.9. The normalized stiffness coefficients k_R and k_H appearing in Eqs. 32 and 33 depend on the characteristics of the soil deposit and on the dimensionless frequency $\omega r/\beta$. This dimensionless frequency must be set equal to $\tilde{\omega}_1 r/\beta$ to calculate $\tilde{\omega}_1$, and to $\tilde{\omega}_1^* r/\beta$ to calculate $\tilde{\omega}_1^*$. For this reason, an iterative process may be required to obtain $\tilde{\omega}_1$ and $\tilde{\omega}_1^*$. For a circular foundation supported on a uniform elastic half-space (ν = 0.33), the foundation stiffness coefficients, at low frequencies, take values of the order of $k_R \sim 4$ and $k_H \sim 5$.

The deviations of the system frequency $\tilde{\omega}_1$ and of the apparent system frequency $\tilde{\omega}_1^*$ from the fixed-base frequency ω_1 are illustrated in Fig. 1. In this Figure the ratios $\tilde{\omega}_1/\omega_1$ and $\tilde{\omega}_1^*/\omega_1$ are presented versus the relative stiffness $\omega_1 r/\beta$ for three values of the slenderness ratio H_1/r. The results presented in Fig. 1 are based on a mass ratio $M_1/\pi r^2 \rho_s H_1=0.2$ and on the foundation stiffness coefficients obtained by Luco and Westmann (1971) for a rigid circular foundation resting on a uniform half-space ($\nu=0.33$). The results presented in Fig. 1 indicate that the system frequencies $\tilde{\omega}_1$ and $\tilde{\omega}_1^*$ can be significantly lower than the fixed-base frequency ω_1. The deviations increase with relative stiffness ($\omega_1 r/\beta$) and with slenderness ratio H_1/r. The apparent system frequency $\tilde{\omega}_1^*$ is slightly higher than the true system frequency $\tilde{\omega}_1$. The difference between $\tilde{\omega}_1$ and $\tilde{\omega}_1^*$ decreases as the slenderness ratio increases.

In many cases the fixed-base natural frequency ω_1 is not known and must be determined from the observed values of $\tilde{\omega}_1$ or $\tilde{\omega}_1^*$. The results presented in Fig. 2 can be utilized for that purpose. In Fig. 2, the ratio $\omega_1/\tilde{\omega}_1$ is presented versus $\tilde{\omega}_1 r/\beta$ for four values of the slenderness ratio. The ratio $\omega_1/\tilde{\omega}_1^*$ versus $\tilde{\omega}_1^* r/\beta$ is also shown in Fig. 2. To utilize the results presented in Fig. 2 it is necessary to have an estimate of the characteristic shear wave velocity in the soil and of the equivalent radius of the foundation. It is interesting to notice that $\tilde{\omega}_1 r/\beta$ and $\tilde{\omega}_1^* r/\beta$ cannot exceed the values

$$\frac{\tilde{\omega}_1 r}{\beta} = \left\{ \left(\frac{M_1}{\pi \rho_s r^2 H_1} \right) \left[\left(\frac{H_1}{r} \right)^3 \frac{\pi}{k_R} + \left(\frac{H_1}{r} \right) \frac{\pi}{k_H} \right] \right\}^{-1/2} \tag{34}$$

$$\frac{\tilde{\omega}_1^* r}{\beta} = \left\{ \left(\frac{M_1}{\pi \rho_s r^2 H_1} \right) \left(\frac{H_1}{r} \right)^3 \frac{\pi}{k_R} \right\}^{-1/2} \tag{35}$$

These limiting values correspond to the system frequencies when the superstructure is rigid.

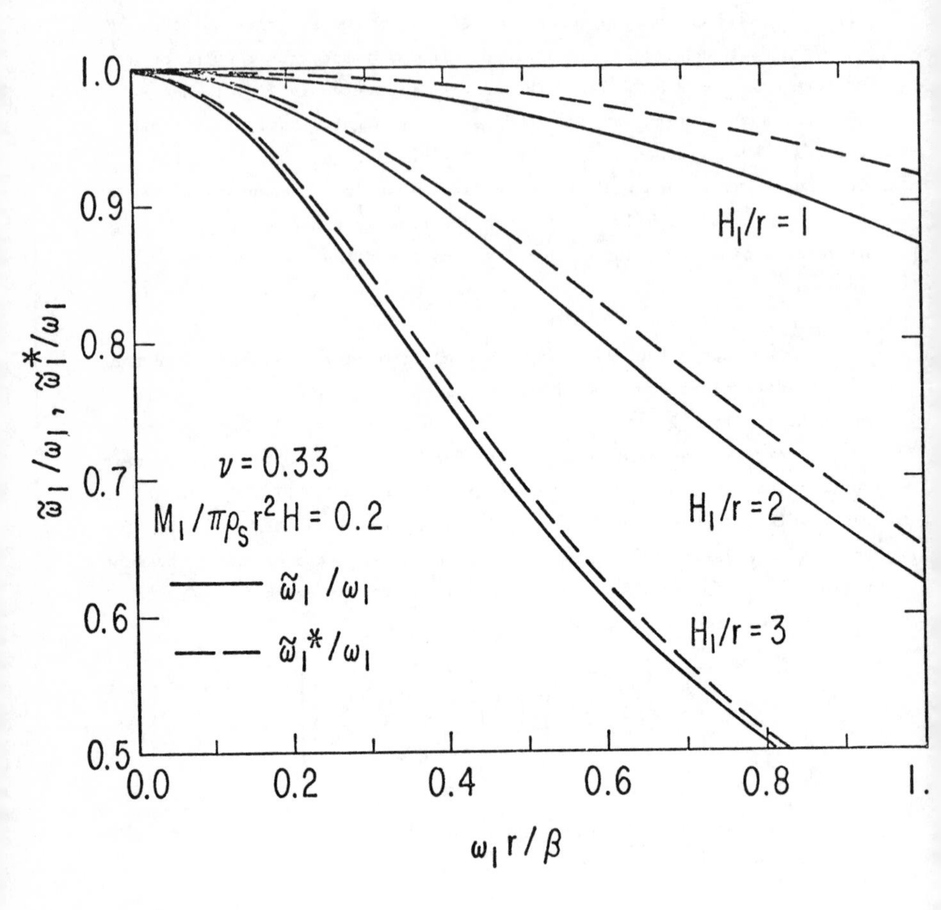

FIG. 1. System frequency and apparent system frequency versus relative stiffness and slenderness ratio.

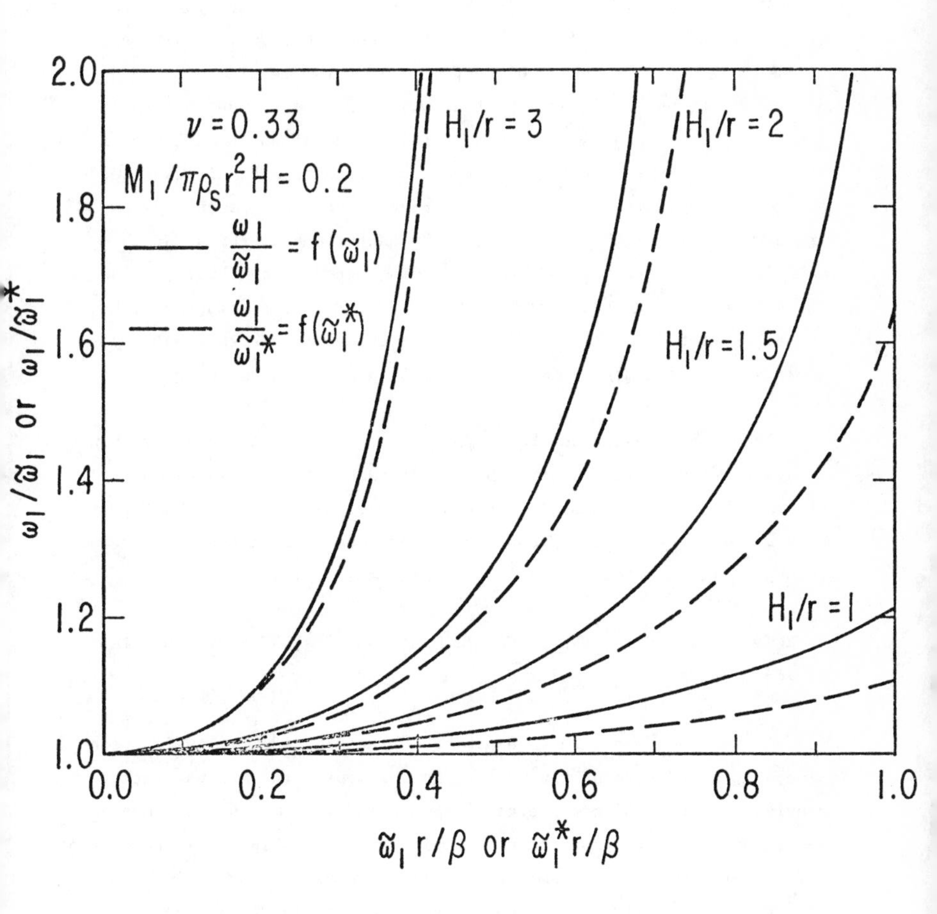

FIG. 2. Fixed-base natural frequency in terms of the system frequency or of the apparent system frequency.

During moderate and strong seismic excitations significant strains are induced in the soil. These strains cause a marked reduction in the effective shear modulus and on the shear wave velocity of the soil. The effects of such changes on the frequency of the soil-structure system are illustrated in Fig. 3. In this Figure, the ratio of the system frequency $\tilde{\omega}_1(\beta')$, when the shear wave velocity has been reduced by 33 percent ($\beta'=0.67\beta$), to the system frequency $\tilde{\omega}_1(\beta)$ for the initial soil conditions is shown versus $\omega_1 r/\beta$ for different values of the slenderness ratio. The results shown in Fig. 3 indicate that a variation in soil properties, as a result of the seismic excitation, may lead to a significant reduction of the system frequency. The variation of the apparent system frequency $\tilde{\omega}_1^*$ is similar to that of $\tilde{\omega}_1$. Reductions of the shear wave velocity of the order of 30 percent can be produced by shear strains in the soil of the order of 0.1 percent. For very stiff structures, Eq. 34 indicates that $\tilde{\omega}_1(\beta')/\tilde{\omega}_1(\beta)=\beta'/\beta$ and the change in system frequency would provide a measure of the variation in soil properties.

The effect described above explains, in part, the temporary changes in system frequencies observed by Udwadia and Trifunac (1974) on the Millikan Library building (50 percent reduction) and on the administration building at the Jet Propulsion Laboratory in Pasadena (20 percent reduction) during the San Fernando earthquake of 1971.

EFFECTS OF SOIL-STRUCTURE INTERACTION ON THE APPARENT SYSTEM DAMPING RATIOS

It has been shown that the common structural identification techniques do not determine the damping in the superstructure but rather provide a measure of the overall system damping $\tilde{\xi}_1(\tilde{\omega}_1)$ or of the apparent system damping $\tilde{\xi}_1^*(\tilde{\omega}_1^*)$. Apparent damping ratios as high as 10 percent of critical for moderate seismic excitation have been reported in several studies (Hart et al., 1973, Iemura and Jennings, 1973, Hart and Vasudevan, 1975). It is of interest to describe the effects of soil-structure interaction on the apparent damping ratios.

In the first place, it should be mentioned that Eq. 19 indicates that the energy dissipation for the complete system does not follow a

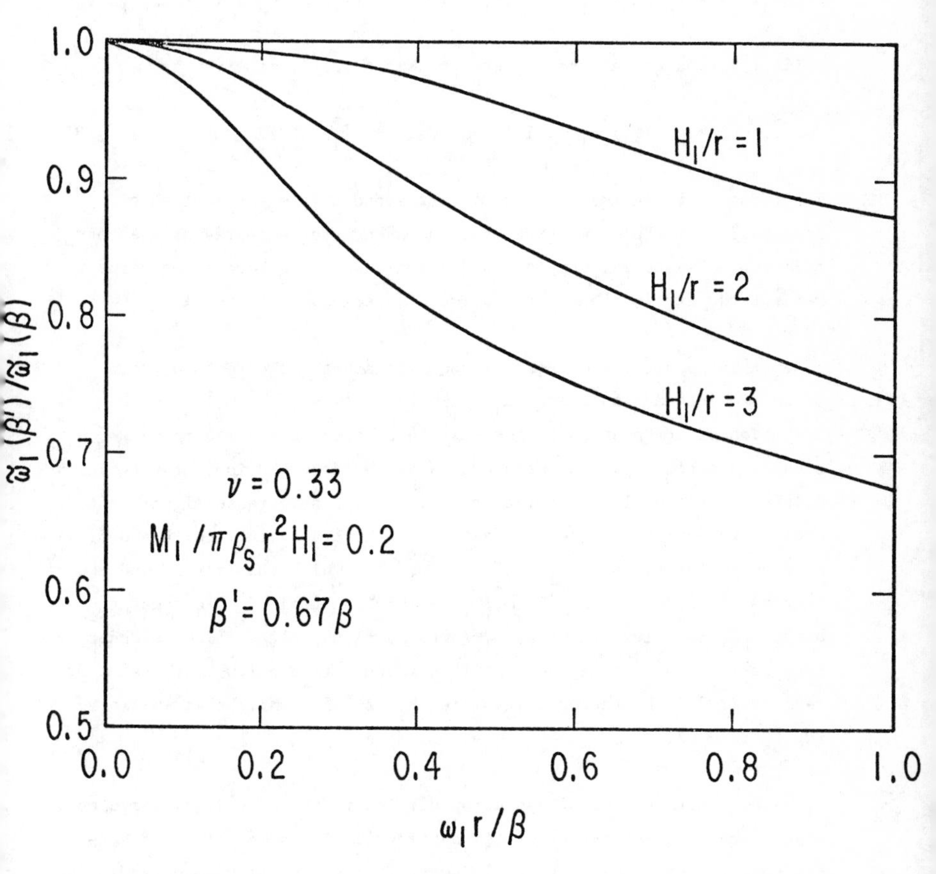

FIG. 3. Effect of a reduction of the soil shear wave velocity on the system frequency.

viscous or hysteretic type of model but a more complete frequency-dependent behavior. The approximate equivalent system damping ratio is given by

$$\tilde{\xi}_1(\tilde{\omega}_1) = (\tilde{\omega}_1/\omega_1)^3\xi_1 + [1-(\tilde{\omega}_1/\omega_1)^2]\xi_s + (\tilde{\omega}_1/\omega_H)^3\xi_H + (\tilde{\omega}_1/\omega_R)^3\xi_R \quad (36)$$

while the apparent equivalent system damping ratio is given by

$$\tilde{\xi}_1^*(\tilde{\omega}_1^*) = (\tilde{\omega}_1^*/\omega_1)^3\xi_1 + [1-(\tilde{\omega}_1^*/\omega_1)^2]\xi_s + (\tilde{\omega}_1^*/\omega_R)^3\xi_R \quad . \quad (37)$$

The first terms in Eqs. 36 and 37 correspond to the contibution of attenuation in the superstructure, the second terms represent the contribution of material damping in the soil, while the remaining terms reflect the attenuation by radiation of energy into the soil.

It is apparent from Eqs. 36 and 37 that as the effects of soil-structure interaction become more important ($\tilde{\omega}_1<\omega_1$, $\tilde{\omega}_1^*<\omega_1$), the contribution of attenuation in the structure to the overall damping decreases drastically. On the other hand, the contributions of material damping in the soil and of the radiation damping increase significantly and become dominant. These effects are illustrated in Figs. 4 and 5. In Fig. 4, the factors $(\tilde{\omega}_1/\omega_1)^3$ and $(\tilde{\omega}_1^*/\omega_1)^3$ which multiply ξ_1 and the factors $[1-(\tilde{\omega}_1/\omega_1)^2]$ and $[1-(\tilde{\omega}_1^*/\omega_1)^2]$ which multiply ξ_s are presented versus the relative stiffness $\omega_1 r/\beta$ for various values of the slenderness ratio. For a relatively stiff structure ($\omega_1 r/\beta=1.0$) and a slender-ratio $H_1/r=2.$, the factor multiplying ξ_1 is 0.25, while that multiplying ξ_s is 0.62.

The contributions of radiation damping to the overall system damping $\tilde{\xi}_1$ and to the apparent system damping $\tilde{\xi}_1^*$ are shown in Fig. 5 versus the relative stiffness $\omega_1 r/\beta$. In general, the contribution of radiation damping to $\tilde{\xi}_1$ is larger than that to $\tilde{\xi}_1^*$. For sufficiently stiff structures (relative to the soil) radiation damping may contribute in excess of one percent of the overall critical damping. For a relative stiffness $\omega_1 r/\beta=2.$, radiation damping may contribute from 5 to 15 percent of the overall critical damping depending on the slenderness ratio

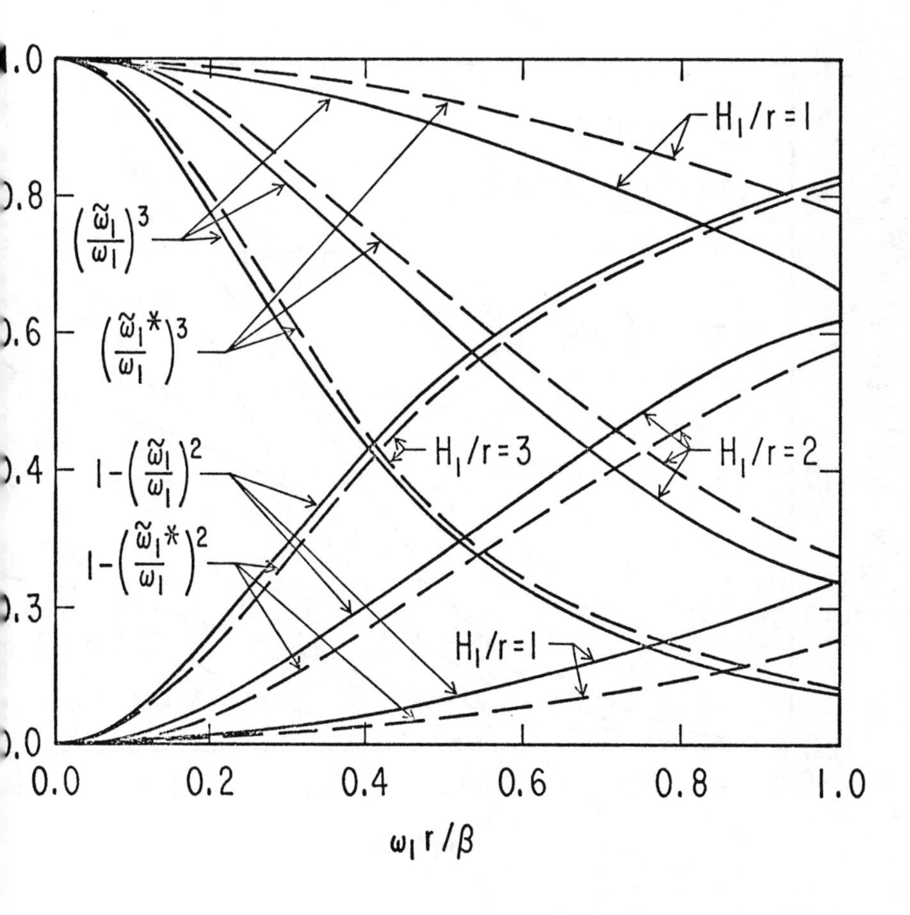

FIG. 4. Participation factors of the structural and material soil damping to the system damping and to the apparent system damping ($M_1/\pi\rho_s r^2 H_1=0.2$, $\nu=0.33$).

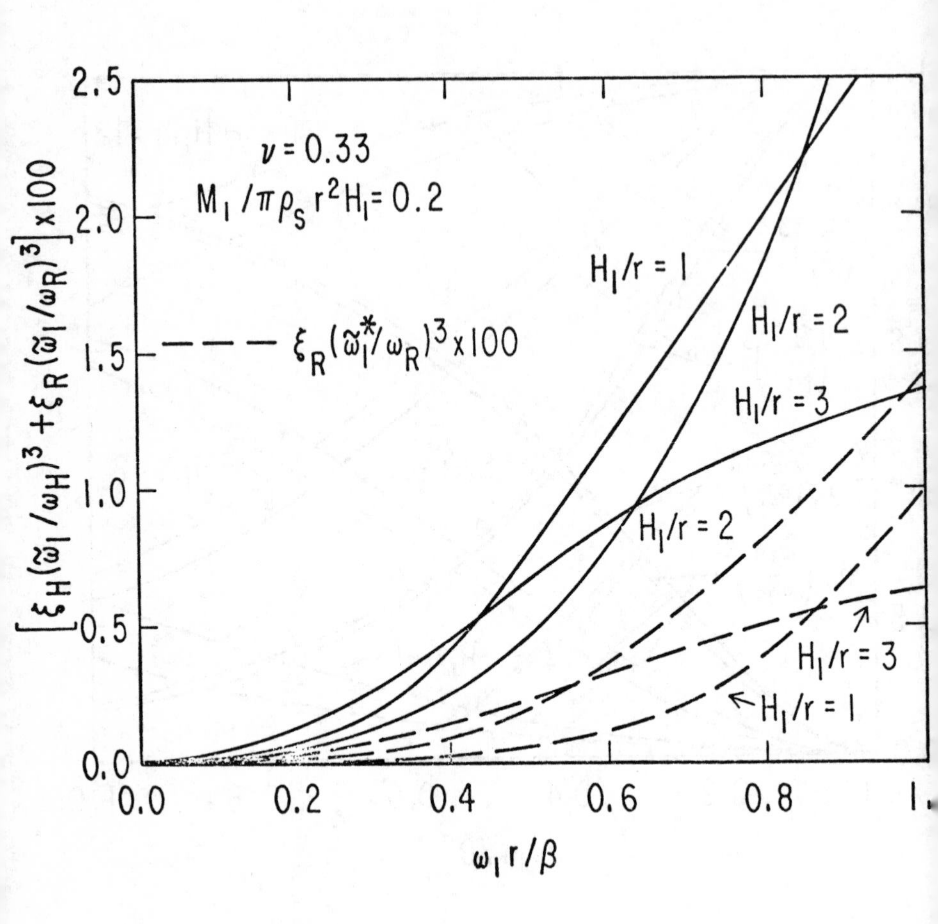

FIG. 5. Contribution of radiation damping to the overall system damping and to the apparent system damping.

(Jennings and Bielak, 1973).

The variation of the system damping ratio $\tilde{\xi}_1$ and of the apparent damping ratio $\tilde{\xi}_1^*$ versus the relative stiffness parameter $\omega_1 r/\beta$ are shown in Fig. 6 for ξ_1=0.05 and for three values of ξ_s ($H_1/r = 2.$, $M_1 \pi \rho_s r^2 H_1 = 0.2$). It can be observed that, in general, $\tilde{\xi}_1^* < \tilde{\xi}_1$. If $\xi_s < \xi_1$, the system damping ratios may be lower than the damping on the structure. This possibility has been noted previously by Jennings and Bielak (1973). On the other hand if $\xi_s > \xi_1$, the system damping ratios for relatively stiff structures can be significantly higher than ξ_1.

It is important to notice that for relatively stiff structures (or for soft soils) the system damping ratio $\tilde{\xi}_1$ and the apparent damping ratio $\tilde{\xi}_1^*$ provide poor measures of the attenuation in the superstructure. Large variations in the value of ξ_1 could have little effect on the values of $\tilde{\xi}_1$ and $\tilde{\xi}_1^*$ and, consequently, it becomes difficult to estimate ξ_1 from observed values of $\tilde{\xi}_1$ or $\tilde{\xi}_1^*$.

During seismic excitation the strains induced in the soil produce not only a reduction in shear wave velocity but also a significant increase in the equivalent hysteretic damping in the soil. The reduction of β coupled with an increase in ξ_s may lead to a large increase in the value of the system damping. Consider an initial situation in which ξ_1=0.05, ξ_s=0.02 and $\omega_1 r/\beta = 0.6$. In this case from Fig. 6 it is found that $\tilde{\xi}_1$=0.04. If the seismic excitation induces shear strains in the soil of the order of 0.1 percent, β would suffer a reduction of 33 percent and ξ_s would increase to 0.10. Under these conditions, $\omega_1 r/\beta \sim 0.9$ and $\tilde{\xi}_1$ would reach a value of 0.09. This mechanism may explain, in part, the large damping values obtained by various studies of the seismic response of structures.

A CASE STUDY: THE MILLIKAN LIBRARY

It has been indicated that the observed variations in apparent modal frequencies and damping ratios could be explained as resulting

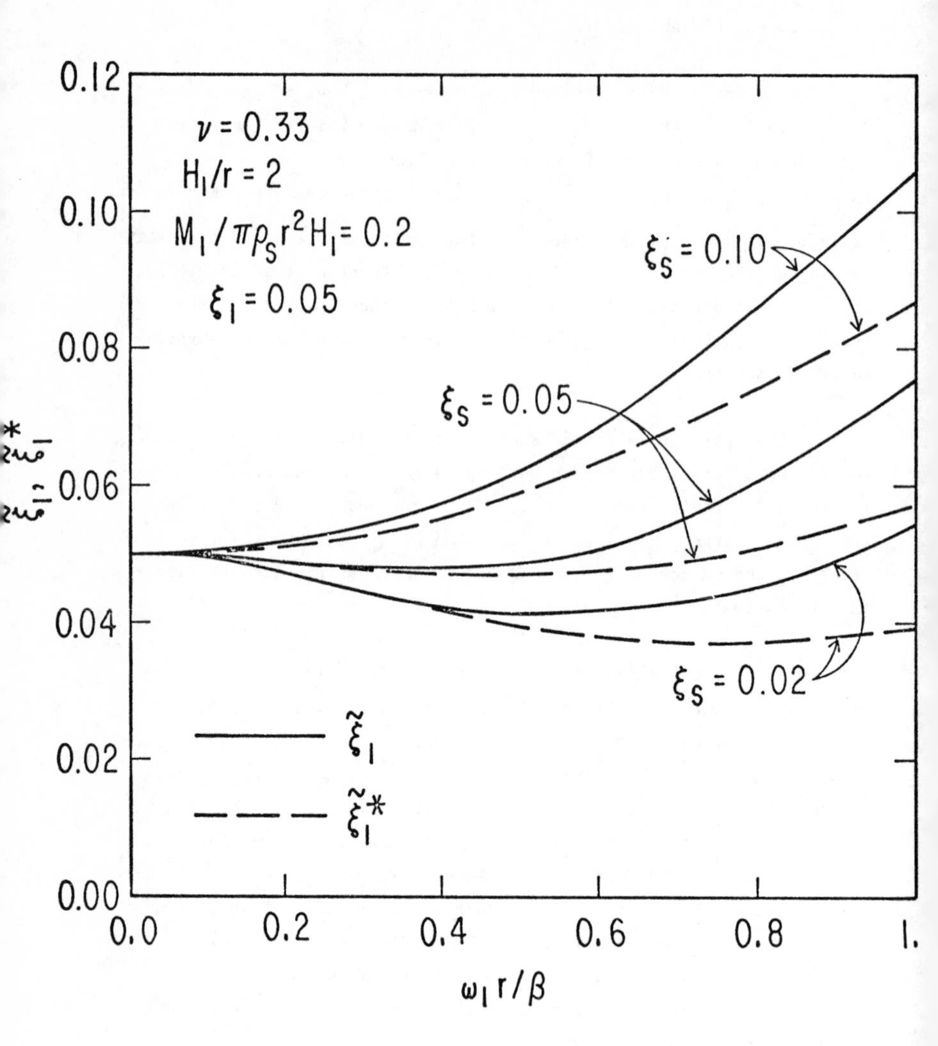

FIG. 6. Effect of the relative stiffness and of the material soil damping on the overall system damping and on the apparent system damping.

from variations in soil properties induced by strong shaking. The information available on the response of the Millikan Library building during the San Fernando earthquake of 1971 provides an opportunity to further explore this possibility.

The Millikan Library is a nine-story reinforced concrete building located on the campus of the California Institute of Technology. The structure has a basement and an enclosed roof area. The typical floor plan covers an area of 21 x 23m (69 x 75 ft) and the height from the basement slab to the roof slab is 43.3 m (142 ft). The majority of the lateral loads in the transverse (N-S) direction are resisted by reinforced shear walls located on the east and west ends of the building. In the longitudinal (E-W) direction the reinforced concrete walls of the central core provide most of the lateral resistance. The north and south-side facades correspond to precast window wall panels. The total weight of the superstructure is estimated at 1.05×10^8 Newtons (23.5×10^6 lbs). The foundation system of the library consists of a central pad 9.75 m (32 ft) wide and 1.22 m (4 ft) deep and of foundation beams 3 x 0.61 m (10 x 2 ft) running in the E-W direction. The equivalent radius of the foundation is 13.7 m (45 ft). The foundation rests on alluvium composed of medium to dense sands mixed with gravel. The shear wave velocity at foundation depth is 382 m/sec (1250 ft/sec).

Forced and ambient vibration tests conducted before the San Fernando earthquake of 1971 (Jennings and Kuroiwa, 1968; Udwadia and Trifunac, 1974) indicate system frequencies $\tilde{f}_1(\tilde{f}_1 = \tilde{\omega}_1/2\pi)$ of the order of 1.92 and 1.48 Hz in the N-S and E-W directions, respectively. Forced vibration tests after the San Fernando earthquake indicate (Foutch et al., 1975, Luco et al., 1980) system frequencies $\tilde{f}_1$ of the order of 1.79 and 1.21 Hz in the N-S and E-W directions, respectively. By use of forced vibration tests conducted in 1975, Luco, Wong and Trifunac (1980) isolated the fixed-base natural frequencies of the superstructure and found values of f_1 of the order of 2.33 and 1.45 Hz in the N-S and E-W directions, respectively. These tests also provided the estimates of $f_R=\omega_R/2\pi$ and $f_H=\omega_H/2\pi$ listed in Table 1. Assuming that the foundation and soil did not experience permanent changes during the San

Table 1. Characteristic frequencies of the Millikan Library before and after the San Fernando Earthquake of 1971 (in Hz).

	Before			After				
	$\tilde{f}_1$	$\tilde{f}_1^*$	f_1	$\tilde{f}_1$	$\tilde{f}_1^*$	f_1	f_R	f_H
N-S	1.92	1.98	2.64	1.79	1.84	2.33	3.01	7.51
E-W	1.48	1.50	2.00	1.21	1.22	1.45	2.26	9.27

Fernando earthquake it is possible to calculate by use of Eq. 18 the fixed-base frequencies before the earthquake. The calculated values for f_1 are 2.64 Hz in the N-S direction and 2.00 Hz in the E-W direction. These results would imply permanent reductions in structural stiffness of 22 and 47 percent in the N-S and E-W directions, respectively (Luco et al., 1980). It is possible that the precast window wall panels which contribute 25 percent of the structural stiffness in the E-W direction may have ceased to act as structural elements (Iemura and Jennings, 1973). Comparison of the results for the torsional response during vibration tests conducted before and after the San Fernando earthquake also indicate a reduction in torsional stiffness of the order of 20 percent (Luco, Wong and Trifunac, 1980) which is consistent with the change in stiffness in the east and west shear walls. The modal characteristics of the Millikan Library before and after the San Fernando earthquake are summarized in Table 1. In this table the values of $\tilde{f}_1^* = \tilde{\omega}_1^*/2\pi$ have been calculated by use of Eq. 23.

In addition to permanent changes described above, the Millikan Library system underwent significant temporary changes during the San Fernando earthquake. Udwadia and Trifunac (1974) using an 8-sec moving window Fourier analysis of the transfer function between the recorded total translation of the base and the total translation at the roof found a marked variation in the apparent system frequency $\tilde{f}_1^*$ in both the N-S and E-W directions. Similar results have been obtained by Iemura and Jennings (1973) and McVerry (1979). The estimates of $\tilde{f}_1^*$ obtained by Udwadia and Trifunac are shown versus time in Fig. 7 (solid lines) for the 80-sec duration of the records. The values of $\tilde{f}_1^*$ calculated from the pre- and post-earthquake conditions (Table 1)

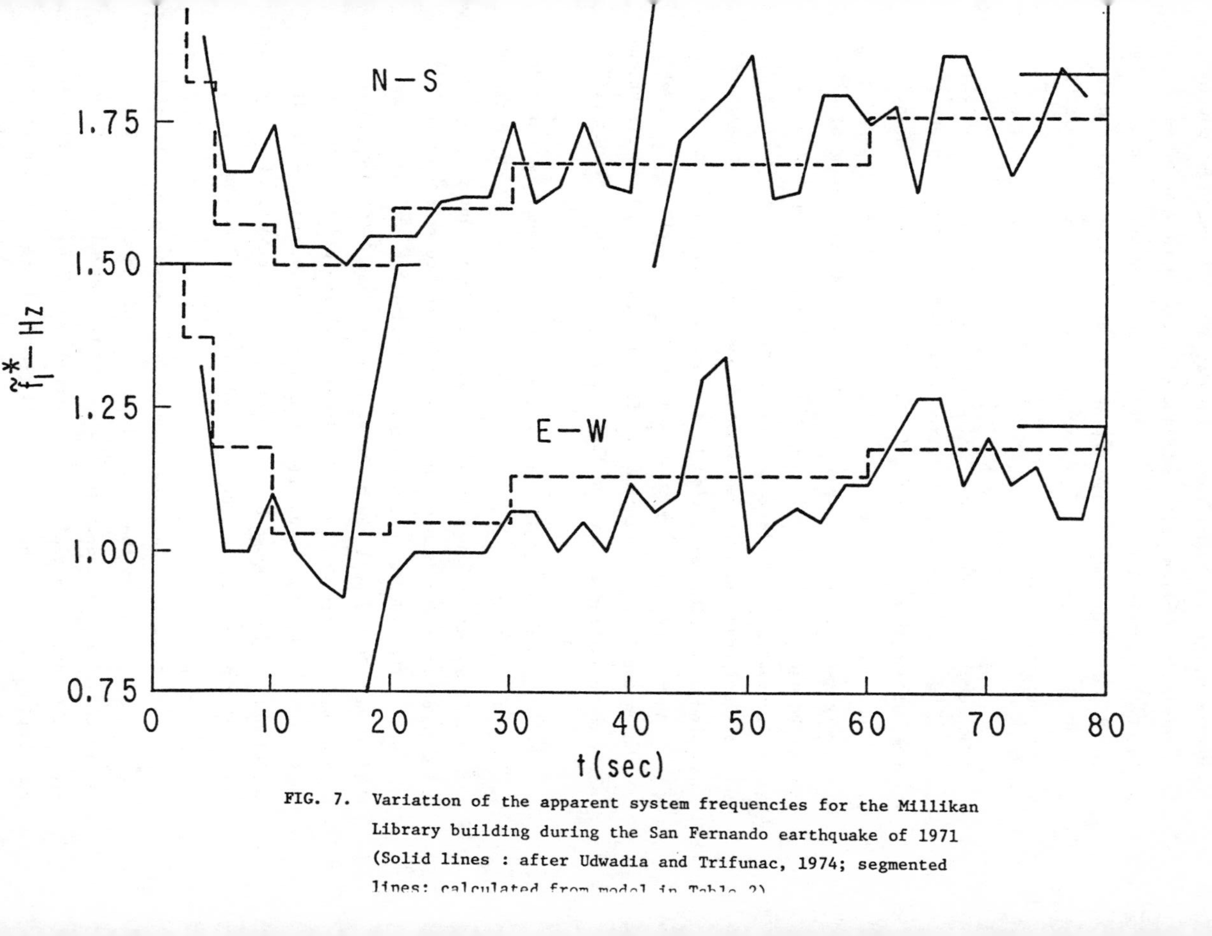

FIG. 7. Variation of the apparent system frequencies for the Millikan Library building during the San Fernando earthquake of 1971 (Solid lines : after Udwadia and Trifunac, 1974; segmented lines: calculated from model in Table 2)

are shown in Fig. 7 by short solid lines at the left and right margins. The variation of apparent system frequencies can be thought of as resulting from structural degradation followed by a harder to explain healing process. An alternative explanation is to assume that the observed changes in apparent frequencies can be associated with variation of soil properties.

The results of Udwadia and Trifunac can be explained by a combination of a permanent reduction in structural stiffness taking place at the time of the strongest shaking (t~10 sec) and a temporary variation of the shear modulus for the soil. The assumed variation of shear modulus (normalized by the low-strain shear modulus before and after the earthquake G_o) is listed in Table 2. The values of the apparent system frequency $\tilde{f}_1^*$ calculated by use of Eq. 23 are listed in Table 2 and are also presented (with segmented lines) in Fig. 7. The calculations are based on the assumption that within each time interval the system can be represented by an equivalent linear model. It is apparent that the calculated values of f_1^*, for both the N-S and E-W directions, follow the same trends as those obtained by Udwadia and Trifunac (1974).

It remains to ascertain whether the assumed changes in shear modulus actually took place during the San Fernando earthquake which caused peak accelerations at the site of the order of 20 percent of gravity. The assumed variation in shear modulus would require shear strains in the soil of the order of 0.2 percent. A preliminary estimate of the shear strain in the soil can be obtained by $\gamma \sim V_{max}/\beta$. Since the recorded peak velocity at foundation level is of the order of 20 cm/sec and β = 382 m/sec, the resulting estimate of γ is 0.05 percent. The additional strains induced by motion of the foundation may increase this value bringing it close to the required value of 0.2 percent.

It has been shown that the assumed variation of shear modulus in the soil could explain the observed behavior of the apparent system frequencies. It remains to study whether the assumed changes in soil properties could also explain the observed variations in apparent system damping. This can be accomplished by use of the standard relations between shear modulus, damping ratio and cyclic strain (Seed and Idriss,

Table 2. Model for the variation of the soil and structural properties during the San Fernando earthquake.

			N-S			E-W		
t sec	G/G_o	ξ_s (%)	f_1 Hz	$\tilde{f}_1^*$ Hz	$\left[1-\left(\frac{\tilde{f}_1^*}{f_1}\right)^2\right]\xi_s$ (%)	f_1 Hz	$\tilde{f}_1^*$ Hz	$\left[1-\left(\frac{\tilde{f}_1^*}{f_1}\right)^2\right]\xi_s$ (%)
0-2.5	1.00	2.0	2.64	1.98	0.88	2.00	1.50	0.88
2.5-5	0.70	4.8	2.64	1.82	2.52	2.00	1.37	2.54
5-10	0.42	10.8	2.64	1.57	6.98	2.00	1.18	6.84
10-20	0.42	10.8	2.33	1.50	6.32	1.45	1.03	5.35
20-30	0.45	10.0	2.33	1.60	4.72	1.45	1.05	4.76
30-60	0.65	5.6	2.33	1.68	2.69	1.45	1.13	2.20
60-80	0.80	3.6	2.33	1.76	1.55	1.45	1.18	1.22
-	1.00	2.0	2.33	1.84	1.02	1.45	1.22	0.81

1970). By eliminating the shear strain, it is possible to obtain a relation between shear modulus and damping ratio. The resulting relations for cohesive and cohesionless soils are shown in Fig. 8 in which G_o represents the low-strain ($\gamma \sim 10^{-6}$) value of the shear modulus.

Given the assumed variation of the shear modulus listed in Table 2 it is possible to determine the corresponding variation of ξ_s by use of the relations presented in Fig. 8. The resulting values of ξ_s for a cohesionless soil are also listed in Table 2. The apparent system damping ratio $\tilde{\xi}_1^*$, given by Eq. 37 and repeated here for convenience, is

$$\tilde{\xi}_1^*(\tilde{\omega}_1^*) = (\tilde{\omega}_1^*/\omega_1)^3\xi_1 + [1-(\tilde{\omega}_1^*/\omega_1)^2]\xi_s + (\tilde{\omega}_1^*/\omega_R)^3\xi_R \quad . \tag{38}$$

In the case of the Millikan Library the contribution of the radiation damping, corresponding to the last term in Eq. 38, is small and can be neglected. The contribution of material attenuation in the soil, given by $[1-(\tilde{\omega}_1^*/\omega_1)^2]\xi_s$, is the dominant term in Eq. 38.

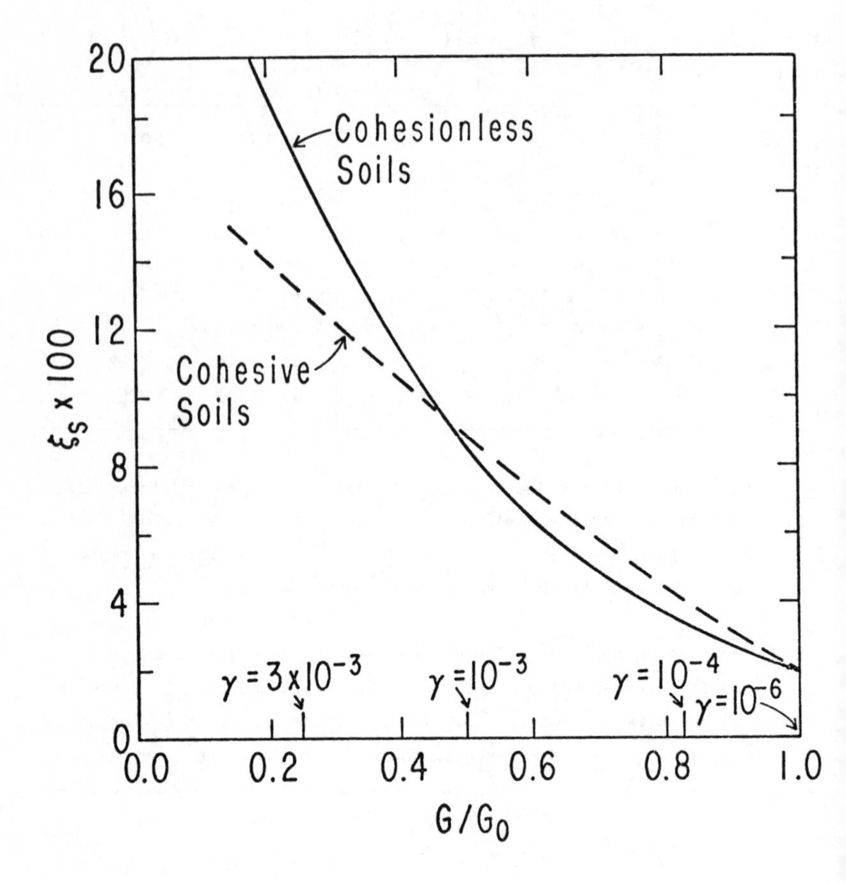

FIG. 8. Relation between soil material damping and reduced shear modulus.

The calculated values of $[1-(\tilde{\omega}_1^*/\omega_1)^2]\xi_s$ are listed in Table 2 and are presented in Fig. 9 (solid lines). The estimates of $\tilde{\xi}_1^*$ obtained by Iemura and Jennings (1973), Udwadia and Marmarelis (1976) and McVerry (1979) by use of different identification techniques are also shown in Fig. 9. It is apparent from Fig. 9 that the contribution of material damping in the soil can account for most of the observed variation of the apparent damping ratio in both the N-S and E-W directions. In the time interval between 2.5 and 30 sec., the factor $(\tilde{\omega}_1^*/\omega_1)^3$ multiplying ξ_1 in Eq. 38 takes values in the range from 0.21 to 0.38. Under these conditions the contribution of damping in the superstructure is small. The contribution of ξ_1 to $\tilde{\xi}_1^*$ becomes more important at the end of the recorded motion.

In summary, it appears that a combination of a permanent change in structural stiffness together with a temporary variation of the soil properties can explain simultaneously most of the observed variation of the apparent system frequencies and apparent damping ratios in both the N-S and E-W directions.

CONCLUSIONS

It has been shown that the typical structural identification techniques which employ strong motion records in an attempt at determining modal structural characteristics lead to results that may not correspond to the modal characteristics of the superstructure. In particular, the apparent modal frequencies and damping ratios obtained depend on the properties of the complete structure-foundation-soil system. Analytic expressions for the apparent modal quantities have been derived by considering the interaction between the structure and the soil. It has been shown that the apparent modal quantities are not equal to the modal frequencies and damping ratios for the complete system. For stiff ($\omega_1 r/\beta>0.5$) and slender structures ($H_1/r>1$) the apparent modal quantities may be significantly different from the modal characteristics of the superstructure. In particular, it has been found that attenuation in the soil by hysteresis has a strong effect on the apparent modal damping.

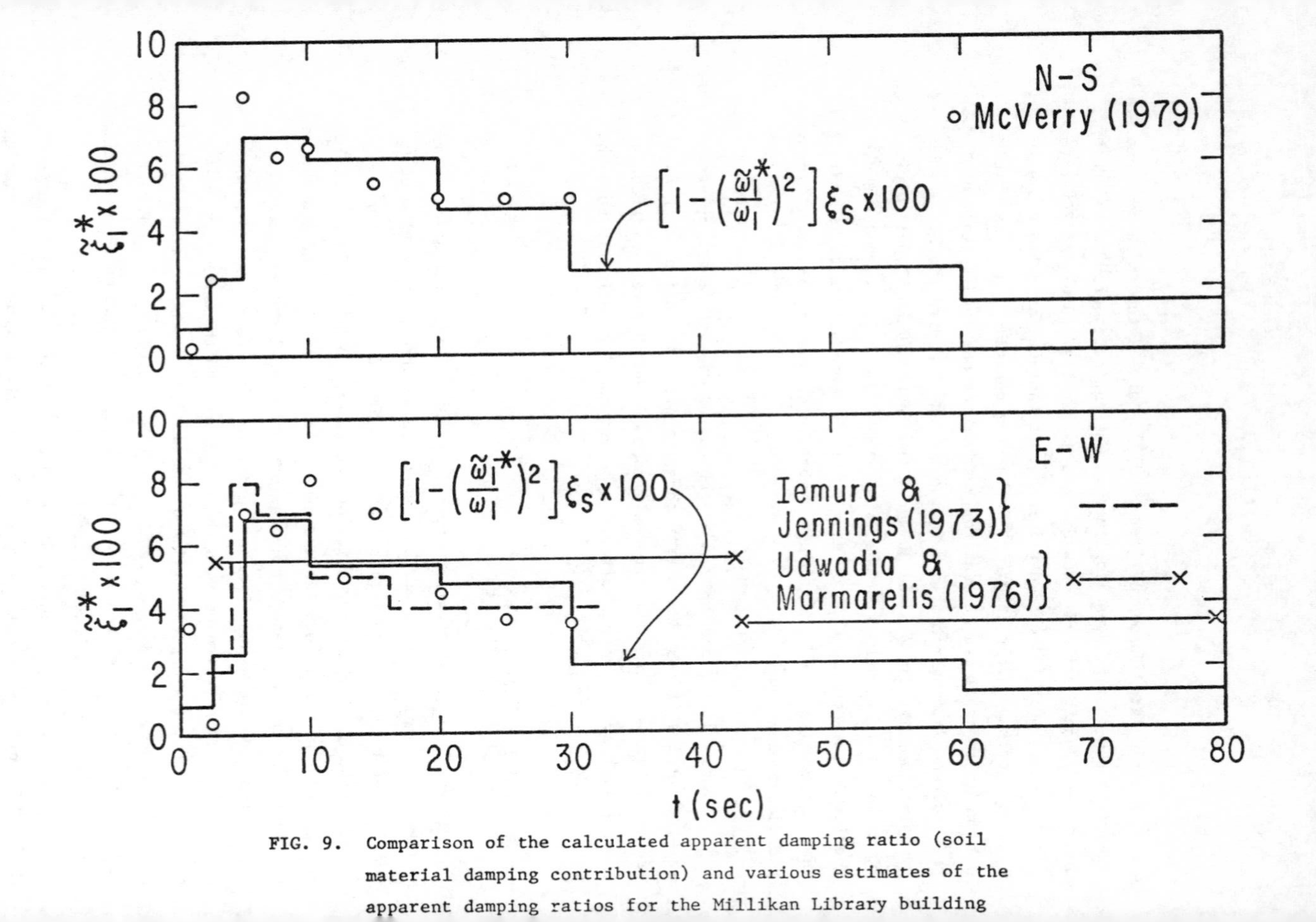

FIG. 9. Comparison of the calculated apparent damping ratio (soil material damping contribution) and various estimates of the apparent damping ratios for the Millikan Library building

When the effects of soil-structure interaction are important, the apparent modal damping is relatively independent of the damping in the superstructure.

The effects of variations of the soil properties on the apparent modal characteristics have been illustrated. It has been shown that the variations of the apparent modal frequencies and modal damping ratios of the Millikan Library building during the San Fernando earthquake could be explained by a combination of a permanent loss of structural stiffness together with temporary variations of the soil properties.

Detailed understanding of the behavior of structures during seismic excitation can only be achieved with more complete instrumentation. If the structural properties are to be isolated, the rocking and torsional motion at foundation level must be recorded.

ACKNOWLEDGEMENT

The work described here has been supported by Grant NSF PFR 79.00006 from the National Science Foundation.

APPENDIX.--REFERENCES

1. Foutch, D. A., Luco, J. E., Trifunac, M. D., and Udwadia, F. E., "Full Scale, Three-Dimensional Tests of Structural Deformations During Forced Excitation of a Nine-Story Reinforced Concrete Building," *Proc. U.S. National Conference on Earthquake Engineering,* Ann Arbor, Michigan, 1975, pp. 206-215.

2. Hart, G. C., Lew, M., and DiJulio, R., "High Rise Buiding Response: Damping and Period Nonlinearities," *Proc. Fifth World Conference on Earthquake Engineering,* Vol. 2, Rome, Italy, June, 1973.

3. Hart, G. C., and Vasudevan, R., "Earthquake Design of Buildings: Damping," *Journal of the Structures Division,* ASCE, Vol. 101, ST 1, Jan., 1975, pp. 11-30.

4. Iemura, H., and Jennings, P. C., "Hysteretic Response of a Nine-Story Reinforced Concrete Building," *International Journal of Earthquake Engineering and Structural Dynamics*, Vol. 3, No. 2, Oct.-Dec., 1974, pp. 183-202.

5. Jennings, P. C., and Kuroiwa, J. H., "Vibration and Soil-Structure Interaction Tests of a Nine-Story Reinforced Concrete Building," *Bulletin of the Seismological Society of America*, Vol. 58, No. 3, 1968, pp. 891-916.

6. Jennings, P. C., and Bielak, J., "Dynamics of Building-Soil Interaction," *Bulletin of the Seismological Society of America*, Vol. 63, 1973, pp. 9-48.

7. Luco, J. E., and Westmann, R. A., "Dynamic Response of Circular Footings," *Journal of the Engineering Mechanics Division*, ASCE, Vol. 97, 1971, pp. 1381-1395.

8. Luco, J. E., Wong, H. L., and Trifunac, M. D., "Soil-Structure Interaction Effects on Forced Vibration Tests," *Report*, Dept. of Civil Engineering, Univ. So. Calif., Los Angeles, Calif., 1980.

9. McVerry, G. H., "Frequency Domain Identification of Structural Models from Earthquake Records," *Report EERL 79-02*, Earthquake Engineering Res. Lab., Calif. Inst. of Tech., Pasadena, Calif, Oct., 1979.

10. Seed, H. B., and Idriss, I. M., "Soil Moduli and Damping Factors for Dynamic Response Analyses," *Report EERC-70-10, Earthquake Engineering Research Center*, Univ. of Calif., Berkeley, Calif, 1970.

11. Udwadia, F. E., and Trifunac, M. D., "Time and Amplitude Dependent Response of Structures," *International Journal of Earthquake Engineering and Structural Dynamics*, Vol. 2, 1974, pp. 359-378.

12. Udwadia, F. E., and Marmarelis, P. Z., "The Identification of Building Structural Systems. I. The Linear Case. II. The Nonlinear Case," *Bulletin of the Seismological Society of America*, Vol. 66, No. 1, Feb., 1976, pp. 121-171.

Figure Captions

FIG. 1. System frequency and apparent system frequency versus relative stiffness and slenderness ratio.

FIG. 2. Fixed-base natural frequency in terms of the system frequency or of the apparent system frequency.

FIG. 3. Effect of a reduction of the soil shear wave velocity on the system frequency.

FIG. 4. Participation factors of the structural and material soil damping to the system damping and to the apparent system damping ($M_1/\pi\rho_s r^2 H_1$=0.2, ν=0.33).

FIG. 5. Contribution of radiation damping to the overall system damping and to the apparent system damping.

FIG. 6. Effect of the relative stiffness and of the material soil damping on the overall system damping and on the apparent system damping.

FIG. 7. Variation of the apparent system frequencies for the Millikan Library building during the San Fernando earthquake of 1971 (Solid lines : after Udwadia and Trifunac, 1974; segmented lines: calculated from model in Table 2).

FIG. 8. Relation between soil material damping and reduced shear modulus.

FIG. 9. Comparison of the calculated apparent damping ratio (soil material damping contribution) and various estimates of the apparent damping ratios for the Millikan Library building during the San Fernando earthquake.

TWO-STEP APPROACH IN SOIL-STRUCTURE INTERACTION: HOW GOOD IS IT?

by

Eduardo Kausel (1)

Introduction

The difficulty in performing soil-structure interaction analyses for complicated structures exhibiting three-dimensional geometry has motivated the frequent use of an approximate procedure that separates the computation into two steps (Fig. 1). In the first step, the three-dimensionality and structural detail is ignored, concentrating efforts instead on computing the response of an "equivalent" two-dimensional (coarse) model of the structure to the seismic environment, typically with finite elements. The resulting motion of the foundation (consisting of both translations and rotations) are then used in the second step as input to a refined three-dimensional model of the structure, assuming the support to be rigid (i.e., no soil-structure interaction). While this technique appears to be attractive for the potential savings in the analyses, it often produces erroneous results because of the inconsistencies in the dynamic properties of the two models used. In particular, spurious amplification peaks may be observed at the natural frequencies of the (detailed) structure that would not have developed if the analysis had been done in a single step. This phenomenon is explored below for the simple but illustrative case of shear beam modeled with springs and lumped masses.

Direct method vs. two-step approach:

Consider a cantilever shear beam (Fig. 2) supported on a single spring-dashpot system representing the flexibility of the soil. A unit harmonic excitation is prescribed at the support of the "springs", i.e., under the soil spring-dashpot system. The motion amplitude observed at a structural location, and in particular at the top of the structure,

(1) Associate Professor of Civil Engineering, M.I.T.

Step 1

Coarse structural model

Detailed soil model

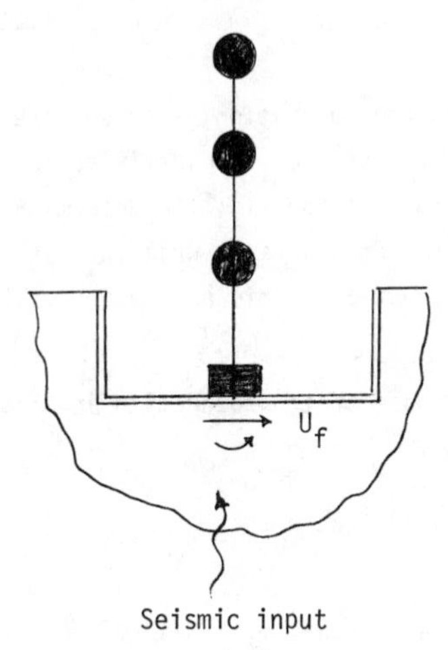

Seismic input

Step 2

Detailed structural model

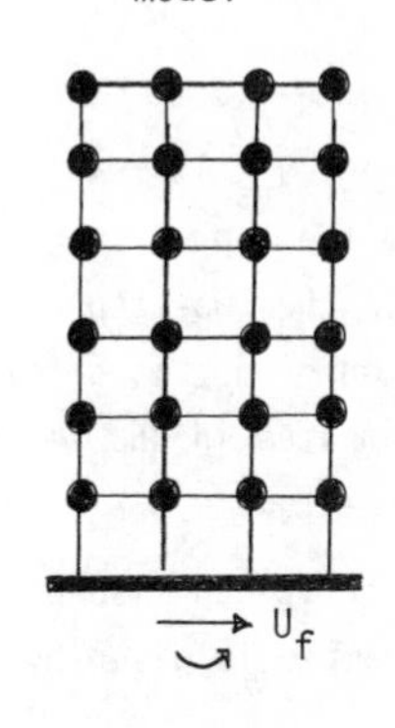

Foundation motion from step 1 as input

Fig. 1

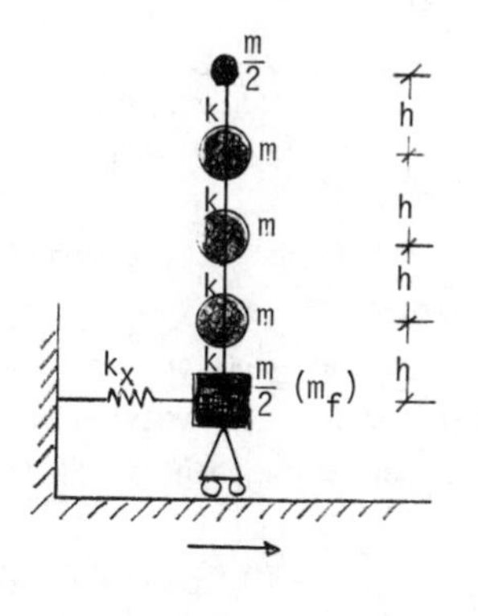

Discrete shear beam

Fig. 2

is referred to as the transfer function (or frequency response function) for that location. If u_{gf} and u_{gs} are the transfer functions from the ground (g) to the foundation (f) and to the top of structure (s) respectively, then it can be shown that

$$u_{gs} = u_{gf} \cdot u_{fs} \tag{1}$$

where u_{fs} is the transfer function from the foundation, assumed fixed, to the top of the structure (Fig. 3). In essence, the justification for the two-step approach is embodied in equation (1); the determination of u_{gf} is obtained in the first step of analysis, while u_{gs} is computed in the second step, using as input a seismic excitation with amplitude u_{gf}.

It can also be shown that the response of the foundation in this example is given by

$$u_{gf} = [k_f - \omega^2(m_f + \sum\gamma_{j1}^2\, a_j)]^{-1}\, k_f\, u_g \tag{2}$$

where $k_f = k + i\omega c$ is the complex soil stiffness, m_f = mass of foundation, γ_{j1} = j^{th} modal participation factor of the structure on fixed base for horizontal base motion $u_g \equiv u_1$ (i.e., in the 1st or horizontal coordinate direction), ω = excitation frequency, and

$$a_j = \frac{\omega_j^2 + 2i\beta_j\, \omega_j\, \omega}{\omega_j^2 - \omega^2 + 2i\beta_j\, \omega_j\, \omega} \tag{3}$$

are the amplification functions for the structural modes (i.e., on fixed base) having natural frequency ω_j and fraction of critical damping β_j. More generally, the equation corresponding to equation (2) for systems of higher complexity involving multiple seismic inputs and/or degrees of freedom per nodal point is of the form

$$U_f = [K_f - \omega^2(M_f + \Gamma^T A\Gamma)]^{-1}\, K_f\, U_g \tag{4}$$

with matrices K_f, M_f, Γ, $A = \text{diag}\,\{a_j\}$ and vectors U_f, U_g substituting for the scalars k_f, m_f, γ_{j1}, a_j, u_{gf}, u_g. The participation factors

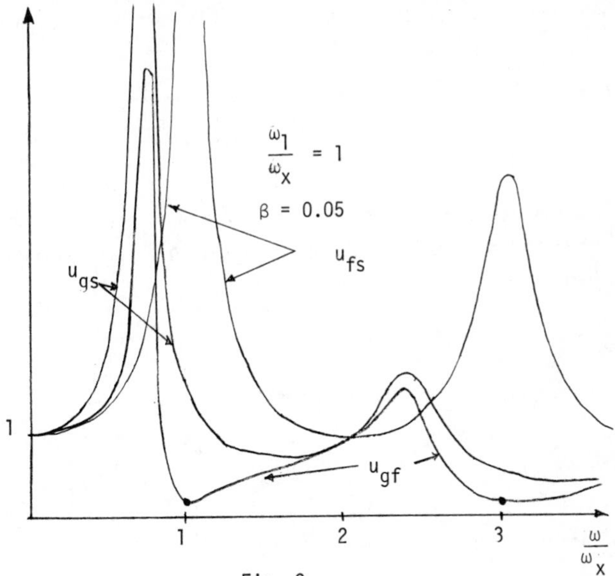
$\frac{\omega_1}{\omega_x} = 1$
β = 0.05
u_{fs}
u_{gs}
u_{gf}
1
1
2
3
$\frac{\omega}{\omega_x}$

Fig. 3

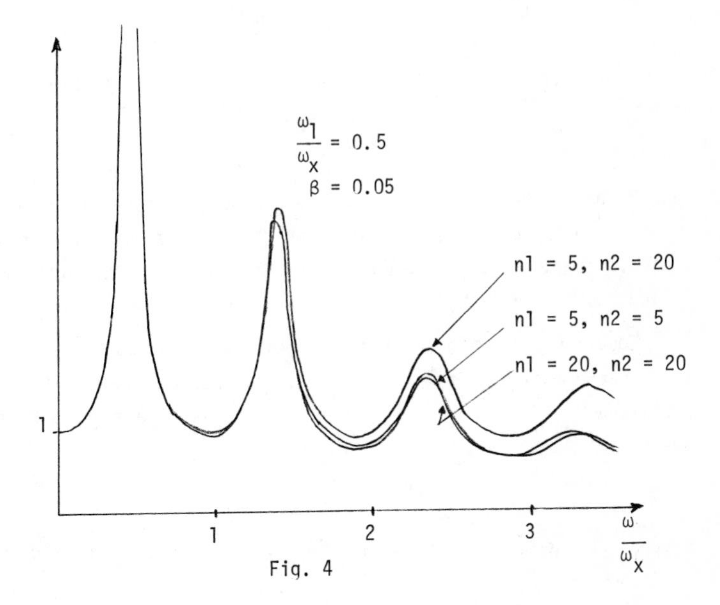
$\frac{\omega_1}{\omega_x} = 0.5$
β = 0.05
n1 = 5, n2 = 20
n1 = 5, n2 = 5
n1 = 20, n2 = 20
1
1
2
3
$\frac{\omega}{\omega_x}$

Fig. 4

matrix Γ has up to six columns, depending on how many seismic components are prescribed under the springs. This matrix can also be written as

$$\Gamma = \begin{Bmatrix} \Gamma_1 \\ \vdots \\ \Gamma_n \end{Bmatrix} = \{\Gamma_j\} \tag{5}$$

with n being the number of degrees of freedom of the structure (i.e., the number of modes on fixed base), and the modal row-vector Γ_j being given by

$$\Gamma_j = \{\gamma_{j1}\ \gamma_{j2} \cdots \gamma_{j6}\} \tag{6}$$

Defining the (6 x 6) virtual modal mass matrices

$$\Delta M_j = \Gamma_j^T \Gamma_j$$

it follows that equation (4) is of the form

$$U_f = [K_f - \omega^2(M_f + \sum \Delta M_j\ a_j)]^{-1}\ K_f\ U_g \tag{7}$$

These virtual mass matrices can be shown to satisfy the relationship

$$\sum \Delta M_j = M_s \tag{8}$$

with M_s being the (6 x 6) mass matrix of the structure, assumed rigid, relative to the coordinate point at which the impedances K_f are attached. Thus, the complete mass matrix for the structure and the foundation would be $M_s + M_f$.

Equation (7) (with equation (2) being a particular case) describes the motion of the foundation for harmonic seismic inputs with amplitude U_g. It can be observed that the effective inertia of the system,

$$M_{eff} = M_f + \sum \Delta M_j\ a_j \tag{9}$$

depends on the amplification factors a_j; thus, it is a function of the excitation frequency of the seismic motion, and the natural frequencies and modal damping values of the structure on fixed base. If the superstructure did not have any damping at all, then at each natural frequency the effective mass would become infinitely large. As a result, the foundation would remain exactly motionless at these frequencies; at other structural locations, the motion would have a finite amplitude which could be interpreted as resulting from the indeterminate product $0 \cdot \infty$(motion of foundation x structural amplification). On the other hand, most real structures have some amount of structural damping of the order of 2% to 7% of critical in each mode. As a result, the effective mass is not infinitely large, although it is still substantial when compared with the actual mass of the structure. Hence, the transfer functions for the motion of the foundation will not display a zero value but a small value at each of the resonant frequencies of the structure. When this "small" value is multiplied times the "large" amplification function for the structure, one obtains the finite response of the structural masses, accounting for interaction, at the natural frequencies of the structure (Fig. 3). If, on the other hand, the structural frequencies are changed in the second step, as they must be if the structural model is changed, then the large amplification peaks for the altered structural model will not coincide with the valleys in the foundation response obtained with the original structural model. As a result, large spurious amplifications will be observed in the response of the hybrid system. Peaks will then be observed not only at the coupled interaction frequencies, but at the frequencies of the structure on fixed base as well. It should be observed that the amplitude of these spurious peaks is controlled by the structural damping only, and not by the effective soil-structure interaction damping (i.e., accounting for radiation). Further, they will develop even for biases in the structural frequencies as small as 5%, as will be shown. The actual bias in the frequencies of a solid block modeled, for example, with finite elements as opposed to those of a refined model with lumped masses and linear members, is likely to be much higher.

Example

To illustrate the effect of changing the structural model in the second step of analysis, consider the lumped mass shear beam, supported on a single translational spring at the base, shown in Fig. 2. For a structure discretized into n masses (n+1 when counting the foundation), the inertia properties are

$$M = mn \quad \text{(total mass)}$$

$$S = \frac{mhn^2}{2} = \frac{MH}{2} \quad \text{(static moment)}$$

$$J = \frac{mh^2n^3}{3}\left(1 + \frac{1}{2n^2}\right) = \frac{MH^2}{3}\left(1 + \frac{1}{2n^2}\right) \quad \left(\begin{matrix}\text{moment of inertia} \\ \text{w.r. to base}\end{matrix}\right)$$

The above formulae indicate that when changing the number of masses in the structural model it is not possible to match simultaneously the static moment (center of mass) and the moment of inertia; nevertheless, the error is small, since $1/2\,n^2 \ll 1$. Also, since this example does not involve rotation, this error is irrelevant.

On the other hand, the frequencies, modal shapes and participation factors for the superstructure on fixed support are

$$\omega_j = 2\sqrt{\frac{k}{m}}\sin\theta_j \quad , \qquad \text{frequencies, rad/sec}$$

$$\text{with} \quad \theta_j = \frac{\pi}{4n}(2j-1) \quad , \qquad j = 1, 2, \ldots n$$

$$\phi_j = \sqrt{\frac{2}{mn}}\begin{Bmatrix} 1 \\ \cos 2\theta_j \\ \cos 4\theta_j \\ \vdots \\ \cos 2(n-1)\theta_j \end{Bmatrix} \quad \text{modal shape}$$

$$\gamma_{j1} = -(-1)^j\sqrt{\frac{m}{2n}}\cot\theta_j \qquad \text{p. factor for base translation}$$

Also, the swaying frequency associated with a rigid structure is

Eduardo Kausel

$$\omega_x = \sqrt{\frac{k_x}{M}}$$

It is convenient at this point to introduce the following notation :

n_1 = number of masses in structural model used in the first step to determine the motion of the foundation.

n_2 = number of masses in structural model used in the second step to determine the motion at the top.

Thus, if n1 = n2, the model is consistent, and the motions computed with the two-step approach coincide with those of a direct evaluation. In general, when the model is "refined" in the second step, the number of masses increases, i.e., $n2 \geq n_1$.

Fig. 4 shows the motion at the top of the structure for $\omega_1/\omega_x = 0.5$, $\beta = 0.05$ (structural damping), and three structural configurations: (n1 = 5, n2 = 5), (n1 = 20, n2 = 20), and (n1 = 5, n2 = 20). Thus, the first two are consistent models, while the last is inconsistent. The stiffness of the "refined" model (i.e., n2 = 20) was adjusted so as to match the fundamental frequency ω_1 of the "coarse" (i.e., n1 = 5) model. It can be observed that the results of the consistent models are in good agreement, while the inconsistent model shows discrepancies at higher frequencies. Thus, the refinement (change!) of the structure in the second step deteriorates rather than improves the computation.

The situation described above is the most favorable possible for the two-step approach with dissimilar structures, since the fundamental frequencies were matched exactly. In actual computations, no such exact match is accomplished, so that even the fundamental frequency may differ between steps. In fact, in some situations it may not even be possible to attempt a match, as when the nature of the structure is altered entirely. An example is the case of a coarse structure modeled first with finite element in two dimensions, and then refined to a structure with lateral and torsional coupling in three dimensions. The effect of a bias in the frequencies is illustrated in Fig. 5. Here, the motion was computed first with a fully consistent model (n1 = 20, n2 = 20) ;

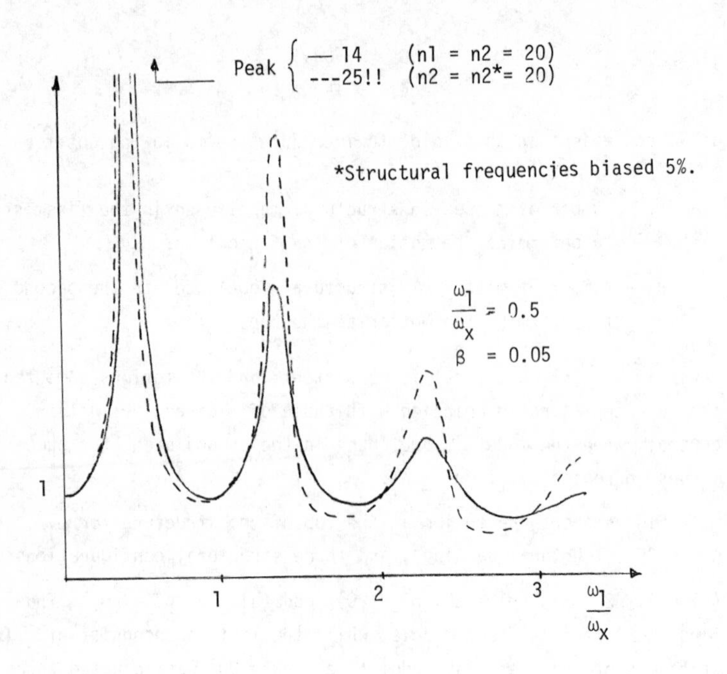
Peak
— 14 (n1 = n2 = 20)
---25!! (n2 = n2*= 20)
*Structural frequencies biased 5%.
ω1/ωx = 0.5
β = 0.05
1
1
2
3
ω1/ωx

Fig. 5

the computation was then repeated with a model having again 20 masses, but changing the stiffness of the structure in the second step so as to bias the frequencies by 5% (all frequencies are multiplied by the factor 0.95). It may be observed that the amplitude of the amplification peaks nearly doubles as a result of the bias. Even worse results are obtained when using models with a different number of masses <u>and</u> frequencies that do not match exactly.

Conclusions:

The results shown here, as well as further numerical experiments for other soil configurations (changing stiffness of soil spring, adding rocking spring, etc.), lead to the following conclusions:

a) The technique of refining the structure in the second step of the two-step approach deteriorates rather than improves the computations. It is then preferable to adhere to the coarse model throughout the analysis. The requirement on the consistency of the structural models may in some cases be as important or even more important than the need to consider soil-structure interaction, unless the structure has considerable damping.

b) For structures that must clearly be modeled in three dimensions, it may be necessary to use the impedance (soil springs) approach in connection with equation (4), or generalizations of two-dimensional finite element codes that are compatible with three-dimensional superstructures (in essence, an implicit impedance approach).

Acknowledgement

The research reported herein was made possible by an NSF Research Initiation Grant: ENG 7908080, "Numerical Procedures for Rigid Plate Foundations."

The Use of Simple Models in Soil Structure Interaction

by J. M. Roesset,* M. A. S. C. E.

Introduction. No method of analysis can be considered exact when attempting to predict the true physical response of a structure subjected to uncertain loads. This is even more so in the area of seismic soil structure interaction, where large uncertainties exist in the estimation of the characteristics of the potential earthquake (including its frequency and wave content) and of the in situ soil properties (including their variation with state and level of strains). Analyses of nuclear power plants are conducted, however, with as much sophistication and rigor as possible, within the limitations imposed by the state of the art knowledge, the availability of specific computer programs and the requirements imposed by codes and regulations (which will introduce inaccuracies in order to guarantee a conservative design). The importance of this type of construction has led naturally to a desire to incorporate in practice results of academic research as soon as they are available, sometimes before they can be properly tested and evaluated, and in some occasions generalizing them to situations for which they were not intended. There has also been, understandably, a great emphasis placed on the development of general computer programs which can be qualified and used as a standard, even if they must be periodically modified to account for new knowledge. The pressure to improve the sophistication and numerical accuracy of the mathematical models has resulted, however, in an unfortunate neglect of parametric studies, which would shed more light on the significance of various parameters, and the derivation of simplified, engineering type procedures which would allow to estimate the importance of various effects. Approximate methods are nevertheless of great value, not only to obtain solutions

*

Professor of Civil Engineering, The University of Texas at Austin which will have often the desired degree of accuracy (particularly when considering all the uncertainties), but also to check the validity of different formulations, to interpret the results of more complex analyses, to estimate the effects of variations in parameters and to identify the key features of the problem that must be correctly modelled.

Two general procedures are used at present for soil structure interaction analyses: a direct approach in which the structure and the soil are modelled together and the complete system is solved in a single step, and a substructure approach in which the problem is solved in three separate steps (determination of the motions of a massless foundation, computation of the dynamic stiffness of the foundation and dynamic analysis of the structure on a foundation with the stiffness computed in the second step and subjected to the motions obtained from the first).

There is only a number of simplifications which can be introduced in direct analyses of the complete soil structure system: use of a two dimensional or a pseudo three dimensional model instead of a true three dimensional representation of the soil; the assumption of vertically propagating waves leading to the specification of a uniform motion at the base of the model, instead of a more general excitation; the use of an equivalent linearization technique, with iterations carried out on the complete model or only on the one dimensional wave propagation problem, instead of a true nonlinear analysis in the time domain; and the use of a crude model of the structure (represented sometimes as a shear block) instead of a more detailed one. These simplifications are intended primarily to reduce the cost of computation; they are justified mainly on the basis of the lack of proven transmitting boundaries in three dimensional cartesian coordinates or in the time domain, uncertainties in the wave content of the design earthquake, and insufficient knowledge on the true nonlinear behavior of the soil.

Approximate procedures are particularly suited for solutions using the substructure approach, where each one of the steps can be investigated and its results interpreted, by more than one method. The flexibility of this approach allows to combine solutions obtained by different procedures as long as they are based on consistent assumptions. The purpose of this paper is to review some of the simplified models that have been developed to date and discuss their potential use to estimate soil structure interaction effects in the seismic design of nuclear power plants.

Simplified models.

The simplest model developed to estimate the effect of the flexibility of the foundation on the dynamic response of a structure consists of a rigid mass supported by a set of springs and dashpots. This model has been used extensively in the study of the vibrations of machine foundations (Whitman and Richart, 1967). For vertical excitation it has a single degree of freedom. Calling M the mass and k_z and c_z the vertical spring and dashpot representing the foundation the natural frequency of the system is given by

$$\omega_z^2 = \frac{k_z}{M} \tag{1}$$

and the corresponding camping at this frequency is

$$D_z = \frac{1}{2} \frac{c_z}{k_z} \omega_z \tag{2}$$

For lateral excitation the system has two dynamic degrees of freedom. Calling then M the mass, I_o the mass moment of inertia with respect to the centroid, h the height of the centroid over the base, u and ϕ the translation and rotation of the centroid, $y = u - h\phi$ the translation of the base and k_x k_p c_x c_ϕ the foundation springs and dashpots one can define the uncoupled frequencies

$$\omega_x^2 = \frac{k_x}{M} \qquad \omega_\phi^2 = \frac{k_\phi}{I_o + Mh^2} \tag{3}$$

The natural frequencies of the system are then given by

$$\omega^2 = \frac{1}{2}\left(1 + \frac{Mh^2}{I_o}\right)(\omega_x^2 + \omega_\phi^2)\left(1 \pm \sqrt{1 + \frac{4\omega_x^2\,\omega_\phi^2}{(\omega_x^2 + \omega_\phi^2)^2 \left(1 + \frac{Mh^2}{I_o}\right)}}\right) \tag{4}$$

and the mode shapes are

$$h\phi_i = \left(1 - \frac{\omega_i^2}{\omega_x^2}\right) U_i \qquad y_i = \frac{\omega_i^2}{\omega_x^2} U_i \qquad i = 1,2 \tag{5}$$

For general cases the two degree of freedom system will not have normal modes in the classical sense once the dashpots c_x c_ϕ are considered. The best approach in this case, if an exact solution is desired, is to obtain the transfer functions of the desired effects and to solve the problem in the frequency domain. Considering, on the other hand, that this is a simplified model, if one is only interested in estimating a fundamental frequency of the system and an effective damping (at that frequency) further approximations can be introduced. Using for instance Dunkerley's rule the first natural frequency can be estimated as

$$\omega_1^2 = \frac{\omega_x^2 \omega_\phi^2}{\omega_x^2 + \omega_\phi^2} \tag{6}$$

a formula which is valid for

$$\omega_1^2 << \frac{1}{2} (\omega_x^2 + \omega_\phi^2)(1 + \frac{Mh^2}{I_o})$$

The effective damping at this frequency is approximately

$$D_1 = \frac{1}{2}\omega_1 \frac{c_x h^2 + c_\phi (1 - \omega_x^2/\omega_1^2)^2}{k_x h^2 + k_\phi (1 - \omega_x^2/\omega_1^2)^2} \tag{7}$$

Expressions (2) and (7) assume an elastic soil. If the soil has an internal damping D_s, of a hysteretic nature, to account for nonlinear behavior, the effective damping would increase by D_s.

Typical nuclear power plants are very stiff structures. When they are resting on relatively soft soils, the above formulae, based on the assumption of a rigid mass, can be sufficiently accurate for practical purposes. Some care must be exercised, however, when using this model to interpret the effects of soil structure interaction, since it would suggest that the flexibility of the foundation results in a peak of the transfer function at the fundamental frequency of the system which would not exist for a rigid foundation (while in reality the effect would be a shift in the frequency at which the peak occurs and in most cases a decrease in its amplitude).

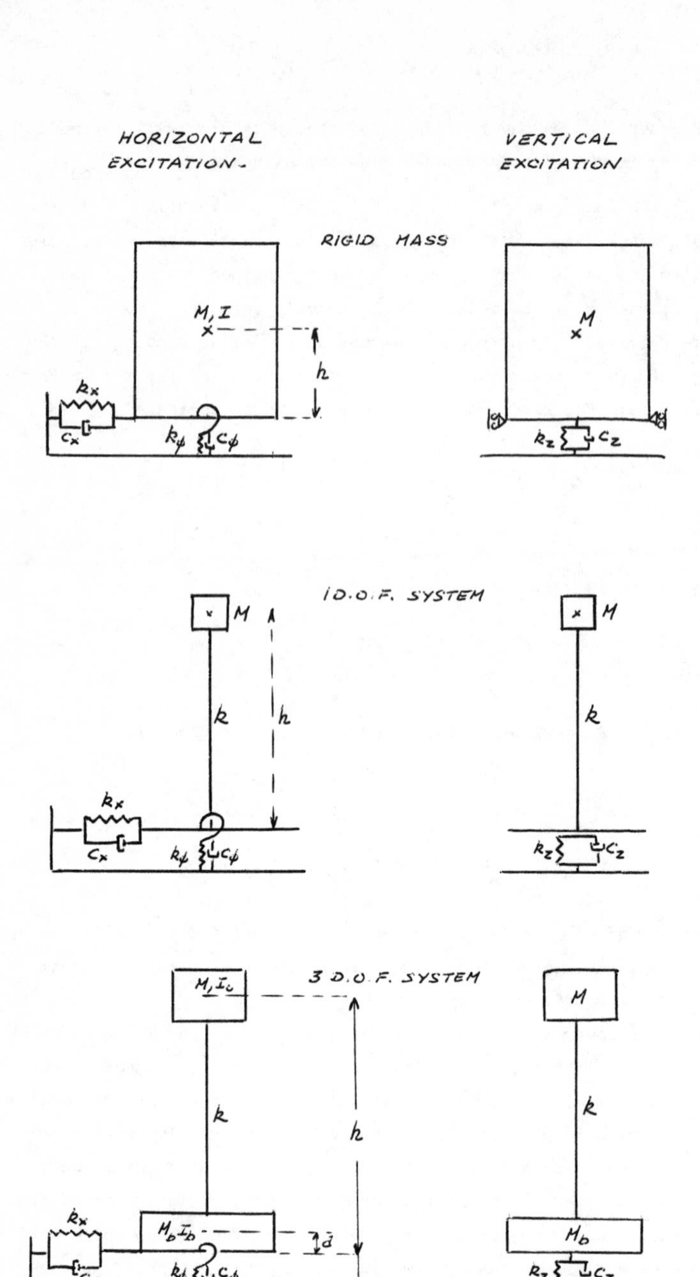

Figure 1. Simple Models.

A better model from a conceptual point of view is the one originally used by Parmelee (1967). The structure is then represented as a single degree of freedom system consisting of a mass M at a height h above the base of the foundation and a spring k. The parameters of this system may be chosen from simple physical considerations, to model the overall behavior of the structure on the basis of an assumed deflected shape, or to represent only its first mode (this implies that a dynamic analysis of the structure on a rigid base has been performed previously, a practice which has some merit). In the simplest form of the model the foundation stiffnesses are represented again by a set of springs and dashpots.

For vertical excitation the systam has again a single dynamic degree of freedom. Its natural frequency is given by

$$\omega_z^2 = \frac{k}{M(1+k/k_z)} = \frac{\omega_o^2}{1+k/k_z} \qquad (8)$$

and the effective damping is approximately

$$D_z = D_{st} \left(\frac{\omega_z}{\omega_o}\right)^2 + \left(D_s + \frac{\omega_z c_z}{2k_z}\right) \left[1-\left(\frac{\omega_z}{\omega_o}\right)^2\right] \qquad (9)$$

where k is the vertical (axial) stiffnessof the structure, ω_o its natural frequency on a rigid base, D_{st} is the structural damping and D_s the internal soil damping.

For horizontal excitation the system has three static degrees of freedom (two translations and a rotation) but two of them can be eliminated from equilibrium considerations if the mass of the foundation is neglected as well as rotatory moments of inertia. Calling then k the lateral stiffness of the structure, ω_o its first natural frequency under lateral vibration and k_x k_ϕ c_x c_ϕ the corresponding foundation springs and dashpots

$$\omega^2 = \frac{\omega_o^2}{1+\frac{k}{k_x}+\frac{kh^2}{k_\phi}} \qquad (10)$$

and

$$D \simeq D_{st} \left(\frac{\omega}{\omega_o}\right)^2 + D_s \left[1-\left(\frac{\omega}{\omega_o}\right)^2\right] + \left(\frac{\omega}{\omega_o}\right)^2 \left[\frac{1}{2}\frac{k}{k_x}\frac{c_x \omega}{k_x} + \frac{1}{2}\frac{kh^2}{k_\phi}\frac{c_\phi \omega}{k_\phi}\right] \qquad (11)$$

Even for the more complex model with three dynamic degrees of freedom, including the mass M_b of the foundation and rotatory moments of inertia, I_o, I_b, the solution can be easily carried out with a calculator. Moreover using approximations such as Dunkerly's rule, calling h the height of the structural mass with respect to the base, d the height of the centroid of the foundation, and

$$\omega_o^2 = \frac{k}{M}, \omega_x^2 = \frac{k_x}{M+M_b}, \omega\phi^2 = \frac{k_\phi}{Mh^2+M_b d^2+I_o I_b}$$

the fundamental frequency can be estimated as

$$\frac{1}{\omega^2} = \frac{1}{\omega_o^2} + \frac{1}{\omega_x^2} + \frac{1}{\omega_\phi^2} \tag{12}$$

These simple models have often been considered as synonimous with the substructure approach. Thus the distinction in regulatory guides between finite element solutions (implying a direct solution) and the lumped spring mass method (implying a substructure analysis). It is clear that the above idealizations are too crude for the detailed dynamic analysis of a nuclear power plant. The structure cannot be reproduced well by a single degree of freedom system due to the various components. Even so it is important to notice that the subsrtucture approach is much more general than a spring mass model and can cover any practical situation within the limitations of a linear analysis. In between the most sophisticated forms of the substructure approach and the simple models with constant springs at the base there is a large number of possible variations which increase the accuracy while increasing only slightly the complexity. Finally while the simplest models are too crude for many cases there are situations in which they provide all the needed accuracy. More importantly they provide a means to estimate the importance of interaction effects; to assess the significance of various parameters and to make decisions on modelling for direct analysis.

Before these models can be used it is necessary, however, to estimate the input motions, the effective soil properties and the foundation stiffnesses.

Determination of the motions at the foundation level

For foundations with small embedment ratio (ratio of the embedment depth to the base dimension of 15 percent or less) present regulations allow to consider directly the design motion at the free surface of the soil deposit as the one occurring at the foundation level. For foundations with deeper embedment it is necessary, however, to compute compatible motions of a massless (normally rigid) foundation accounting for the geometry of the excavation.

Morray (1977) conducted a series of parametric studies for circular foundations embedded in a finite layer of soil with uniform properties and obtained the transfer functions from the surface to the foundation level assuming a train of vertically propagating shear waves. From these studies he suggested as an approximation a transfer function for the horizontal motion given by

$$F_u(\Omega) = \begin{cases} \cos \frac{\pi}{2} \frac{f}{f_1} & \text{for } f \geq 0.7 f_1 \\ 0.453 & \text{for } f > 0.7 f_1 \end{cases} \quad (13)$$

where f_1 is the fundamental frequency of the embedment layer and $\Omega = 2\pi f$.

There is in addition a rotation of the foundation. The transfer function from the surface motion to the rotation is approximately

$$F_\phi(\Omega) = \begin{cases} \frac{0.257}{R} \left(1 - \cos \frac{\pi}{2} \frac{f}{f_1}\right) & \text{for } f \geq f_1 \\ \frac{0.257}{R} & \text{for } f > f_1 \end{cases} \quad (14)$$

These simple formulae are intended to provide smoothed, average type effects. The true transfer functions will exhibit oscillations around these values. Similar studies for strip footings and for rectangular foundations considering not only vertically propagating shear waves but also trains of body waves at other angles yield the same kind of results but show that the lower bound for the translation (0.453 in Morray's case) increases slightly with increasing internal soil damping and for other wave trains.

Present regulations impose a limit on the amount of reduction of the translational motion which can occur with embedment. Imposing such a limitation on the transfer function makes sense in order to account for uncertainties in the types of waves and the soil properties. The simplified expressions above are particularly suited for this kind of considerations. At the present time the limitations are being imposed, however, on the response spectra. This would require obtaining similar formulae for the ratio of response spectra instead of the transfer function. More importantly present limitations are supposed to apply to the translational motion that would occur at the foundation level in the free field (ignoring the geometry of the excavation) and make no mention of the rotational component. This is not a logical approach.

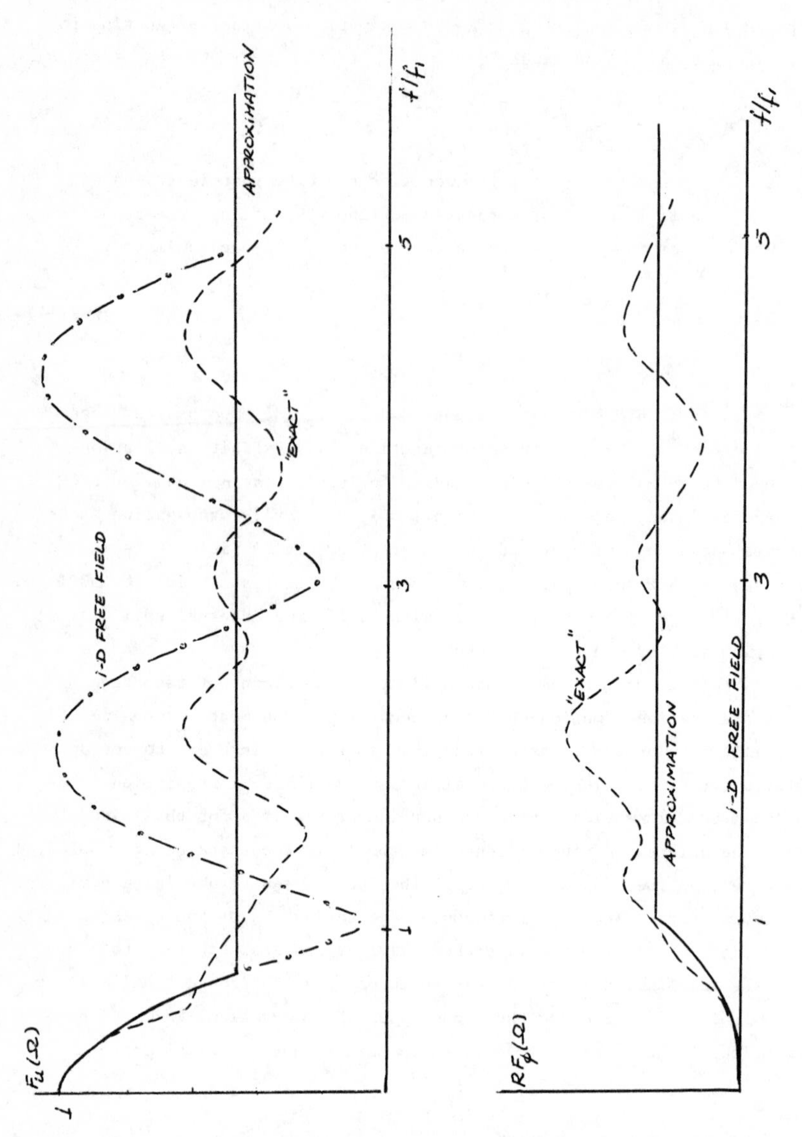

Figure 2. Transfer functions of foundation motions.

A procedure like that suggested by Morray, with further verification and possible improvements through more extensive parametric studies, including variable soil profiles, nonlinear soil behavior and various kinds of waves, and determining expressions for both the transfer functions and the ratio of response spectra would provide a simple means to obtain solutions which are reasonably approximate and which can account for some of the uncertainties. Until additional and more comprehensive studies are performed expressions (13) and (14) can be used for preliminary estimates of the input motions for embedded foundations, applying them as transfer functions or even as ratio of response spectra.

The equations of motion for the rigid mass model accounting for the combination of a horizontal and a rocking excitation become

$$M\ddot{y} + Mh\ddot{\phi} + k_x y = -M(\ddot{u}_G + h\ddot{\phi}_G)$$

$$Mh\ddot{y} + (I + Mh^2)\ddot{\phi} + k_\phi \phi = -Mh\ddot{u}_G - (I + Mh^2)\ddot{\phi}_G \qquad (15)$$

where the translation of the base is given by $y + u_G$, the rotation is $\phi + \phi_G$ and the absolute displacement of the centroid is $y + u_G + h(\phi + \phi_G)$.

Using on the other hand the single degree of freedom model the equation of motion is

$$M(1 + \frac{k}{k_x} + \frac{kh^2}{k_\phi})\ \ddot{y} + ky = -M(\ddot{u}_G + h\ddot{\phi}_G) \qquad (16)$$

where y represents the distorsion of the structural spring. The motion of the foundation (including the structure) consists then of a translation $k/k_x y + u_G$ and a rotation $\phi_G + ykh^2/k\phi$. The absolute displacement of the mass if $u_G + h\phi_G + y(1 + k/k_x + kh^2/k\phi)$

This indicates than when the rotatory mass moment of inertia can be neglected the base rotation can be accounted for by using an "effective" base motion $u_G + h\phi_G$. Using Morray's results the transfer function for this motion is

$$F(\Omega) = \begin{cases} 0.257\,\frac{h}{R} + (1 - 0.257\,\frac{h}{R})\cos\frac{\pi}{2}\frac{f}{f1} & \text{for } f \le 0.7\,f1 \\ 0.453 + 0.257\,\frac{h}{R}\ (1 - \cos\frac{\P}{2}\frac{f}{f1}) & \text{for } 0.7f \le f \le f1 \\ 0.453 + 0.257\,\frac{h}{R} & \text{for } f < f \end{cases} \qquad (17)$$

As a first, crude, approximation one can evaluate this function at the fundamental frequency of the soil structure system to assess the effect of the reduction in motion due to embedment (so called kinematic interaction effects) on the acceleration of the mass or the base shear. This function is always less than one but close to unity for low frequencies. For high frequencies the value is $0.453 + 0.257 \frac{h}{R}$, or since the ratio $\frac{h}{R}$ is typically between 0.5 and 1 for nuclear power plants (reactor building), between 0.68 and 0.80 (reductions of 20 to 32%).

Estimation of Soil Properties

To determine the foundation stiffness as well as the natural frequency of the embedment layer for the previous formulae, and the natural frequency of the total soil stratum when dealing with a soil deposit of finite depth it is necessary to determine the soil properties consistent with the expected levels of strain. Great uncertainties exist not only in the determination of in situ properties but also in the way nonlinear soil behavior is handled in most direct analyses (through an equivalent linearization). Accepting, however, the validity of the iterative procedure, at least from a qualitative point of view, it has been shown that for moderate to large seismic excitations (where nonlinear effects would be important) results are very similar conducting the iterations on the complete two or three dimensional model (including the structure) or using the equivalent soil properties (shear modulus and damping) resulting from one dimensional soil amplification studies in the free field (as those conducted as a preliminary step in direct solutions to obtain a compatible base motion).

One dimensional amplification studies as performed by the program SHAKE or other similar ones are very simple and economical. Even so, the uncertainties involved in the soil characteristics and the accuracy of the iterative linearization scheme are such that even simpler procedures may be justified. Jakub (1977) conducted a series of parametric studies for uniform soil profiles and for a soil deposit where the shear modulus increased with depth, assuming in all cases a Ramberg-Osgood model for the soil. The constitutive equation was of the form

$$\frac{\gamma}{\gamma_y} = \frac{\tau}{\tau_y}\left(1 + \alpha \left|\frac{\tau}{\tau_y}\right|^{r-1}\right) \tag{18}$$

The effective shear modulus corresponding to a shear stress τ is then given by

$$G = G_o \frac{1}{1 + \alpha \left|\frac{\tau}{\tau_y}\right|^{r-1}} \tag{19}$$

and the internal soil damping is

$$D_s = \frac{2}{\pi} \alpha \frac{r-1}{r+1} \frac{G}{G_o} \left|\frac{\tau}{\tau_y}\right|^{r-1} \tag{20}$$

where $G_o = \tau_y / \gamma_y$ is the shear modulus for levels of strain from 10^{-5} to 10^{-6}. From his results Jakub obtained curves relating $\tau/\rho az$ versus Z/H where τ_m is the maximum shear stress at depth Z, ρ is the mass density of the soil, a is the peak ground acceleration at the free surface and H is a characteristic depth (function of the shear wave velocity of the soil at the desired depth, the predominant frequency of the earthquake and the soil parameters τ_y, α in addition to ρ and a).

While the actual curves derived by Jakub need further verification through additional parametric studies, considering a larger number of earthquakes and soil profiles, the procedure seems interesting for typical soil profiles (for unusual conditions individual studies would always be required). It allows in particular to estimate with simple hand computations the maximum shear stress τ_m at a number of points of the soil profile. Assuming then a characteristic stress $\tau = \frac{2}{3}\tau_n$ one can compute from equations (19) and (20) effective values of shear modulus and damping.

Determination of the foundation stiffnesses.

In the simplest form of the approximate models described earlier the foundation is reproduced by a set of springs and dashpots. The corresponding constants $k_z k_x k_\phi c_z c_x c_\phi$ are typically obtained from available solutions for circular foundations on the surface of an elastic half space as obtained for instance by Veletsas et al (1971, 1974).

Then $$k_x = \frac{8GR}{2-\upsilon} \qquad k_\phi = \frac{8GR^3}{3(1-\upsilon)} \qquad k_z = \frac{4GR}{1-\upsilon} \tag{21}$$

and $$c_x \simeq \frac{4.5}{2-\upsilon}\rho c_s R^2 \qquad c_z \simeq \frac{4}{1-\upsilon}\rho c_s R^2 \qquad c_\phi \simeq \frac{0.9}{1-\upsilon}\rho c_s R^4 \tag{22}$$

where $G = \rho c_s^2$ is the shear modulus of the soil, ρ its mass density, c_s its shear wave velocity, υ its Poisson's ratio and R is the radius of the foundation.

Several improvements can be introduced in this model. Such are:

- to account for the frequency variation of the stiffness coefficients.
- to account for the nonuniformity of the soil profile (variation of properties with depth)
- to account for the shape of the foundation
- to account for the effects of embedment.

1. When dealing with a surface foundation or a deep and relatively uniform soil deposit which could be considered as a half space, the variation of the stiffness coefficients with frequency has been tabulated by Veletsos et al. A solution of a 1, 2 or 3 degree of freedom system in the frequency domain is straightforward and inexpensive. It can be easily carried out with a calculator. Accounting in this case for the frequency variation of the stiffness functions offers no particular difficulty.

The terms k_x c_x and c_z are almost independent of frequency and can be considered constant without any serious error. The terms k_z k_ϕ and c_ϕ vary strongly with frequency, particularly in the low and moderate frequency range. Some earlier work recommended the use of an added soil mass (which would produce a dynamic stiffness of the form $k-m\Omega^2$) to ccount for the frequency variation of the k terms. This approach will produce a good approximation for low values of frequency but will cause serious errors for larger frequencies. Moreover the concept of an added soil mass is physically incorrect. If it is desired to obtain a solution, or an estimate of the solution, by hand it is possible to perform a few cycles of iteration, starting with the static values, or the values corresponding to an estimated frequency; the natural frequency of the system is computed using formulae (6), (10), or (12); the values of the stiffness coefficients at this frequency are obtained from the tables or figures presented by Veletsos et al and the process is repeated until the variation in results is smaller than a desired tolerance. Alternatively one can use simplified approximate expressions for k_z k_ϕ and c_ϕ as a function of frequency, suggested by Veletsos and by Kausel.

2. It is clear that soil deposits will rarely have uniform properties, particularly when considering the variation of modulus and damping with level of strain. Jakub (1977) conducted a limited number of parametric studies on strip footings (two dimensional, plane strain model) using the same soil profiles of his amplification studies and the soil properties computed for different earthquake levels. From these studies he concluded that a reasonable approximation to the foundation stiffnesses could be obtained assuming a uniform soil deposit with the properties of the soil at a depth of 0.6 B if the soil deposit were originally uniform or at a depth of 0.4 B if the original, low level of strain modulus increased with depth. B is the half width of the footing. Considering the uncertainties in the soil properties it may be sufficient for practical purposes to use in all cases the adjusted properties at a depth of 0.5 B or 0.5 R for a circular foundation.

While the approach needs further verification it is consistent with many approximate methods in static geotechnical engineering problems. Combined with the procedure suggested to estimate strain compatible soil properties it provides an extremely easy solution which accounts at least qualitatively for the key features of the problem.

A more important consideration related to the layering characteristics of the soil deposit is the possible existence of a finite soil stratum, resting on much harder rock-like material by opposition to a half space.

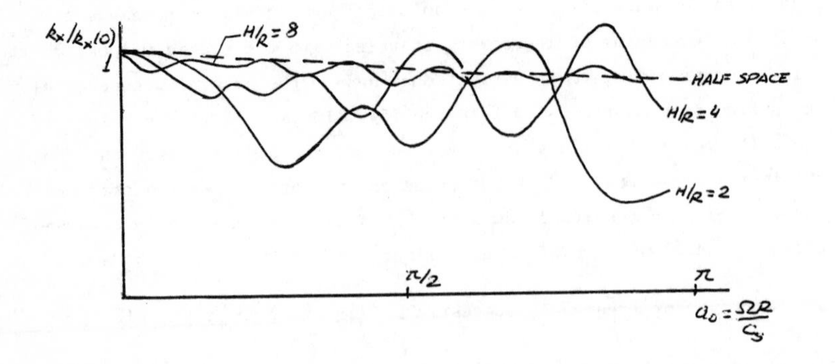

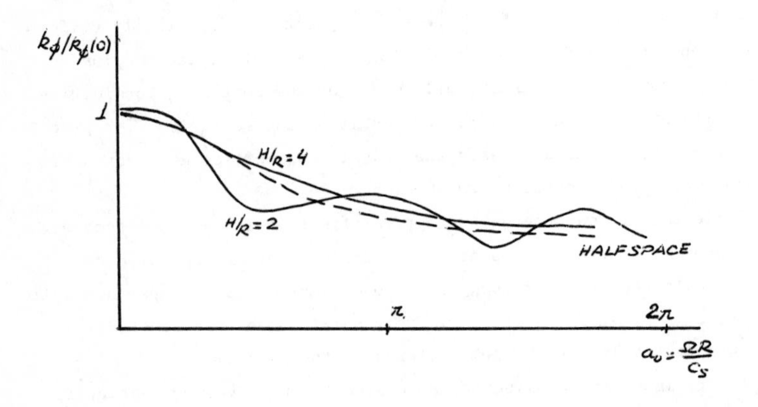

Figure 3. Effect of layer depth on stiffness coefficients.

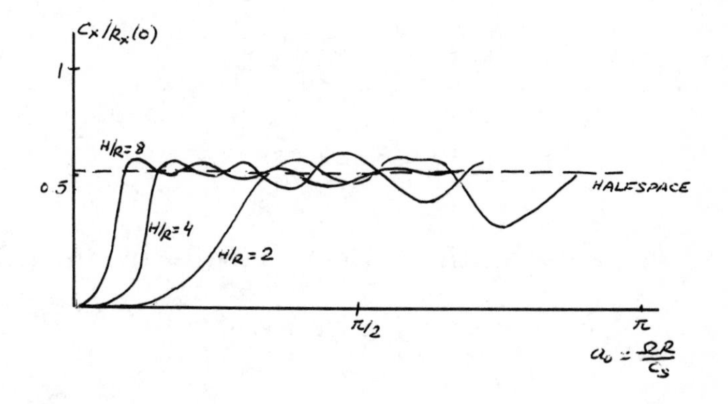

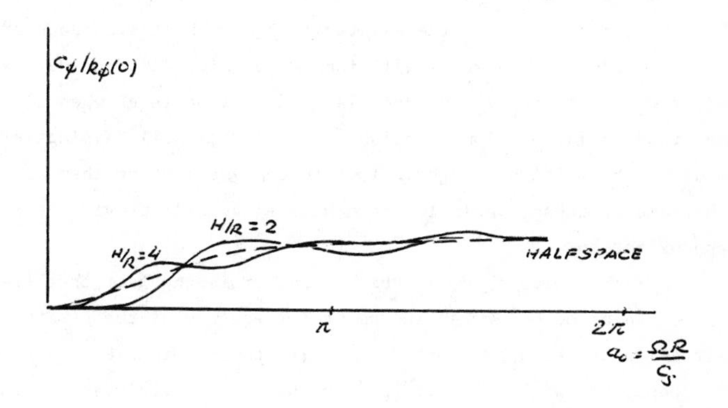

Figure 4. Effect of layer depth on damping coefficients.

The main effects in this case are:

-an increase in the values of the static stiffnesses. From parametric studies with circular foundations resting on the surface of a homogeneous soil stratum of variable depth, Kausel (1974) suggested the formulae

$$k_x = \frac{8GR}{2-\upsilon} \left(1 + \frac{1}{2}\frac{R}{H}\right)$$

$$k_\phi = \frac{8GR^3}{3(1-\upsilon)} \left(1 + \frac{1}{6}\frac{R}{H}\right) \tag{23}$$

For the vertical stiffness Kausel and Ushijima (1979) suggested

$$k_z = \frac{4GR}{1-\upsilon} \left(1 + 1.28 \frac{R}{H}\right) \tag{24}$$

More studies of this kind are needed to consider typical soil profiles where the properties vary with depth. In the meantime the above formulae can be used as reasonable approximations

-a change in the frequency variation of the stiffness coefficients. For the terms k_x, k_ϕ, k_z the existence of a much stiffer boundary at a finite depth will cause oscillations associated with the natural frequencies of the stratum. THe oscillations are not large when dealing with moderate to large seismic motions and internal soil dampings of the order of 5 to 10% or larger. In this case neglecting them and assuming the same frequency variation as a half space will provide an acceptable approximation.

For the terms c_x c_ϕ c_z the key factor is the fact that there will be no radiation below the fundamental frequency of the stratum. This situation can be approximated by making these terms zero below the layer's frequency and with the half space values for larger frequencies (a solution which should be on the conservative side). This will produce reasonable results unless the natural frequency of the combined system is very close to that of the layer. In this case a more refined analysis would be warranted.

3. The foundations of nuclear power plants will not be always perfectly circular slabs. In many cases they may be polygonal or even rectangular mats. It should be noticed that even the more sophisticated analyses using the direct approach will ignore this effect and that in fact the approximation provided bya circular mat will be generally better than that of a strip footing (two dimensional or pseudo three dimensional solution). For foundations whose dimensions are nearly equal in two orthogonal directions using the solution for an equivalentcircular mat with the same area will be acceptable. For rectangular foundations it is common practice to derive equivalent radii for horizontal and vertical excitation on the basis of the area, and for rotational motions equating the appropriate moment of inertia. It is generally accepted that these solutions will be reasonable for aspect ratios of up to 4:1 and this belief is confirmed by comparative studies conducted for both static values and the dynamic coefficients as a function of the frequency.

4. A relatively large number of studies have been conducted to determine the static and dynamic stiffnesses of embedded foundations and the results are available in the literature.

Novak (1973), using Baranov's equations, determined spring constants and dashpots to reproduce the soil surrounding the foundation. In this approach the embedment is treated as a Winkler foundation with distributed or lumped horizontal and vertical springs and dashpots applied to the lateral walls. The springs and dashpots are functions of frequency including therefore inertial and radiation effects. (For very low frequencies results must be extrapolated since the solution is not valid for the static case).

While it involves the approximation inherent in a Winkler foundation, Novak's solution is particularly attractive because of its flexibility. For each individual case the effect of embedment can be evaluated by adding to the dynamic stiffnesses of a surface foundation the contribution of the lateral springs and dashpots. It is possible in this form to consider foundations of artitrary shape as well as foundations which are embedded only over a portion of their perimeter. It appears that further studies should be conducted using this approach to compare results with those of other formulations and to investigate the effect of partial embedment. In addition to the expressions for the lateral springs,

Novak obtained solutions for a number of cases and presented curves which have been extensively used in practice, as well as simple formulae for the effect of embedment or the static stiffnesses.

Elsabee (1977) conducted a series of parametric studies with circular foundations embedded in a homogeneous soil stratum. From the results of these studies he extended Kausel's formulae to account for embedment

$$k_{xx} = \frac{8GR}{2-\upsilon} \left(1 + \frac{1}{2}\frac{R}{H}\right) \left(1 + \frac{2}{3}\frac{E}{R}\right) \left(1 + \frac{5}{4}\frac{E}{H}\right)$$

$$k_{\phi\phi} = \frac{8GR^3}{3(1-\upsilon)} \left(1 + \frac{1}{6}\frac{R}{G}\right) \left(1 + 2\frac{E}{R}\right) \left(1 + 0.7\frac{E}{H}\right) \tag{25}$$

where E is the embedment depth and H the layer thickness. There is in addition a coupling term

$$k_{x\phi} = k_{\phi x} = \left(0.4\frac{E}{R} - 0.03\right)Rk_{xx} \tag{26}$$

These expressions will provide a good approximation for values of $\frac{R}{H} \leq \frac{1}{2}$ and $\frac{E}{R} \leq 1$. For foundations with deeper embedment the increase in the stiffnesses will be larger than that predicted by the formulae.

Elsabee recommended to use the same frequency variation of these stiffness coefficients as for a surface foundation, an approximation which is reasonable, although the increase in stiffness tends to disappear with increasing frequencies (the approximation deteriorating therefore in the high frequency range). In the case of a finite soil stratum the terms c_x and c_ϕ will still be essentially zero below the fundamental frequency of the soil layer. Above this frequency, or over the complete frequency range for a half space, they will be larger than those of a surface foundation. Elsabee noted, however, that the values of these coefficients are affected by the conditions of the backfill and suggested, as a conservative measure, to use the results of a surface foundation. He did not attempt therefore to obtain expressions for these terms as a function of the embedment ratio. Studies by Scaletti (1977) indicate, however, that when separation occurs between the sidewalls of the foundation and the backfill a frictional loss of energy takes place that compensates the decrease in radiation damping. Accounting for the complete

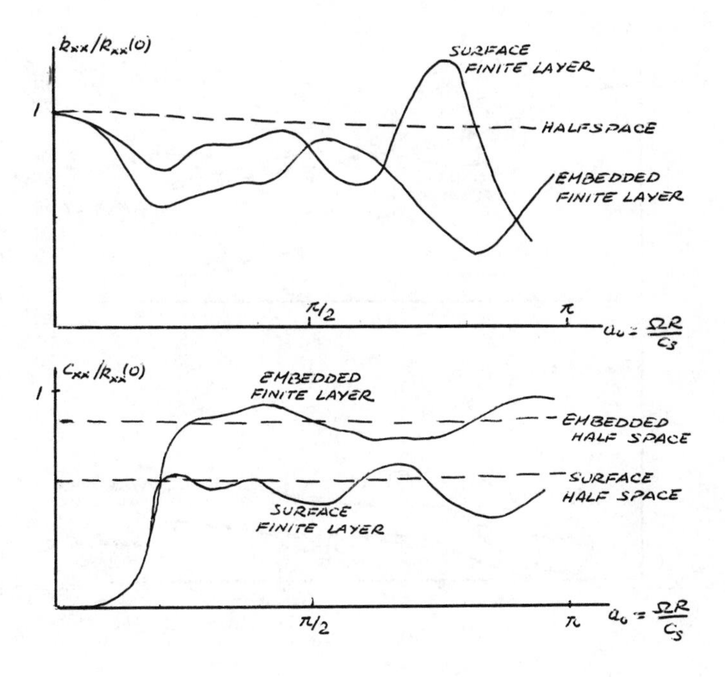

Figure 5. Effect of embedment on horizontal stiffness coefficients.

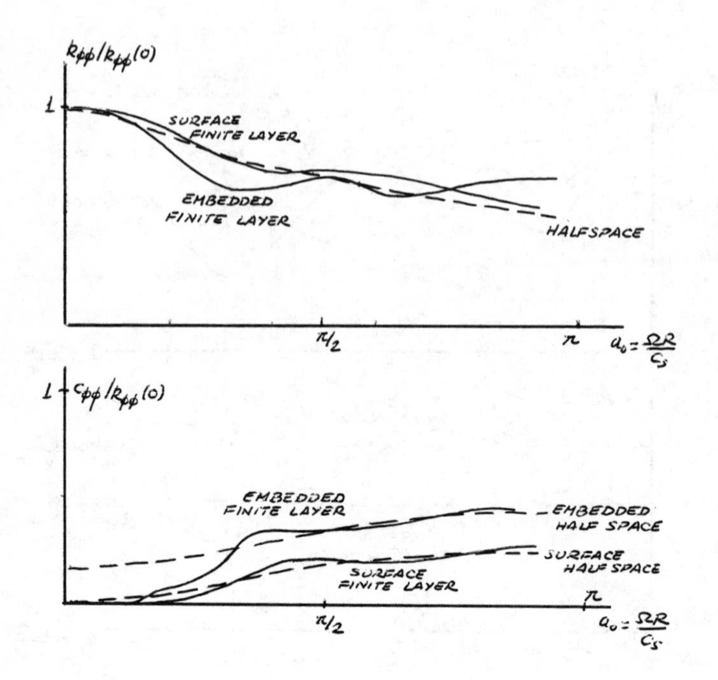

Figure 6. Effect of embedment on rocking stiffness coefficients.

increase in the c terms due to embedment would then be reasonable. Particularly significant is the increase in the term c_ϕ for low frequencies in the case of a half space (or a very deep soil deposit).

To account for the coupling term $k_{x\phi}$ with the one degree of freedom model it is sufficient to use for k_x k_ϕ the expressions

$$k_x = \frac{k_{xx}k_{\phi\phi} - k_{x\phi}^2}{k_{\phi\phi} - hk_{x\phi}} \tag{27}$$

$$k_\phi = \frac{h(k_{xx}k_{\phi\phi} - k_{x\phi}^2)}{hk_{xx} - k_{x\phi}}$$

(In reality these computations should be performed with the complex quantities $k + i\Omega c$. One should then obtain the real and imaginary part divided by Ω of the results).

For the rigid mass or the more accurate three degree of freedom system there is no problem in considering a foundation stiffness matrix with the coupling terms $k_{x\phi}$. One can alternatively use only the springs k_{xx} $k_{\phi\phi}$ by placing them at a distance from the base of the foundations equal to the ratio of $k_{x\phi}$ to k_{xx}, which according to formula (26) would be approximately 0.4 E - 0.03 R.

Following the work of Elsabee, Kausel and Ushijima derived expressions for the vertical stiffness of an embedded circular foundation of the form

$$k_z = \frac{4GR}{1-\upsilon}\left(1 + 1.28\frac{R}{H}\right)\left(1 + 0.47\frac{E}{R}\right)\left\{1 + \left(0.85 - 0.28\frac{E}{R}\right)\frac{E/H}{1.-E/H}\right\} \tag{28}$$

Similar studies were conducted by Jakub (1977) for strip footings (plane Strain, two dimensional models) and by Dominguez (1978) for rectangular foundations. Dominguez pointed out that for embedded rectangular foundations the definition of an equivalent radius, in order to extrapolate results for circular foundations, becomes more difficult since both the area (or moment of the inertia) of the slab and its perimeter (area of the sidewalls in contact with the backfill) are involved.

Use of simple models. In the preceeding pages a number of approximate procedures, applicable to the various phases of a soil structure interaction analysis using the substructure approach, have been presented. No attempt has been made, however, to conduct an exhaustive survey of all the procedures developed in recent years. Further refinements are possible to increase the accuracy of simple models without a large increase in cost of computation, but, as mentioned above, a substantial amount of work is still needed to verify, generalize and improve approximate methods.

While some approximations have a degree of accuracy comparable, or even superior, to that of more complex procedures, the emphasis in this paper has been on methods which can be used to obtain meaningful estimates of key response parameters by hand computations. These methods are of value to verify and interpret results of more complex analyses, to estimate the effect of uncertainties (sensitivity studies) and to decide on modelling issues. So for instance:

1. The degree of accuracy and the general applicability of two dimensional or pseudo three dimensional models has been a subject of controversy for several years. It appears that the transfer functions for the foundation motions (translation and rotation) are not very different for 2D models than for circular foundations. Jakub (1977) obtained, on the other hand, expressions for the stiffnesses of embedded strip footings similar to those discussed for circular foundations. When dealing with a horizontal earthquake excitation it would be desirable to match the stiffness functions k_x k_ϕ c_x c_ϕ over the complete range of frequencies of interest. There are only, however, two parameters one can play with: the width of the footing and the thickness of the soil slice. By selecting appropriately the width and thickness one can match two of the four functions at a specific frequency. It is important therefore to estimate first the frequency at which the match should be achieved (the fundamental frequency of the soil structure system), to identify the two functions which affect most the system's response and to assess the effect that errors in the other two can have in the final results.

2. When performing a direct analysis of the complete system a discrete model based on a finite element or finite differences representation of the soil is normally used. An important question is then to define a finite domain and appropriate boundary conditions at the edges.

If a soil deposit of finite depth is resting on much stiffer, rock-like material, the bottom boundary should be placed at the interface between the soil and the rock. If the fundamental frequency of the soil layer is larger than the natural frequency of the soil structure system there will be very little or no radiation damping and the lateral boundaries can be simple boundaries placed at a sufficient distance to reproduce properly static effects. On the other hand if the natural frequency of the system is above that of the layer radiation may play an important effect in the response. For linear analyses in the frequency domain consistent boundaries (Kausel-1974) can be placed at the edge of the foundation providing excellent results; viscous or simple boundaries must be placed, however, at a sufficient distance to guarantee that reflected waves have very small amplitudes when reaching back the structure.

For a very deep soil deposit with uniform properties or whose properties vary smoothly with depth a half space solution might be more realistic. The bottom boundary will introduce then artificial errors which must be minimized. It should be placed at a depth of the order typically of 4 radii, and one should check in addition that the natural frequency of the system is above that of the fictitious layer created by the mathematical model. Lateral boundaries should be again carefully selected.

3. If the solution is to be carried out in the frequency domain an appropriate increment of frequency must be chosen for the evaluation of the transfer functions. The size of this increment will be affected by the natural frequency of the system and the effective damping, in order to ensure that the peak of the transfer functions at this frequency is properly reproduced.

4. The main sources of uncertainties in soil structure interaction analyses are the characteristics of the design earthquake and the soil properties. It is now required to change the latter by ± 50%. Even so a larger number of variations may be advisable in some cases. Simple models allow to estimate in an economic way whether further studies are needed and what the possible effects of variations in parameters are.

REFERENCES

1. Dominguez J. (1978). "Dynamic Stiffness of Rectangular Foundations." Research Report R78-20. Civil Engineering Department. M.I.T. August.

2. Elsabee F. and Morray J. P. (1977). "Dynamic Behavior of Embedded Foundations." Research Report R77-33. Civil Engineering Department. September.

3. Jakub M. (1977). "Nonlinear Stiffness of Foundations." Research Report R77-35. Civil Engineering Department. M.I.T. September.

4. Jakub M. (1977). "Dynamic Stiffness of Foundations: 2-D vs 3-D Solutions." Research Report R77-36. Civil Engineering Department. M.I.T. October.

5. Kausel E. (1974). "Forced Vibrations of Circular Foundations on Layered Soils." Research Report R74-11. Civil Engineering Department. M.I.T. January.

6. Kausel E. and Ushijima R. (1979). "Vertical and Torsional Stiffness of Cylindrical Footings." Research Reprot R79-6. Civil Engineering Department. M.I.T. February.

7. Novak M. (1973). "Vibrations of Embedded Footings and Structures." ASCE National Structural Engineering Meeting. San Francisco, California. Preprint 2029. April.

8. Parmelee R. A. (1967). "Building Foundation Interaction Effects." Journal of the Engineering Mechanics Division. ASCE. Vol 93. No. EM2. April.

9. Scaletti H. (1977). "Nonlinear Effects in Soil Structure Interaction." Ph.D. Dissertation. Civil Engineering Department. M.I.T. September.

10. Veletsos A. S. and Wei Y. T. (1971). "Lateral and Rocking Vibrations of Footings." Journal of the Soil Mechanics and Foundations Division. ASCE. Vol 97. No. Smg. September

11. Veletsos A. S. and Verbic B. (1974). "Basic Response Functions for Elastic Foundations." Journal of the Engineering Mechanics Division ASCE No. EM2. April.

12. Whitman R. V. and Richart F. E. (1967). "Design Procedures for Dynamically Loaded Foundations." Journal of the Soil Mechanics and Foundations Division. ASCE. Vo. 93. No. SM6. Novmeber

FOUNDATION VIBRATIONS WITH SOIL DAMPING

by

John Lysmer, M.ASCE[1]

Introduction

The major topic discussed in this paper is how to evaluate the dynamic stiffness of a footing on a damped subgrade from a presumed available solution for the same footing on an undamped subgrade. The fundamental work on this problem was done by Veletsos and Verbic (1973) who showed how the correspondence principle of the theory of visco-elasticity can be used to solve this problem.

The work described herein employs the same method with slightly different assumptions and leads to a much simpler and potentially more accurate method for evaluating the effects of soil damping.

In order to develop this new method it was considered necessary first to discuss in some detail the definition of some measures of soil damping, to review the notation used in the theory of foundation vibrations and to introduce some new notations for the purpose of this paper. The new information is the simple formulas expressed by Eqs. (36) - (43). These formulas are applicable to very general cases such as embedded footings on layered foundations.

Undamped Foundation Vibrations

Harmonic vibrations of rigid footings on an undamped half space was one of the earliest topics studied in the field of soil dynamics. A very complete exposition of this topic has been given by Richart et al. (1970). Solutions are now available for all modes of vibration of a circular footing founded at the surface and for many other cases such as footings of arbitrary shape, embedded footings, and footings on layered half spaces. It is not the intention here to discuss the development and limitations of these solutions but rather to discuss the

[1]Professor of Civil Engineering, University of California, Berkeley, CA.

forms in which these solutions have been presented in the literature and to introduce some notation to be used in later parts of the present paper. For this purpose it will suffice to discuss the special case of vertical vibrations of a symmetric massless footing.

A central parameter in all presentations is the dimensionless frequency ratio, a_o, defined by

$$a_o = \omega R\sqrt{\rho/G} = \omega R/V_s = 2\pi R/\lambda_s \quad (1)$$

where ω is the circular frequency of excitation and R is a typical dimension of the footing (usually the radius or the side dimension). ρ, G and V_s are the mass density, shear modulus and shear wave velocity, respectively, of the soil and λ_s is the wavelength of shear waves. For a layered foundation G, ρ, V_s and λ_s may refer to average values or to the top layer. The last expression of Eq. (1) shows that a_o expresses how large the footing is compared to the length of waves which propagate in the soil.

Virtually all solutions in the literature express the solutions according to the notation of the complex response method, i.e.

$$\text{Exciting force:} \quad p(t) = P \cdot e^{i\omega t} \quad (2)$$

$$\text{Displacement:} \quad u(t) = U \cdot e^{i\omega t} \quad (3)$$

where P and U are the complex amplitudes of force and displacement, respectively.

Prior to about 1970 the relationship between P and U was expressed by

$$G \cdot R \cdot U = P(f_1 + if_2) \quad (4)$$

or

$$K \cdot U = P(F_1 + F_2) \quad (5)$$

where K is the static spring constant and f_1, f_2, F_1 and F_2 are real dimensionless functions of a_o, Poisson's ratio and the geometry of the problem. The static spring constant is, in general, proportional to G. For the special case of vertical vibrations of a circular footing on a perfect half space K has the value $4GR/(1-\nu)$, where ν is Poisson's ratio.

Numerous solutions have been published, see Richart et al. (1970), in terms of graphs of f_1 and f_2 or F_1 and F_2 as functions of a_o. The latter functions have the advantage that they are essentially independent of Poisson's ratio in the low frequency range.

Solutions published since about 1970 usually express the relationship between P and U in a form similar to the direct stiffness method, i.e.,

$$\tilde{K}_o U = P \tag{6}$$

In this formulation $\tilde{K}_o$ is the complex stiffness (also called compliance)

$$\tilde{K}_o = K_o(\omega) + i\omega C_o(\omega) = K(k_o(a_o) + ia_o \cdot c_o(a_o)) \tag{7}$$

where $K_o(\omega)$ and $C_o(\omega)$ are the equivalent spring and dashpot coefficients for the foundation and $k_o(a_o)$ and $c_o(a_o)$ are real dimensionless frequency response functions of a_o, Poisson's ratio and the geometry of the problem. A simple calculation will show that k_o and c_o are related to F_1 and F_2 through

$$k_o = \frac{F_1}{F_1^2 + F_2^2} \quad ; \quad c_o = \frac{-F_2/a_o}{F_1^2 + F_2^2} \tag{8}$$

or inversely

$$F_1 = \frac{k_o}{k_o^2 + a_o^2 c_o^2}, \quad F_2 = \frac{-a_o c_o}{k_o^2 + a_o^2 c_o^2} \tag{9}$$

The major advantage of the modern notation, Eqs. (6) and (7), is that published graphs of the functions $k_o(a_o)$ and $c_o(a_o)$ enable the designer to compute the equivalent spring and dashpot coefficients from the simple formulas

$$K_o(\omega) = K \cdot k_o(a_o) \tag{10}$$

$$C_o(\omega) = K \cdot \frac{R}{V_s} c_o(a_o) \tag{11}$$

which follow directly from Eq. (7). The old notation requires conversion through Eqs. (9).

The Constant Hysteretic Material

In order to introduce soil damping it will be assumed that the foundation consists of viscoelastic materials. For the special case of harmonic excitation of a given frequency, ω, it can be shown, Bland (1960), that the complex stress, σ_{ij}, and strain, γ_{ij}, amplitudes satisfy the relation

$$\begin{Bmatrix} \sigma_x \\ \sigma_y \\ \sigma_z \\ \tau_{xy} \\ \tau_{yz} \\ \tau_{zx} \end{Bmatrix} = \begin{bmatrix} \tilde{M} & (\tilde{M}-2\tilde{G}) & (\tilde{M}-2\tilde{G}) & 0 & 0 & 0 \\ (\tilde{M}-2\tilde{G}) & \tilde{M} & (\tilde{M}-2\tilde{G}) & 0 & 0 & 0 \\ (\tilde{M}-2\tilde{G}) & (\tilde{M}-2\tilde{G}) & \tilde{M} & 0 & 0 & 0 \\ 0 & 0 & 0 & \tilde{G} & 0 & 0 \\ 0 & 0 & 0 & 0 & \tilde{G} & 0 \\ 0 & 0 & 0 & 0 & 0 & \tilde{G} \end{bmatrix} \begin{Bmatrix} \varepsilon_x \\ \varepsilon_y \\ \varepsilon_z \\ \gamma_{xy} \\ \gamma_{yz} \\ \gamma_{zx} \end{Bmatrix} \quad (12)$$

which is identical to Hooke's law except that the constrained modulus, $\tilde{M}$, and shear modulus, $\tilde{G}$, are complex numbers. In general, these moduli are functions of frequency. A material for which $\tilde{G}$ and $\tilde{M}$ are frequency independent is called a constant hysteretic material. Such materials will when subjected to cyclic tests exhibit elliptical hysteresis loops; the shape and slope of which are independent of frequency. This behavior is in reasonable agreement with observed behavior of soils and only constant hysteretic materials will be considered in the following.

Since $\tilde{G}$ and $\tilde{M}$ are complex constants they can be expressed in the forms

$$\tilde{G} = G(1 + i\tan\delta_s) \quad (13)$$

$$\tilde{M} = M(1 + i\tan\delta_p) \quad (14)$$

where G and M are the real shear and constrained moduli, respectively.

The angles δ_s and δ_p are known as Loss angles. δ_s is associated with the attenuation of S-waves and δ_p with the attenuation of P-waves. Although there is in fact evidence that $\delta_p < \delta_s$ for many soils, the available data is inconclusive and it is usually assumed that $\delta_s = \delta_p = \delta$. This assumption will be made in the following.

The Correspondence Principle

As shown by Bland (1960) the similarity of Eq. (12) to Hooke's law implies that, if a closed-form undamped solution, $u(t) = U(\omega,G,M) \cdot \exp(i\omega t)$

is available to a harmonic response problem, a damped solution can be obtained simply by replacing G and M in the function U by the complex quantities $\tilde{G}$ and $\tilde{M}$. This method, which is known as the correspondence principle, has been used by Veletsos and Verbic (1973), to evaluate the effect of soil damping on foundation vibrations. It should be observed that the method requires that the closed-form solution for U can be evaluated for complex arguments.

Modal Damping

The loss angle defined above is the measure used by pure theoreticians for damping in a constant hysteretic material. However, most available data on soil damping have been reported (Seed and Idriss (1970), Hardin and Drnevich (1972)), in terms of another measure of damping: the modal damping, β. This is also the measure used in most engineering analyses. It is therefore appropriate to discuss the definition of this measure and its relation to the loss angle.

The concept of modal damping used in the analysis of multi-degree-of-freedom systems has its origin in the theory of the simple damped oscillator which for the special case of harmonic excitation has the equation of motion

$$m\ddot{u} + c\dot{u} + ku = P \cdot e^{i\omega t} \tag{15}$$

which should require no further explanations.

The modal damping, β, is defined by

$$\beta = \frac{c}{2\sqrt{km}} \tag{16}$$

and is also referred to as the fraction of critical damping or simply the damping ratio.

The relationship between displacement and force amplitudes for the simple damped oscillator is

$$P = U \cdot H(\omega) \tag{17}$$

where $H(\omega)$ is the transfer function

$$H(\omega) = k + i\omega c - \omega^2 m \tag{18}$$

It is interesting in this connection to consider the response of the complex oscillator

$$m\ddot{u} + k^*u = P \cdot e^{i\omega t} \tag{19}$$

in which the real spring and dashpot in Eq. (15) have been replaced by a complex spring k^*. By simple substitution into Eq. (18) the transfer function for this complex oscillator is found to be

$$H^*(\omega) = k^* - \omega^2 m \tag{20}$$

which with the choice

$$k^* = k[1 - 2\beta^2 + 2i\beta\sqrt{1 - \beta^2}\,] \tag{21}$$

will lead to the identity $|H(\omega)| = |H^*(\omega)|$. Hence the two oscillators, Eqs. (15) and (19) will respond with exactly the same amplitude at all frequencies. A small phase difference of the order

$$\Delta\phi \simeq \frac{2\beta}{1 + (\frac{\omega}{\omega_o})} \text{ [radians] ; } \quad \omega_o = \sqrt{k/m} \tag{22}$$

will, as shown by Lysmer (1973), exist between the two responses. However, since β is usually small this difference is of no consequence for applications.

The above substitution of a complex stiffness to account for damping is of course similar to the correspondence principle and has led to the practice of accounting for damping in finite element analyses (e.g., Lysmer et al. (1974) and Lysmer et al. (1975)) by forming the element stiffness matrices from the complex moduli

$$G^* = G(1 - 2\beta^2 + 2i\beta\sqrt{1 - \beta^2}\,) \simeq G(1 + 2i\beta) \tag{23}$$

$$M^* = M(1 - 2\beta^2 + 2i\beta\sqrt{1 - \beta^2}\,) \simeq M(1 + 2i\beta) \tag{24}$$

A simple comparison between Eqs. (13) and (23) will show that the relationship between δ and β is

$$\beta = \sin(\delta/2) \simeq \delta/2 \tag{25}$$

and that

$$G^* = G\exp(i\delta) = \tilde{G}\cos\delta \simeq \tilde{G} \tag{26}$$

$$M^* = M\exp(i\delta) = \tilde{M}\cos\delta \simeq \tilde{M} \tag{27}$$

In the above equations the approximation involves neglecting terms of order $o(\beta^2)$.

Thus, as long as Eq. (25) is observed, there is therefore little difference between the theory of viscoelasticity (loss angle) and the current method of complex response finite element analysis (modal damping). It should also be mentioned here that, for uniform damping the latter method gives results which are, for practical purposes, identical to those obtained by classic modal analysis.

Damped Foundation Vibrations

The solution for a footing on a viscous foundation can in full analogy with Eqs. (6) and (7) be expressed in the form

$$\tilde{K}_\beta U = P \tag{28}$$

where

$$\tilde{K}_\beta = K(k_\beta(a_o) + ia_o c_\beta(a_o)) \tag{29}$$

Observing that K is proportional to G, and neglecting terms of order $o(\beta^2)$, the complex stiffness can also, according to the correspondence principle, be evaluated from

$$\tilde{K}_\beta = K \cdot (1 + 2i\beta)(k_o(a_\beta) + ia_\beta \cdot c_o(a_\beta)) \tag{30}$$

where

$$a_\beta = \omega R \sqrt{\frac{\rho}{G^*}} \simeq a_o(1 - i\beta) \tag{31}$$

and k_o and c_o are the frequency response functions defined for the undamped case by Eq. (7). In principle, one should now be able to evaluate the functions $k_\beta(a_o)$ and $c_\beta(a_o)$ in Eq. (29) simply by separation of the real and imaginary parts of Eq. (30) and comparison with Eq. (29). However, $k_o(a_\beta)$ and $c_o(a_\beta)$ are now complex-valued functions of a complex argument and nobody has so far described how these functions vary in the complex plane.

Veletsos and Verbic (1973 and 1974), who were the first to suggest the use of the correspondence principle in foundation vibrations, computed the real response functions, $k_o(a_o)$ and $c_o(a_o)$, for several cases involving a rigid circular footing on an undamped half space. They also introduced convenient approximations of the form

$$k_o(a_o) = A_1(a_o)/B_1(a_o) \tag{32}$$

$$c_o(a_o) = A_2(a_o)/B_2(a_o) \tag{33}$$

where A_1, B_1, A_2, B_2 are simple polynomials in a_o which were fitted to the exact undamped solutions. Some of these and their approximations are shown in Figs. 1 and 2 for the cases of vertical and rocking vibrations, respectively.

In order to obtain solutions for the damped case Veletsos and Verbic (1973) then assumed that the polynomial approximations, Eqs. (32) and (33) were also valid for complex arguments. This assumption is, in the opinion of the writer, questionable. Since the approximate expressions were fitted only along the real axis of the complex plane they cannot be assumed to show how k_o and c_o vary for an imaginary increment which, according to Eq. (31), is what we need to determine. It is therefore proposed that Veletsos' and Verbic's assumption be replaced by the equally valid, or invalid, assumption.

$$k_o(a_\beta) \simeq k_o(a_o) \quad , \quad \text{real} \tag{34}$$

$$c_o(a_\beta) \simeq c_o(a_o) \quad , \quad \text{real} \tag{35}$$

which can be justified by the fact that β is small and thus the increment in the argument, $a_\beta - a_o \simeq -i\beta a_o$, is small. With these simpler assumptions the separation of the real and imaginary parts of Eq. (30) can be achieved and we find, after neglecting terms of the order $o(\beta^2)$, the following simple expressions

$$\tilde{K}_\beta = K(k_o(a_o)(1 + 2i\beta) + ia_oc_o(a_o)(1 + i\beta)) \tag{36}$$

and

$$k_\beta(a_o) = k_o(a_o)(1 - \eta\beta) \quad = k_o(a_o) - \beta a_oc_o(a_o) \tag{37}$$

$$c_\beta(a_o) = c_o(a_o)(1 + 2\beta/\eta) = c_o(a_o) + 2\beta k_o(a_o)/a_o \tag{38}$$

where

$$\eta = a_o \frac{c_o(a_o)}{k_o(a_o)} \quad , \qquad \text{real} \tag{39}$$

from which the behavior of a footing on a damped foundation can be predicted from a given solution for the undamped case.

By using Eqs. (7), (10) and (11) and their equivalents for the damped case the above formulas can also be written

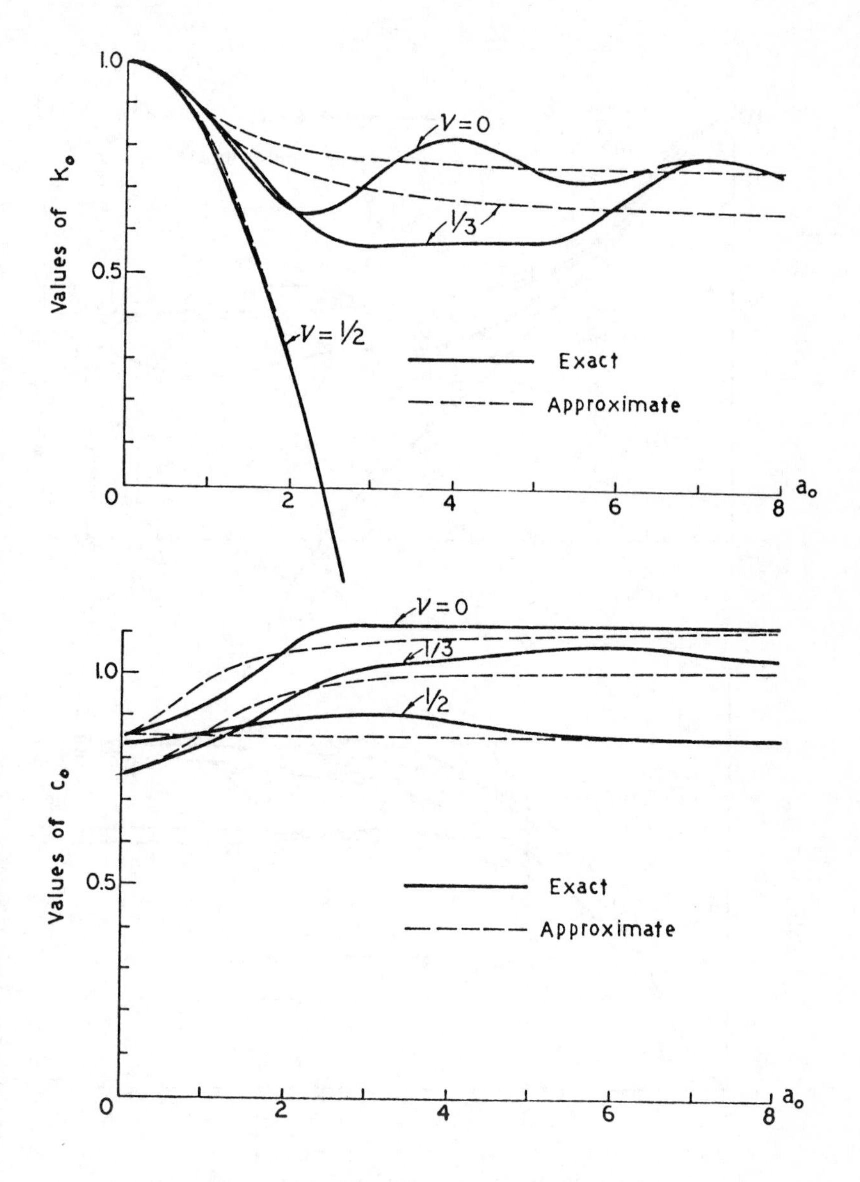

FIG. 1 UNDAMPED SOLUTIONS FOR VERTICAL MODE

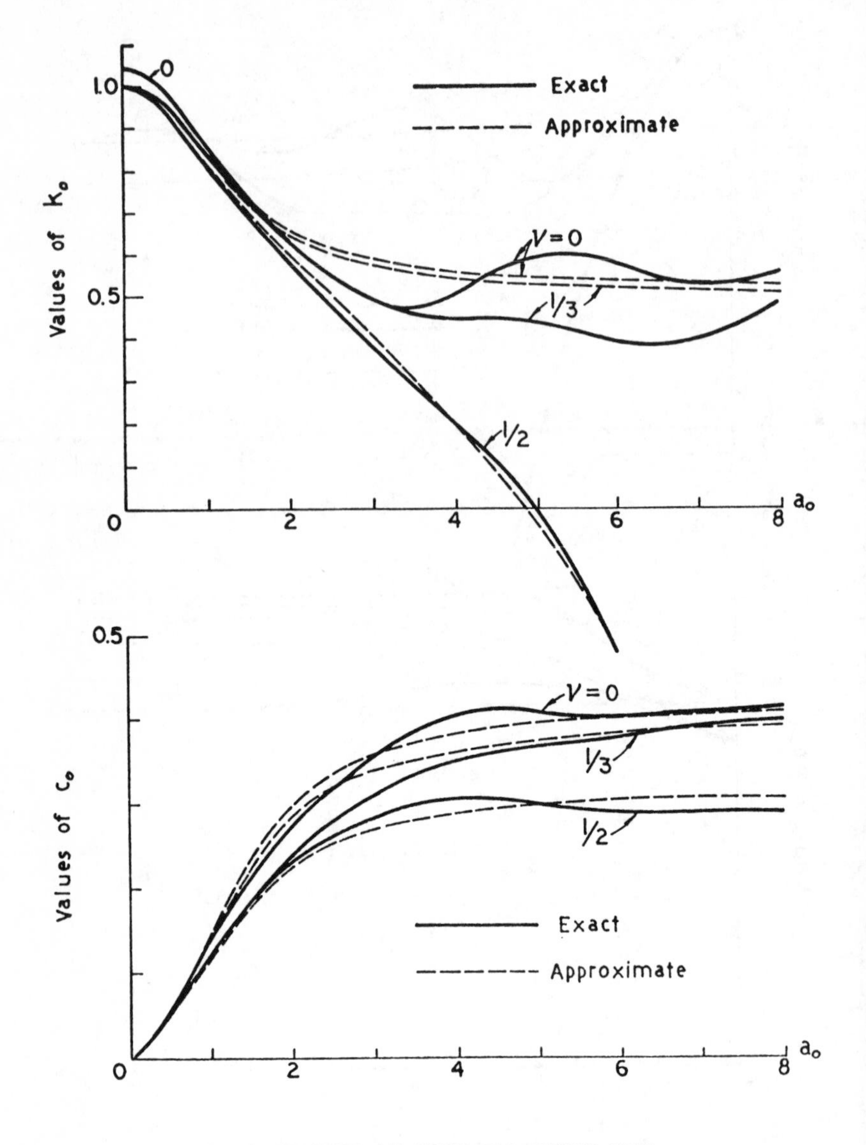

FIG. 2 UNDAMPED SOLUTIONS FOR ROCKING MODE

$$\tilde{K}_\beta(\omega) = K_o(\omega)(1+2i\beta) + i\omega C_o(\omega)(1+i\beta) \tag{40}$$

$$K_\beta(\omega) = K_o(\omega)(1-\eta\beta) = K_o(\omega) - \beta\omega C_o(\omega) \tag{41}$$

$$C_\beta(\omega) = C_o(\omega)(1+2\beta/\eta) = C_o(\omega) + 2\beta K_o(\omega)/\omega \tag{42}$$

where

$$\eta(\omega) = \omega \cdot \frac{C_o(\omega)}{K_o(\omega)} \tag{43}$$

which avoids much of the complicated notation introduced above.

Equations (36) to (43) are applicable to all modes of uncoupled vibration and to embedded footings on layered foundations with the same damping in all layers as long as the assumptions in Eqs. (34) and (35) are satisfied and $k_o(a_o)$ and $c_o(a_o)$ do not vanich. The latter condition is often a sign that the frequency response functions vary strongly for an imaginary increment of the argument. The same is the case in frequency regions where the undamped frequency response functions tend to oscillate. Thus, the formulas should be used with caution in regions where the (undamped) frequency response functions are not smooth. This would occur in the case of a layered foundation with strong stiffness contrast between layers.

Physically, the form of Eq. (40) can be explained in terms of the simple model shown in Fig. 3(a). In this model the spring represents the stiffness of the foundation and the semi-infinite bar represents an element which can provide radiation damping. As shown by Lysmer and Kuhlemeyer (1969) the model shown in Fig. 3(a) behaves exactly like the model shown in Fig. 3(b). This latter model has the complex stiffness

$$\tilde{K}_o = K + i\omega\sqrt{M\rho} \tag{44}$$

Hence, by application of the correspondence principle and neglecting terms of order $o(\beta^2)$

$$\tilde{K}_\beta = K(1+2i\beta) + i\omega\sqrt{M\rho}\,(1+i\beta) \tag{45}$$

which is in complete agreement with Eq. (40). Other examples, with transmitting boundaries to account for radiation damping, can be constructed and they all confirm the form of Eq. (40); i.e., the term of $\tilde{K}_o$ which is due to radiation damping is proportional to $\sqrt{M}$ (or $\sqrt{G}$) and should thus be multiplied by the factor $(1+i\beta)$ to account for soil damping.

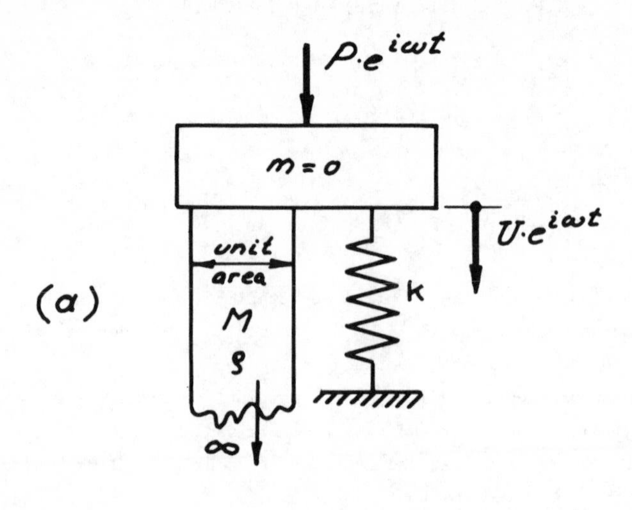

$$P = (k + i\omega\sqrt{\rho M}) \cdot U$$

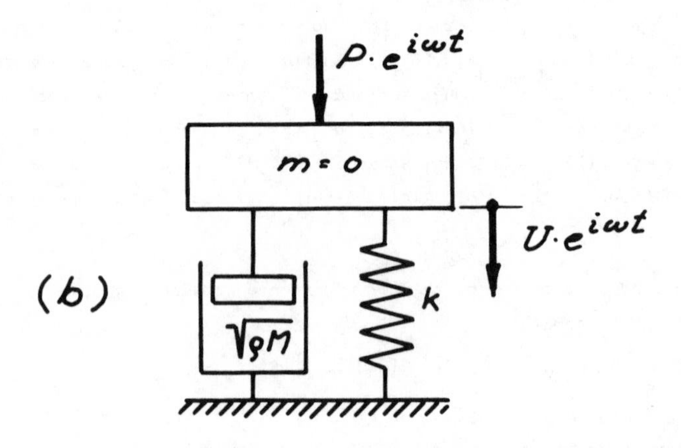

FIG. 3 SIMPLIFIED MODELS FOR FOUNDATION VIBRATIONS

It should be observed that Eq. (36) is different from the equation

$$\tilde{K} = K(k_o(a_o) + ia_o c(a_o))(1 + 2i\beta) \tag{46}$$

which has appeared in several publications, e.g. Kausel et al. (1978). The use of this expression will overestimate the effect of damping on the equivalent spring constant, i.e., a factor of 2 will be introduced in the last term of Eqs. (37) and (41). The equivalent dashpot, Eqs. (38) and (42) remains unchanged.

Examples

In order to illustrate the use of the proposed formulas, Eqs. (36) through (39), two examples are presented herein. Both involve a rigid circular footing on the surface of a perfect half space with Poisson's ratio equal to 1/3. The undamped solutions chosen are those presented by Veletsos and Verbic (1974). The results are compared with the damped solutions presented by Veletsos and Verbic (1973) using a loss angle defined by $\tan\delta = 0.3$, corresponding to $\beta = 14.5\%$, see Eq. (25), which is the order of soil damping which might be used in a strong-motion seismic soil-structure interaction analysis. This is actually the case which is of most interest in connection with the topic discussed in this paper. For normal machine vibration problems soil damping is not particularly important because the strains in and thus the damping ratio for the soil are usually small.

The first case involves vertical vibrations for which Veletsos and Verbic (1974) suggest the undamped solutions shown in Fig. 1 and the approximate formulas

$$k_o(a_o) = 1 - 0.35\,\frac{(0.8a_o)^2}{1 + (0.8a_o)^2} \tag{47}$$

$$c_o(a_o) = 0.75 + 0.28\,\frac{(0.8a_o)^2}{1 + (0.8a_o)^2} \tag{48}$$

The second case involves rocking vibrations for which Veletsos and Verbic (1974) suggest the undamped solutions shown in Fig. 2 and the approximate formulas

$$k_o(a_o) = 1 - 0.5 \frac{(0.8a_o)^2}{1 + (0.8a_o)^2} \tag{49}$$

$$c_o(a_o) = 0.4 \frac{(0.8a_o)^2}{1 + (0.8a_o)^2} \tag{50}$$

The results of the study are shown in Figs. 4-7. Each figure shows five frequency response functions:

Curve 1 is the "exact" undamped solution presented by Veletsos and Verbic, Figs. 1 and 2.

Curve 2 is the "approximate" undamped solution computed from Eqs. (47) - (50).

Curve 3 is a damped solution computed by the method presented herein, Eqs. (37) - (39), from the "exact" undamped solution, Curve 1.

Curve 4 is a damped solution computed by the method presented herein from the "approximate" undamped solution, Curve 2.

Curve 5 is the damped solution computed from the "approximate" undamped solution using the method suggested by Veletsos and Verbic (1973).

Since the figures are similar the conclusions can be drawn from Fig. 4.

1. A comparison between curves 4 and 5 shows that the method proposed by Veletsos and Verbic (1973) and that proposed herein give essentially identical results when starting from the same undamped solution. Hence, if for no other reasons, the simpler method should be used.
2. A comparison between curves 3 and 4 shows that the errors introduced by the necessity of approximating the frequency response functions when using the method by Veletsos and Verbic (1973) can be quite appreciable. The method presented herein does not have this problem since it can operate directly from the exact undamped solution.

Summary

In summary, a new simple and potentially more accurate method has been presented for evaluating the effect of soil damping on foundation vibrations from a knowledge of how the same foundation behaves with no

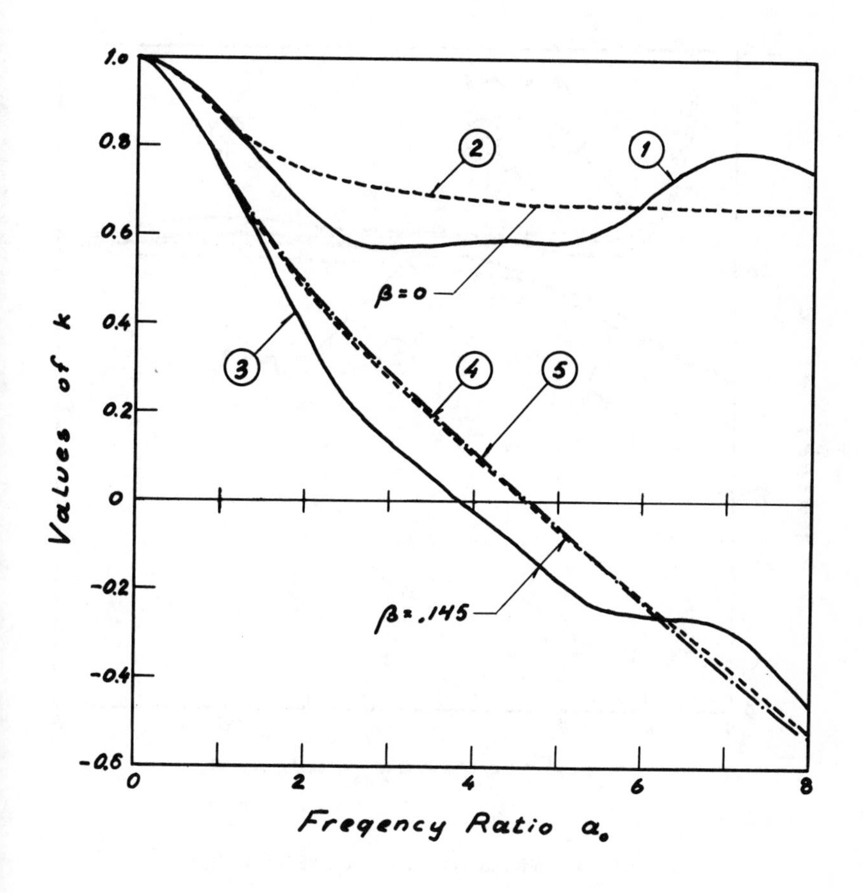

FIG. 4 STIFFNESS FUNCTIONS FOR VERTICAL MODE

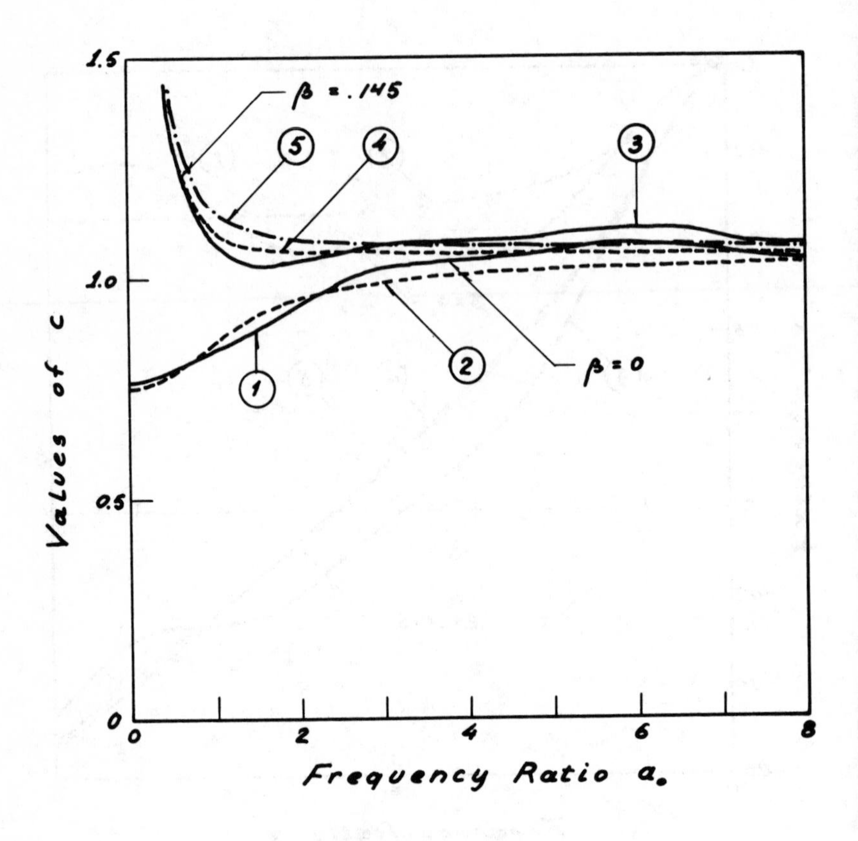

FIG. 5 DAMPING FUNCTIONS FOR VERTICAL MODE

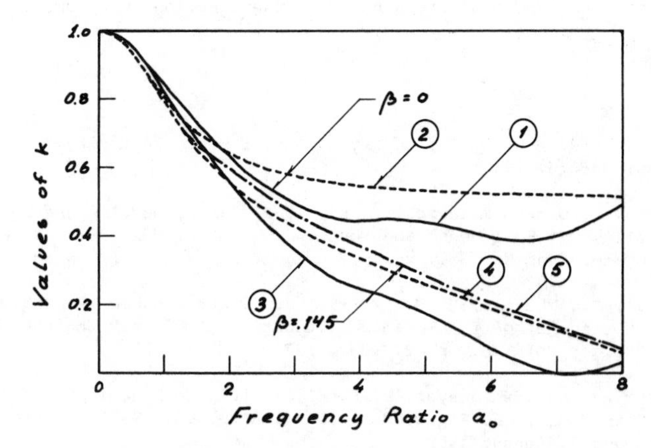

FIG. 6 STIFFNESS FUNCTIONS FOR ROCKING MODE

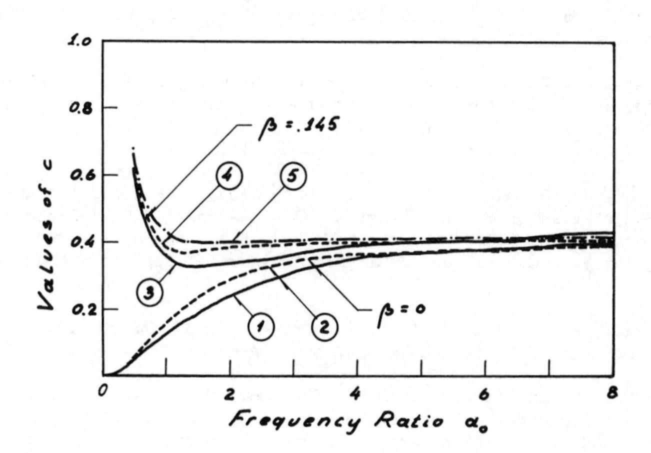

FIG. 7 DAMPING FUNCTIONS FOR ROCKING MODE

soil damping. The assumptions involved have been clearly stated and the form of the resulting expressions have been given a physical explanation. The method is not exact but should give good results in frequency regions where the undamped frequency response functions vary smoothly with frequency.

References

1. Bland, D. R. (1960) "The Theory of Linear Viscoelasticity," Pergamon Press, Oxford, 1960.

2. Hardin, B. O. and Drnevich, V. P. (1972) "Shear Modulus and Damping in Soils," J. Soil Mech. and Found. Div., ASCE, Vol. 98. No. SM6, pp. 603-624 and No. SM7, pp. 667-692, 1972.

3. Kausel, E., Whitman, R. V., Morray, J. P. and Elsabee, F. (1978) "The Spring Method for Embedded Foundations," Nuclear Engineering and Design, Vol. 48, pp. 377-392, 1978.

4. Lysmer, J. and Kuhlemeyer, R. L. (1969) "Finite Dynamic Model for Infinite Media," J. Engrg. Mech. Div., ASCE, Vol. 95, No. EM4, pp. 859-877, August 1969.

5. Lysmer, J. (1973) "Modal Damping and Complex Stiffness," Lecture Notes, Department of Civil Engineering, University of California, Berkeley, 1973.

6. Lysmer, J., Udaka, T., Seed, H. B. and Hwang, R. (1974) "LUSH - A Computer Program for Complex Response Analysis of Soil-Structure Systems," Report No. EERC 74-4, Earthquake Engineering Center, University of California, Berkeley, April 1974.

7. Lysmer, J., Udaka, T., Tsai, C.-F. and Seed, H. B. (1975) "FLUSH - A Computer Program for Approximate 3-D Analysis of Soil-Structure Interaction Problems," Report No. EERC 75-30, Earthquake Engineering Research Center, University of California, Berkeley, November 1975.

8. Richart, F. E., Jr., Hall, J. R., Jr. and Woods, R. D. (1970) "Vibrations of Soils and Foundations," Prentice-Hall, Inc., 1970.

9. Seed, H. B. and Idriss, I. M. (1970) "Soil Moduli and Damping Factors for Dynamic Response Analysis," Report No. EERC 70-10, Earthquake Engineering Research Center, University of California, Berkeley, 1970.

10. Veletsos, A. S. and Verbic, B. (1973) "Vibration of Viscoelastic Foundations," Earthquake Engineering and Structural Dynamics, Vol. 2, pp. 87-102, 1973.

11. Veletsos, A. S. and Verbic, B. (1974) "Basic Response Functions for Elastic Foundations," J. Engrg. Mech. Div., ASCE, Vol. 100, No. EM2, pp. 189-202, April 1974.

EXPLICIT INTEGRATION METHOD FOR NONLINEAR SOIL-STRUCTURE INTERACTION[a]

By David K. Vaughan[1] and Jeremy Isenberg[2], M.ASCE

Explicit integration techniques have been incorporated in finite element codes such as TRANAL (Baylor et al., 1979) and finite difference codes such as STEALTH (Hofmann, 1976). They have been used to simulate soil-structure interaction in the SIMQUAKE series of field tests, where model nuclear containment structures were subjected to strong ground shaking. Two-dimensional and, in the case of TRANAL, three-dimensional nonlinear effects are included in the simulation. The present paper describes how these explicit methods are being applied to simulate the response of 1/12 and 1/8 scale containment structures subjected to ground shaking from time-phased detonation of planar explosive arrays (Isenberg, et al., 1978a,b; Vaughan et al., 1979).

The SIMQUAKE series of field tests, sponsored by the Electric Power Research Institute, was designed to study fundamental soil-structure interaction effects, such as rocking, which occur during earthquakes. Fig. 1 shows the explosive arrays, free field instrumentation and the 1/8 scale test structure in SIMQUAKE II. The back array was detonated first, followed by the front array 1.2 sec later. The measured ground shaking in the vicinity of the 1/8 scale structure, 200 feet from the front array, is presented in Fig. 2 in terms of vertical and horizontal velocity-time histories at selected points along the centerline of the test bed. These motions agree closely with those obtained by scaling and synthesizing motions measured in SIMQUAKE I, which signifies that the ground shaking is reproducible.

---

[a]This work was sponsored by the Electric Power Research Institute, Palo Alto, CA.

[1]Senior Research Engineer, Weidlinger Associates, Menlo Park, CA.
[2]Partner, Weidlinger Associates, Menlo Park, CA.

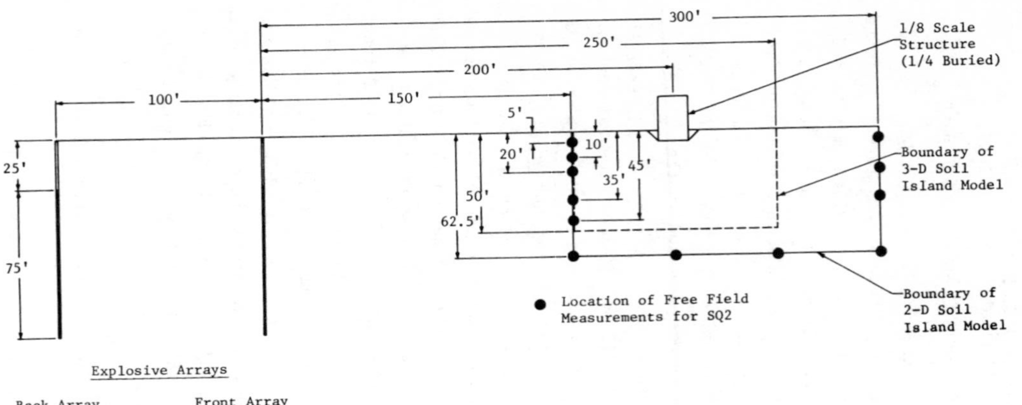

Explosive Arrays

Back Array	Front Array
12 Holes	16 Holes
(18" diameter,	(12" diameter,
spaced 16'8" apart)	spaced 12'6" apart)
94200 lb ANFO	57600 lb ANFO
Detonation at $t=t_o$	Detonation at $t=t_o+1.2$ sec

Fig. 1. Elevation view of SIMQUAKE II including 1/8 scale containment structure.

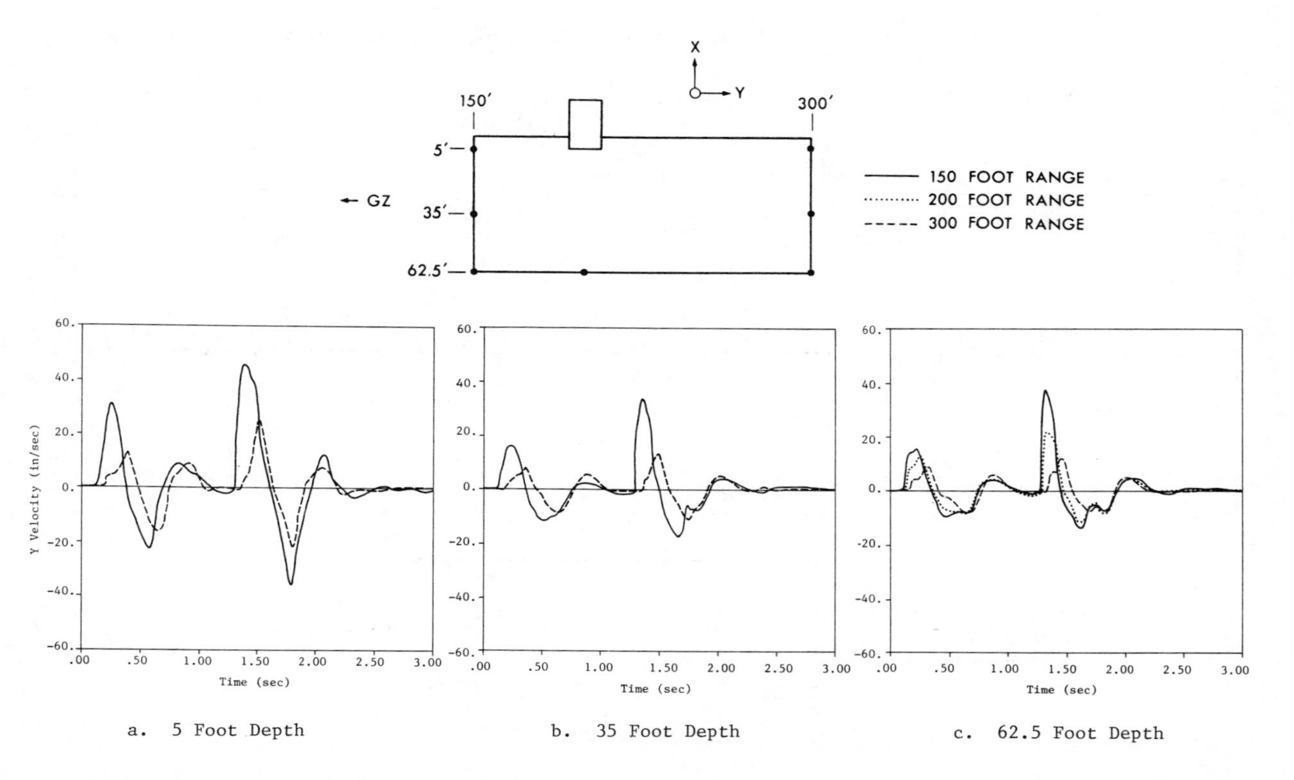

a. 5 Foot Depth

b. 35 Foot Depth

c. 62.5 Foot Depth

Fig. 2a. Selected horizontal velocity-time histories used as input to posttest analysis of SIMQUAKE II.

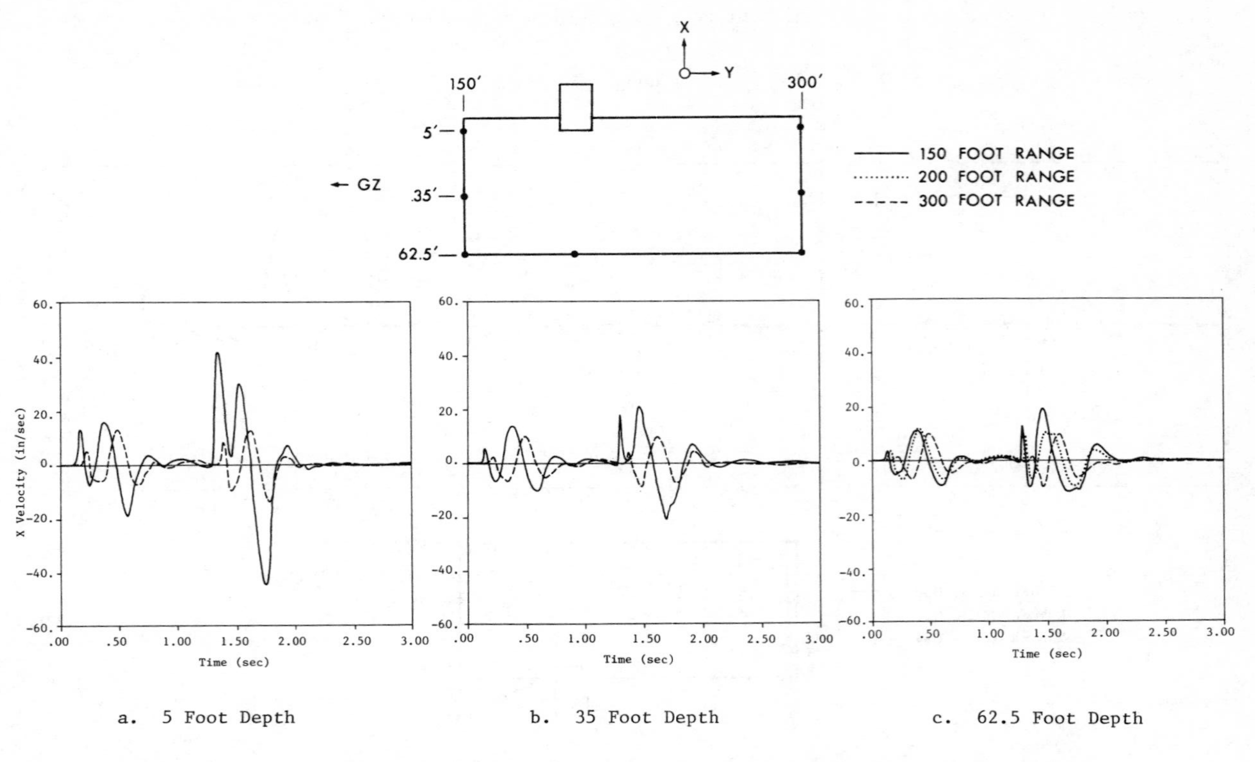

a. 5 Foot Depth

b. 35 Foot Depth

c. 62.5 Foot Depth

Fig. 2b. Selected vertical velocity-time histories used as input to posttest analysis of SIMQUAKE II.

The main new effect demonstrated by the explosive field tests and by the forced vibration tests which accompanied them is the strong dependence of natural frequency of rocking on the amplitude of rocking. The forced vibration tests show that, when the peak horizontal acceleration of a shaker placed at the top of the 1/8 scale structure is increased from .01g to .12g, the resonant rocking frequency decreases from about 10.8 Hz to about 3.6 Hz (Howard et al., to be published).

The problems posed for simulation include three-dimensional geometry, spatial gradients of ground motion input, nonlinear soil properties of the continuum type and nonlinear debonding-rebonding at the soil-structure interface. Explicit methods have the following advantages:

1. Efficiency — TRANAL is capable of calculating the response of a 3D nonlinear soil-structure model for 2.5 seconds (20 seconds full scale) in approximately 3 hours of CDC 7600 time; analysis of a corresponding 2D model requires about 2/3 hour CDC 7600 time. The current 2D version of seismic STEALTH requires about 1½ hours of CDC 7600 time. The relative efficiency of TRANAL is due to its subcycling option.
2. General Input and Geometry — Ground motion input is prescribed in terms of phased velocity-time histories at points along the boundaries of an imaginary soil island containing the structure. Since phase and amplitude vary with range and depth, it is necessary to input a different motion at every subsurface boundary point; it follows that a time-marching, forward integration technique, such as an explicit method, is required. Although the test is designed to eliminate torsion of the structures, other three-dimensional effects such as radiation damping must be investigated with a three-dimensional simulation.
3. General Nonlinearity — Cyclic hysteresis in bulk and shear is required to propagate the input ground motions through the soil island to the structure. The cap model with kinematic hardening shear yield surface and isotropic hardening cap adequately represents nonlinear continuum properties of the site insofar as they are known from laboratory experiments. This model, available in both TRANAL and STEALTH

also represents volumetric compaction which occurs in soil beneath the corners and along the sides of the structure; this effect plays a dominant role in reducing the support afforded by the underlying soil to the structure during and after strong shaking. In addition, TRANAL and STEALTH represent the highly nonlinear effects of cavitation and rebonding at the soil-structure interface by means of limiting interface tension in continuum elements adjacent to the structure (TRANAL) or nodal splitting (STEALTH); see Fig. 3.

Besides providing basic insight into soil-structure interaction under strong ground shaking, the data gathered in the SIMQUAKE test series provide a valuable means of evaluating analytic methods. Since the amplitude dependence of the rocking frequency and related effects is the most significant finding in the test series, attention is focused on simulating these effects as is described below.

Cap model representations (Isenberg, et al., 1978a; DiMaggio et al., 1971) of free field and backfill soil properties were developed on the basis of a suite of laboratory experiments conducted on reconstituted (Mazanti et al., 1970; Kelly et al., 1977) and undisturbed (Chaney et al., 1979) samples. The generalized cap model is shown in Fig. 4. Cross-hole measurements of seismic wavespeeds were used to construct the geologic profile and amend the cap model parameters governing low amplitude wavespeeds. The idealized site profile and a comparison of cyclic shear stress-strain measurements with the model are shown in Fig. 5.

The site model was evaluated by comparing ground motions calculated in the interior of the soil island with corresponding physical measurements. Since the soil island is small and measurement locations in the soil are few, it is necessary to rely on locations which are close to the structure and are to some degree affected by soil-structure response. Fig. 6 compares velocity-time histories calculated without a structure present with measured velocity-time histories. The agreement is judged to be favorable; contamination of the measurement by waves scattered by the structure appears to be negligible.

Analytic simulation of the SIMQUAKE events was performed using soil-structure models of the type shown in Fig. 7. Physical records of

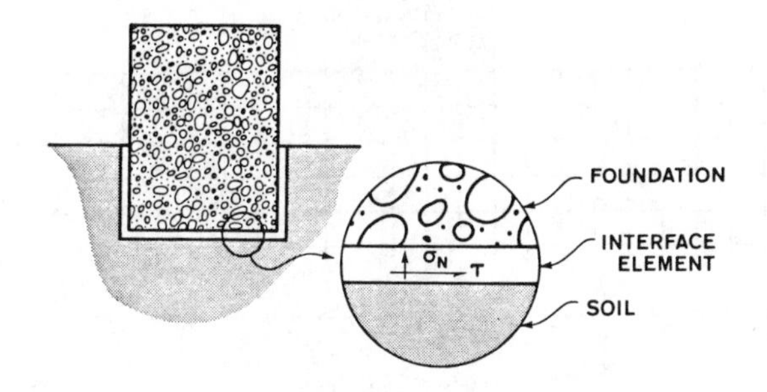

a. Structural foundation surrounded by interface elements

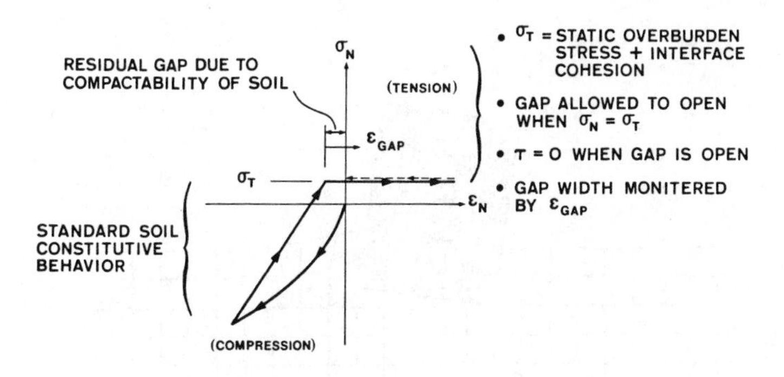

b. Uniaxial stress-strain behavior of interface elements for cycle of compressive and tensile straining

Fig. 3a. Continuum approach to modeling of nonlinear soil-structure interface.

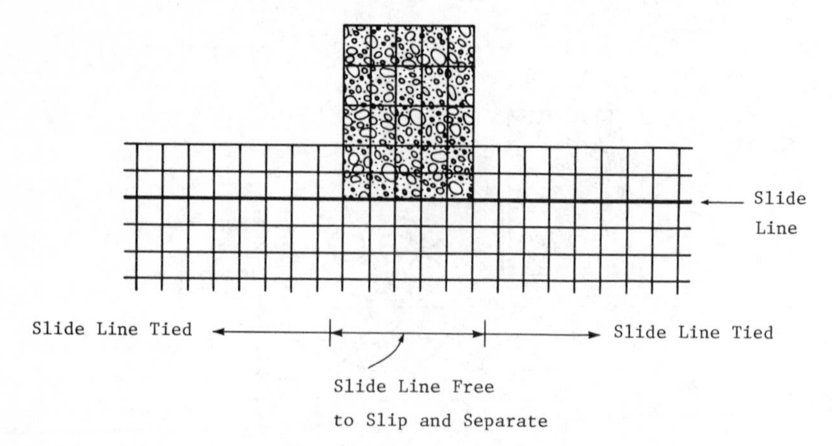

a. Undisturbed configuration of embedded structure with slide-line interface along bottom of structure

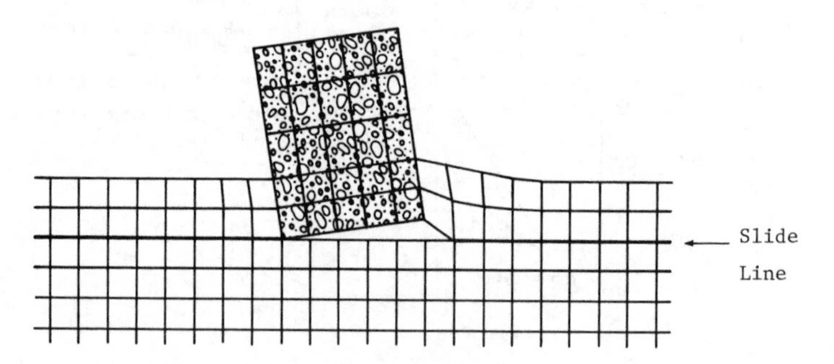

b. As structure rocks, separation (node splitting) may occur at slide line below structure

Fig. 3b. Slide-line (node splitting) approach to modeling of nonlinear soil-structure interface.

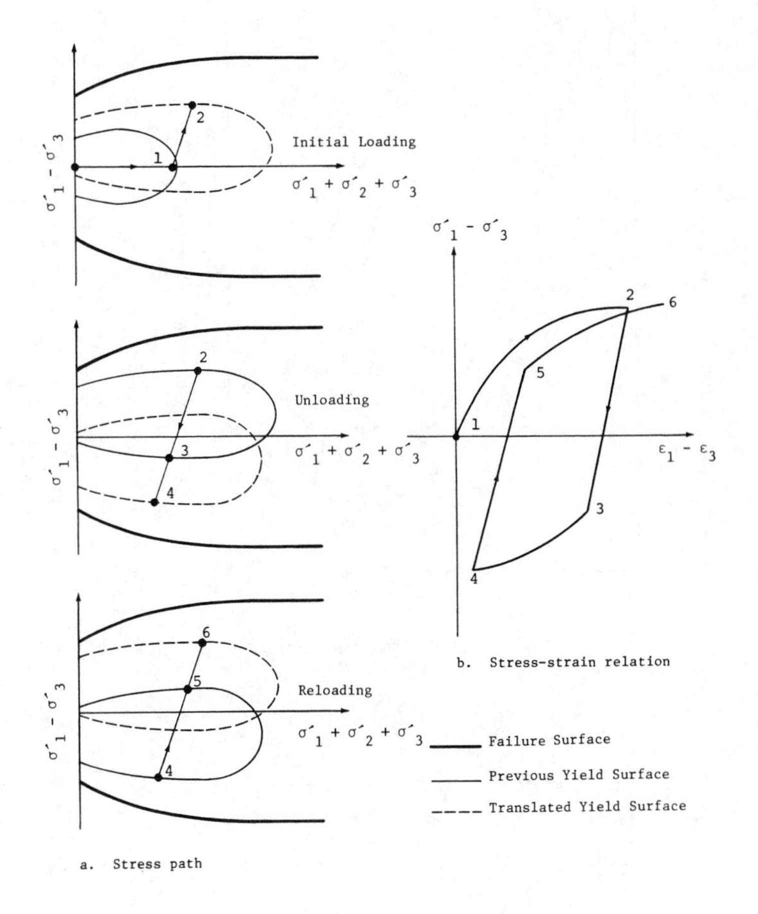

Fig. 4. Role of kinematic hardening in producing hysteresis loops for cyclic triaxial loading.

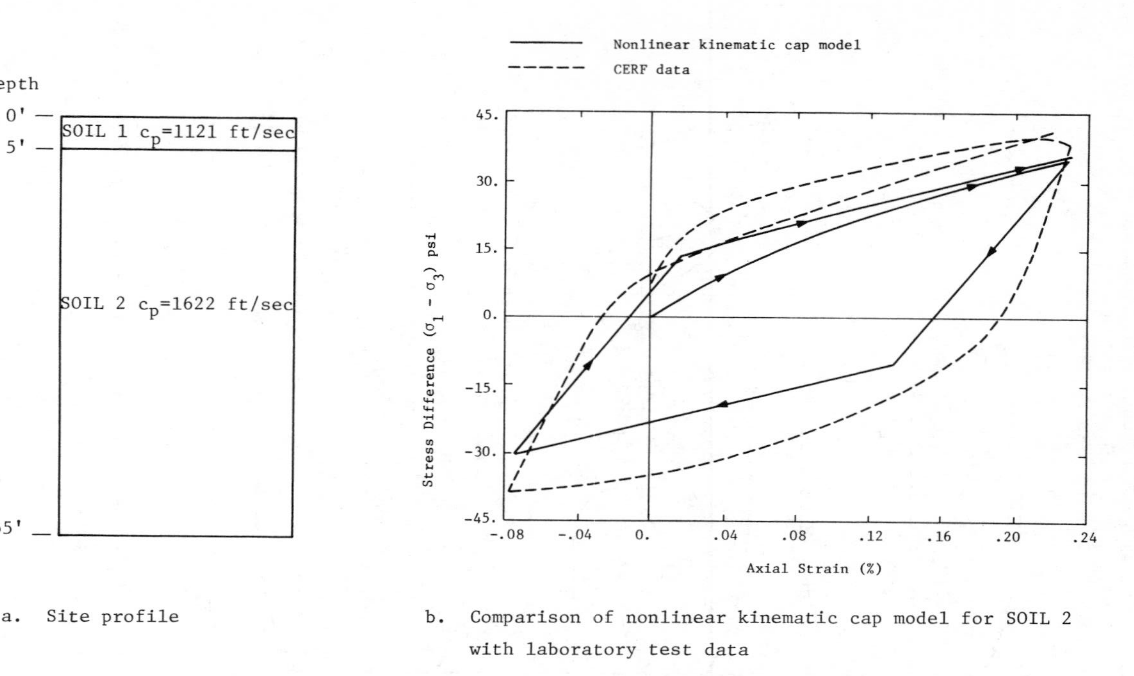

a. Site profile

b. Comparison of nonlinear kinematic cap model for SOIL 2 with laboratory test data

Fig. 5. SIMQUAKE II site profile and comparison of nonlinear kinematic cap model for SIMQUAKE site model with cyclic triaxial test data for soil from the site of SIMQUAKE.

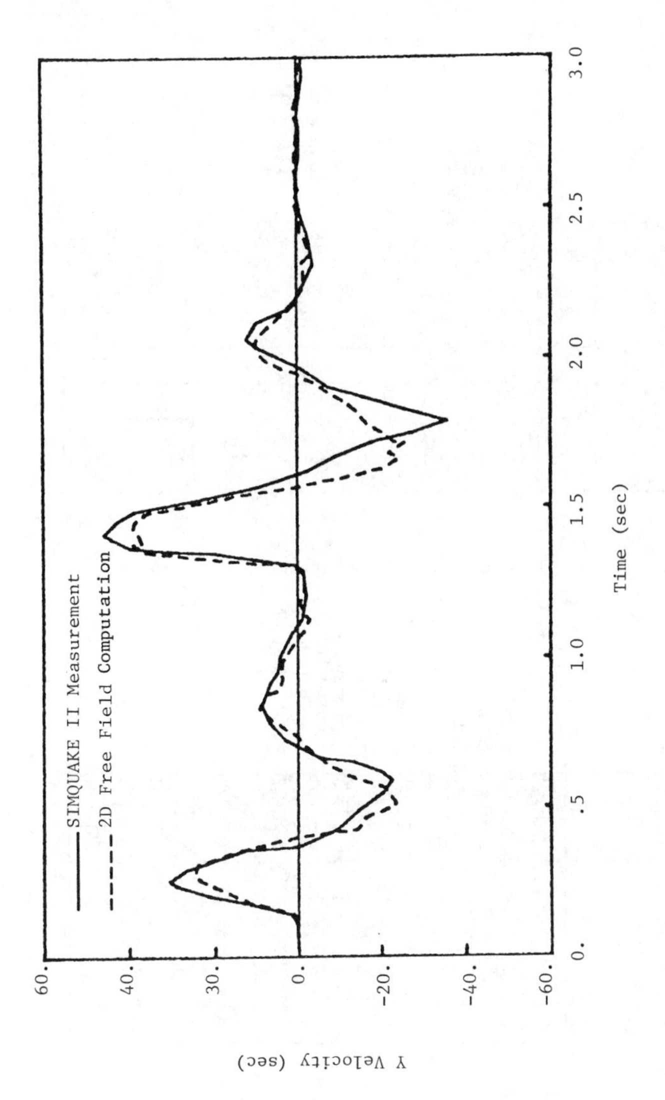

Fig. 6. Free field motion in interior of soil island, calculated vs. measured.

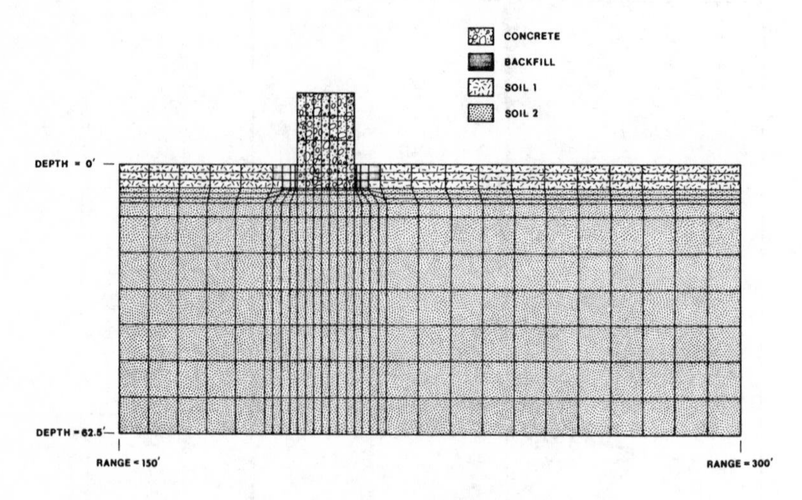

a. Two-dimensional TRANAL grid

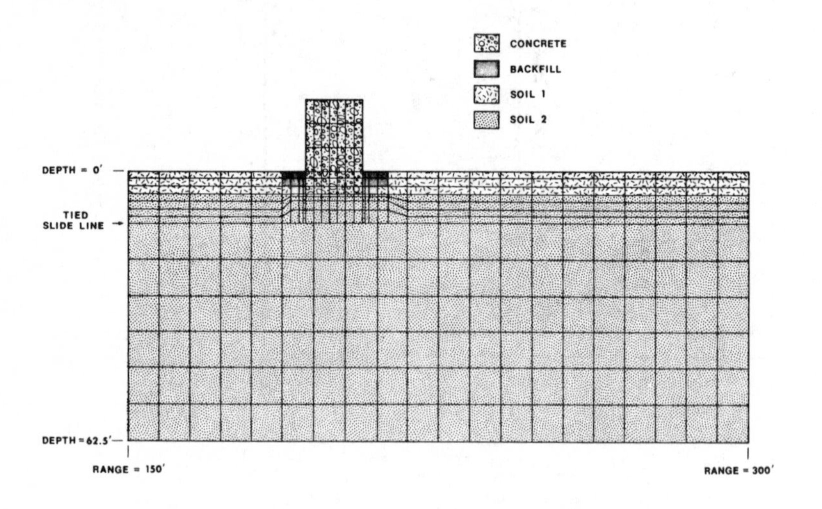

b. Two-dimensional STEALTH grid

Fig. 7a. Two-dimensional TRANAL and STEALTH mesh used in posttest analysis of SIMQUAKE II.

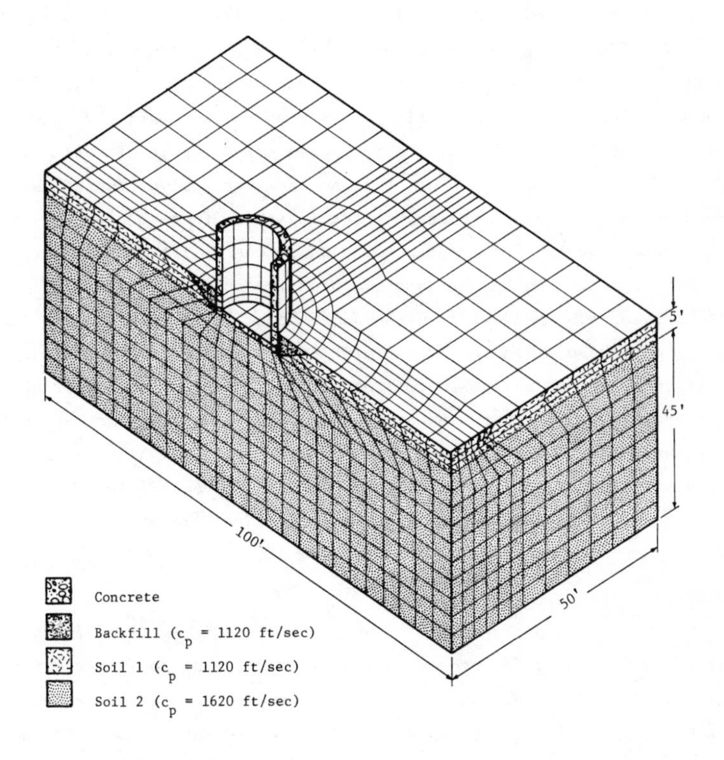

Fig. 7b. 3-Dimensional TRANAL soil island model used for SIMQUAKE II pretest calculation of 1/8 scale model response.

stresses normal to the soil-structure interface show that, when the structure rocks, the soil is beaten back on the sides, thus reducing the resistance to rocking for the duration of strong shaking; Fig. 8 illustrates that this effect is simulated by the analytic model. In addition, physical records show that the edges of the basemat alternately lift off and contact the underlying soil. Analytic simulation of this effect, Fig. 9, indicates that volumetric compaction of the soil occurs where the corner of the structure rocks down. Fig. 10 shows the sequence of foundation compaction as calculated by a two-dimensional TRANAL model of the 1/8 scale structure in SIMQUAKE II. This effect, which complements beating back the soil on the sides, reduces the rocking frequency by reducing the effective diameter of the base support. The accuracy with which the uplift and cavitation effects beneath the basemat are computed is a sensitive function of the spatial discretization in the adjacent soil. In the present study, the analytic solution converges to a stable value of rocking frequency when there are seven constant strain elements across a diameter of the structure. Using fewer elements results in overestimating the rocking frequency; for example, using three elements across a diameter in a 2D simulation of the 1/8 scale structure in SIMQUAKE II led to a rocking frequency of about 6 Hz, whereas using seven elements led to about 2.5 Hz.

Summary

This study illustrates the value of explicit integration finite element and finite difference methods in investigating nonlinear effects in soil-structure interaction. Although there is no direct evidence from strong motion earthquakes (the PG&E reactor at Humboldt Bay is too deeply embedded to have exhibited in the 1975 Ferndale earthquake rocking effects of the type shown here), the results of forced vibration testing before and after SIMQUAKE confirm the SIMQUAKE findings that resonant rocking frequency varies inversely with rocking amplitude.

The results of the analytic simulations confirm and illuminate the experimental data; for example, the analysis shows that bulk hysteresis or compaction of the soil adjacent to and beneath the structure dominates the natural frequency of rocking. Measuring this site property and including it in the analytic model appears to be advisable in cases involving significant rocking of a foundation in compressible soil. To simulate these effects, analytic methods capable of representing discontinuous nonlinear behavior at the soil-structure interface as well as representing the continuum properties of the site are required. Explicit methods appear to offer a useful alternative for such simulation.

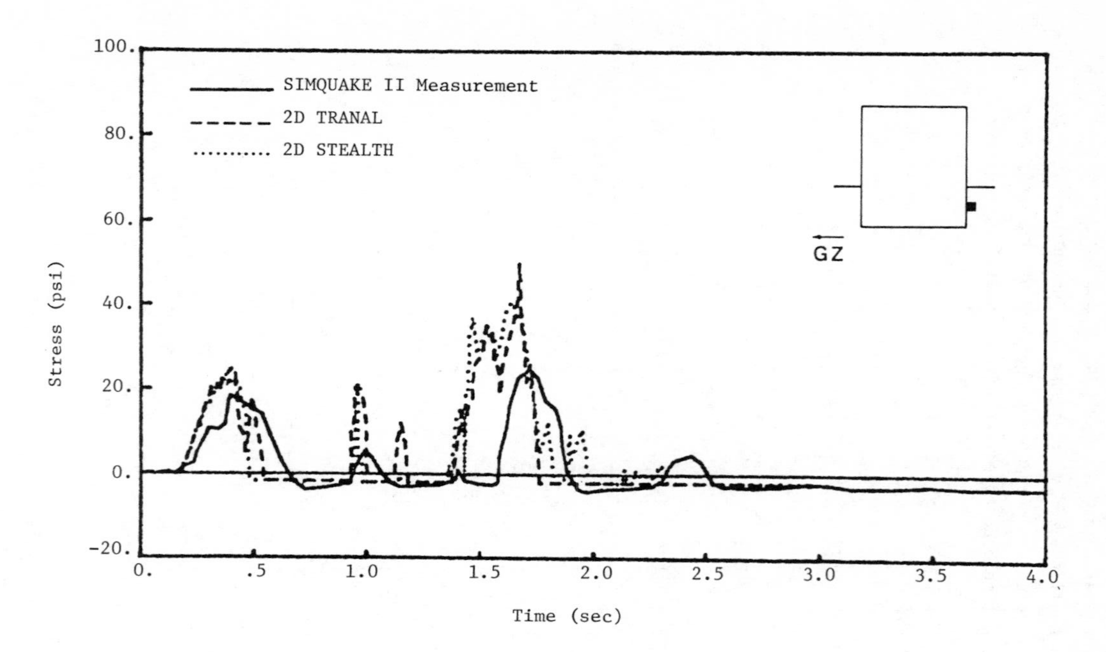

Fig. 8. Horizontal interface stress-time history at downstream face of 1/8 scale structure for SIMQUAKE II, calculated vs. measured.

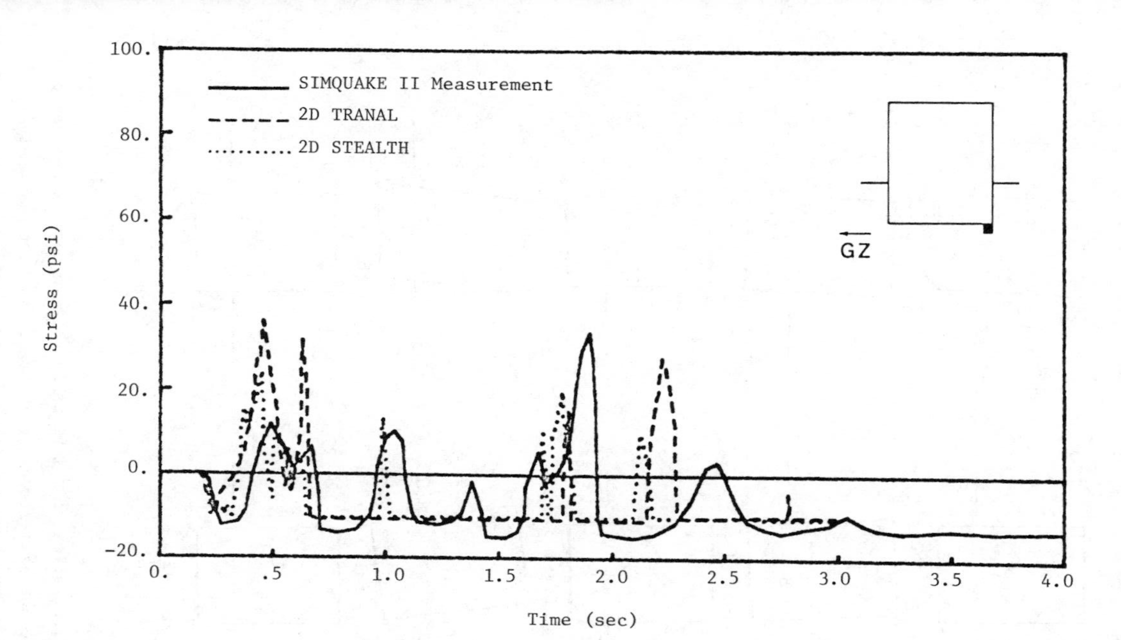

Fig. 9. Vertical interface stress-time history beneath downstream edge of 1/8 scale structure for SIMQUAKE II, calculated vs. measured.

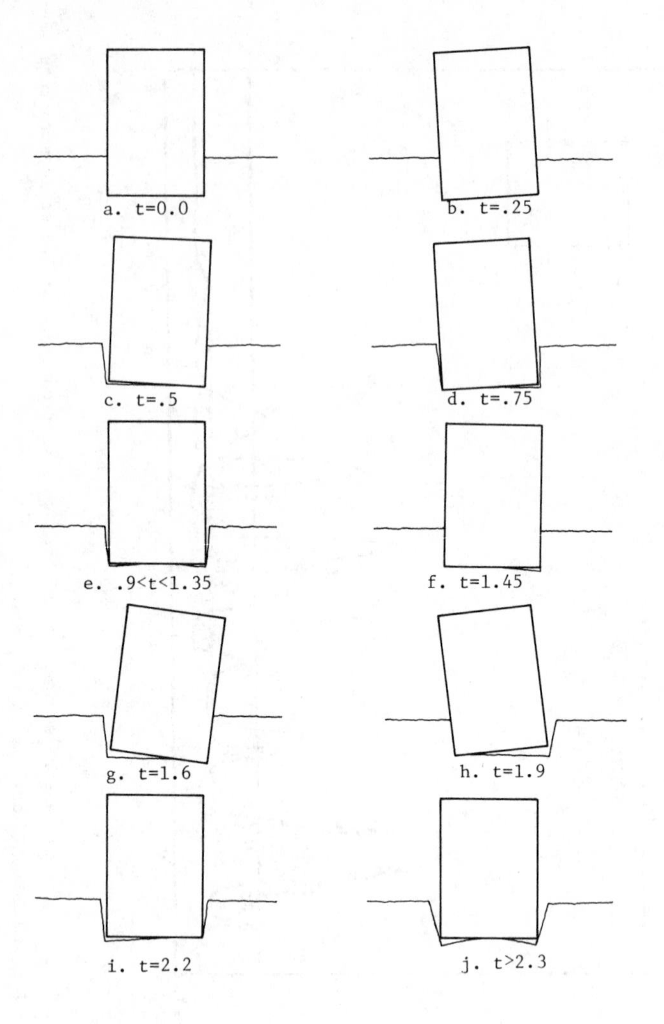

Fig. 10. Illustration of relationship between rocking and cavitation at soil-structure interface in posttest analysis of SIMQUAKE II (t = time in seconds after first detonation).

APPENDIX — REFERENCES

Baylor, J. L., J. P. Wright and C. F. Chung, "TRANAL User's Guide, Part I," Final Report by Weidlinger Associates, NY, for Defense Nuclear Agency, March 1979.

Chaney, R. C. and C. España, "Laboratory Testing Program—Development of Material Parameters for SIMQUAKE Site Soils for Soil-Structure Interaction Studies," FUGRO, Inc., Long Beach, CA, Final Report to Electric Power Research Institute, August 1979.

DiMaggio, F. L. and I. Sandler, "Material Model for Granular Soils," J. of Eng. Mech., Proc., ASCE, Vol. No. 97, No. EM3, 1971, pp. 935-950.

Hofmann, R., "STEALTH—A Lagrange Explicit Finite-Difference Code for Solids, Structural and Thermohydraulic Analysis," EPRI NP-260, Science Applications, Inc., August 1976.

Howard, G., P. Ibáñez and C. Smith, Final Report by ANCO Engineers for Electric Power Research Institute (to be published).

Isenberg, J. and D. K. Vaughan, "Three-Dimensional Nonlinear Analysis of Soil-Structure Interaction in a Nuclear Power Plant Containment Structure," Proc. 2nd Int. Conf. on Microzonation, San Francisco, Vol. II, November 1978, pp. 911-920.

Isenberg, J., D. K. Vaughan and I. S. Sandler, "Nonlinear Soil-Structure Interaction," Weidlinger Associates, Final Report to Electric Power Research Institute, NP-945, December 1978.

Kelly, M. F. and G. T. Baird, "Material Properties of SIMQUAKE I Test Site," Letter Report by Wang Civil Engineering Research Facility, University of New Mexico for Electric Power Research Institute, April 1977.

Mazanti, B. B. and C. N. Holland, "Study of Soil Behavior Under High Pressure, Report 1, Response of Two Recompacted Soils to Various States of Stress," USAEWES, Vicksburg, MS, February 1970.

APPENDIX — REFERENCES (Continued)

Vaughan, D. K., I. S. Sandler, D. Rubin, J. Isenberg and H. Nikooyeh, "Data Report of a Pretest Analysis of Soil-Structure Interaction and Structural Response in Low-Amplitude Explosive Testing (50 KG) of the Heissdampfreaktor (HDR)," Contract Report by Weidlinger Associates for Lawrence Livermore Laboratory, November 1979.

IMPEDANCE APPROACH FOR SEISMIC SSI ANALYSIS

By Wayne W. H. Chen[1], M. ASCE, and M. Chatterjee[2]

INTRODUCTION

In recent years, considerable efforts have been made to compare the seismic soil-structure interaction analysis results obtained using the impedance approach and the finite element method (Agrawal et al, 1973; NUREG/CR-0693, 1979; Seed et al, 1975). Most of these comparison studies are not appropriate since the impedance approach uses inappropriate foundation impedances and does not take into account foundation input motion in the analysis. However, it should be noted that both methods should yield similar results if they are formulated correctly to solve the same problem.

This paper presents an impedance approach for complete seismic soil-structure interaction analysis and comparison studies with the direct finite element method. Important factors affecting seismic soil-structure interaction analysis such as foundation embedment, soil layering, backfill material, side soil separation, superstructure modelling, impedance functions, input motions, and soil material damping are investigated. The current developments in the impedance approach have eliminated uncertainties in defining foundation impedance functions and input motions under varying conditions of embedment.

---

[1] Research and Engineering, Bechtel National Inc., San Francisco, California

[2] Bechtel Power Corporation, San Francisco, California

SEISMIC SOIL-STRUCTURE INTERACTION ANALYSIS METHOD

Analysis of soil-structure interaction effects during earthquakes for nuclear power plant structures is usually made either by the direct finite element method or by the impedance approach. The general procedure for making a complete finite element soil-structure interaction analysis is given by Lysmer et al (1975). The advantage of the direct finite element method is that the nonlinear soil properties and other types of material behavior can be approximately included in the analysis. However, inaccurate solutions and incorrect conclusions can result when discretization of the time and space variables, such as the finite extent of the spatial grid and finiteness of the time sample over which a solution can be economically computed, are not carefully considered.

In the impedance approach, the complete soil-structural analysis problem can be separated into the following five basic problems (Fig. 1): 1) determination of the free-field motion in absence of the foundation, 2) evaluation of the response of the rigid and massless foundation to the free-field motion excitation, 3) evaluation of the base forces and moments that the superstructure exerts on the foundation expressed in terms of the foundation motion, 4) evaluation of the forces that the foundation exerts on the soil in terms of the foundation motion by considering the equation of motion of the foundation including its mass, and 5) determination of the foundation motion caused by the forces that the foundation exerts on the soil (Wong, 1975). The advantage of the substructure approach is that the most appropriate solution for each subproblem can be used in the analysis. The approach allows engineers to have a better understanding of the physical behavior of each subproblem. The procedure described by Lee and Wesley (1971) is usually used to model the three-dimensional behavior of the superstructure.

OVERVIEW OF IMPEDANCE APPROACH

In recent years, the use of impedance approach for seismic soil-structure interaction analysis has been the subject of many studies. In the past, the impedance approach appears to have gained less general acceptance in spite of its computational efficiency and more flexibility in model alternation in relation to the direct finite element method. This is due largely to the limited number of foundation shapes that can be analyzed acceptably. Also, it is difficult to use the impedance approach to assess the frequency-dependent foundation impedance functions (radiative process) and foundation input motions (scattering process) for structures embedded in layered viscoelastic media. An assessment of developments in the use of impedance approach for seismic soil-structure interaction follows.

Several different methods have been developed to determine the impedance matrix and input motion for the foundation-soil system. For example, substructure deletion method was developed by Dasgupta et al (1977) to compute foundation impedance matrices. However, the method is limited to two-dimensional foundations on uniform elastic medium. Explicit finite difference method has been used to study the soil-structure interaction (Report EPRI NP-1091, 1979). In particular, foundation impedance functions and input notions are determined by using finite differences in two dimensions to solve numerically the elastodynamic field equation in the supporting medium. Again, this method is restricted to a two-dimensional structure founded on an elastic half-space. The boundary element method has been applied to the determination of the dynamic stiffness of rectangular foundations rests on the surface or embedded in an elastic half-space (Dominquez, 1978a). This method has also been used to study the motion of rigid massless square foundations under various types of seismic (Dominquez, 1978b). Further studies are still needed to extend this method to investigate the dynamic response of arbitrarily shaped foundations embedded in layered viscoelastic medium. Day (1977) has obtained excellent agreement with analytical method by using the finite element method to evaluate both compliances and scattering.

Integral equation approach is probably the most well-developed method to analyze the effects of soil-structure interaction on the seismic response of structures. In the early studies, the complexity of the mixed boundary value problem was avoided by assuming particular stress distribution under the foundation (Kobori et al, 1967). The mathematical formulation of the mixed boundary value problem was reduced by Wei (1971) to the solution of a set of Fredholm integral equations by assuming relaxed boundary conditions under the foundations. By use of the correspondence principle, Veletsos and Verbic (1973) first consider the dynamic response of a circular foundation placed on a uniform viscoelastic half-space. Recently, a general analytical method was formulated by Wong and Luco (1978) to study the dynamic response of arbitrarily shaped rigid surface foundation subjected to external load and any type of seismic excitation. The mixed boundary value problem is solved by representing the displacements in terms of integrals of the discrete Green's functions for a uniform viscoelastic half-space. When this approach is combined with the Green's functions for an embedded, layered, viscoelastic half-space, the complete soil-structure interaction problem for structures embedded in a layered viscoelastic medium can be solved.

These analytical methods have been documented in a computer program package entitled CLASSI (Wong and Luco, 1979). In case of very complicated soil profile such as a layered medium with backfill material, the computer program PONDORA developed by Day (1977) can be used to obtain the foundation impedance functions and input motions. The method is based on a finite element model in which the model boundaries are eliminated by solving the problem in the time domain. The advantages of the intergral equations approach involving Green's functions are that it incorporates material damping as well as radiation damping into the soil, experiences no frequency limitation, and approximates the boundary conditions very closely by numerically solving the integral equation. However, at present stage of development, its computer cost is only comparable to Day's finite element approach (1977).

FACTORS AFFECTING SEISMIC SOIL-STRUCTURE INTERACTION ANALYSIS

This section describes the effects of important parameters on the results of seismic soil-structure interaction analysis. The impedance functions and input motions were obtained using either integral equations approach involving Green's functions or Day's finite element approach. The computer program CLASSI was used to calculate the structural response.

(a) Foundation Embedment and Soil Layering

The effect of foundation embedment on the impedance function of a rigid foundation has been investigated using the computer programs CLASSI and LUCON (Chen and Chatterjee, 1979a). The LUCON program calculates the response on a surface foundation placed on a layered viscoelastic medium. As Figure 2 shows, the surface foundation has lower rocking foundation impedance function. Figure 3 shows the numerical values obtained for horizontal input motion as functions of the dimensionless frequency a_o. For a surface foundation, the horizontal input motion would simply be equal to the free-field amplitude at all frequencies. However, Figure 3 indicates that the amplitude of horizontal input motion for an embedded foundation is reduced, especially at the high frequency range as a result of foundation embedment. The effect of soil layering introduces a marked frequency dependence of the stiffness and radiation damping coefficients.

(b) Backfill Material

Impedance functions for the foundation embedded in uniform mudstone (V_s = 3,280 ft/sec) and the foundation embedded in mudstone with backfill sand (V_s = 600 ft/sec) were obtained. As can be seen from Figure 4, the presence of sand backfill significantly reduces the foundation impedance.

(c) Side Soil Separation

The effect of lateral separation has been investigated by Apsel (1979) for various degree of contact between a cylindrical foundation and the surrounding soil. As shown in Figure 5, the stiffness undergoes a significant reduction in amplitude at all dimensionless frequency as the percentage of lateral separation is increased. The reduction in amplitude can be attributed directly to the reduced contact area on which the soil resists the movements of the embedded foundation.

(d) Superstructure Modelling

To investigate the effect of the details of structural modeling on the total foundation motion, the following three fixed base lumped mass building models are used.

i. Lumped mass stick model A - this model uses a simplified nine-mass lumped mass Reactor Building stick model.

ii. Lumped mass stick model B - this model is the same as the lumped mass stick model A, except that the stiffnesses of the members below grade are increased 100 times to simulate structural rigidity as assumed for the building in computing the foundation impedance.

iii. Lumped mass stick model C - this Reactor Building model consists of twenty-one lumped masses with the additional masses used to represent the building internals.

It was found that similar foundation motions are obtained independent of the rigidity assumptions below the ground surface. The inclusion of internals does not affect the foundation motions.

(e) Impedance Function and Input Motions

The use of embedded input motion and free-field input motion has a significant effect on the foundation motion. As indicated in Figure 6, the use of free-field input motion gives much more conservative horizontal foundation motion as compared to that obtained using actual embedded input motion. However, the effect of the use of actual embedded impedance functions or flat impedance functions on the foundation motion is not significant.

(f) Soil Material Damping

Apsel (1979) investigated in detail the effect of including material damping in the soil model on the foundation impedance functions. It was found that the material damping, in general, tends to reduce the stiffness coefficients at high frequencies and increase the radiation damping coefficients at low frequencies. The effect of soil material damping on the structural response was investigated by Chen and Chatterjee (1979b). The effect of the use of different soil damping values on the structural response was found to be insignificant.

COMPARISON STUDIES

Comparison seismic soil-structure interaction analyses have been performed using the impedance approach (CLASSI) and the direct finite element method (FLUSH) for two BWR nuclear power plants (Chen and Chatterjee, 1979a, 1979b; Chen et al, 1980). In the first case, both the horizontal and vertical seismic soil-structure interaction analyses were performed. The nuclear power plant analyzed is deeply embedded in layered soil media and is 279 feet square in plan, 249.3 feet high, and 147.6 feet embedded. The design earthquake time history is the first twelve seconds of Taft 1952 EW specified at the surface of the plant free-field. The surface motion is to be so scaled that the design peak accelerations, 0.306 g, is obtained at the base of the free-field soil column. In the free field a Banjin sand layer (V_S = 984 fps) exists from the surface (El. 17.4 ft) to El. -2.3 ft. The soft mudstone (V_S = 984 fps) layer then extends to El. -48.2 ft, after which there is hard mudstone (V_S = 2,198 fps) for the remainder of the soil deposit. The reactor building is founded in the hard mudstone. Backfill material (80 ft wide) is placed in the excavation on the sides of the building down to El. -69.5 ft. Strain-dependent soil properties are used for backfill sand, Banjin sand, and mudstone.

In the second case, only the horizontal seismic soil-structure interaction analysis was performed. The nuclear power plant analyzed is embedded in soil media and is 265.7 ft square in plan, 213.3 ft high and 65.5 ft embedded. The design earthquake is a 30 sec artificial time history with a peak acceleration of 0.30 g. In the free field, a fine sand layer (V_s = 656 ft/sec) extends from the surface to elevation -13 ft. The mudstone layer (V_s = 3,280 ft/sec) then extends to elevation -183 ft. Backfill material (V_s = 656 ft/sec) is placed in the excavation of the sides of the reactor building down to elevation -65.5 ft. The standard Seed and Idriss strain-dependent shear modulus and damping curves are used in the analysis for the backfill fine sand (Seed and Idriss, 1970). The properties of the mudstone is assumed to be elastic.

In Figures 7, 8, 9 the horizontal 5% spectra obtained from the CLASSI analysis are compared with those obtained from FLUSH. At the top of the Reactor Building, the FLUSH analysis predicts a maximum spectral acceleration approximately 30% lower than that predicted by CLASSI. It should be noted that the specified free-field motions are treated directly as the excitation in the impedance approach, thus eliminating the deconvolution calculations and the related assumptions required in the direct finite element method.

The differences in the spectral accelerations between the FLUSH and CLASSI analyses can be attributed to three causes: 1) material damping is included in the CLASSI solution only in an approximate fashion, 2) the foundation is assumed to be rigid in the CLASSI analysis, and 3) the two-dimensional characteristics of the FLUSH model may reduce the spectral response. It has been found that the effect of soil material damping on the structural response obtained from the impedance approach is not significant. Treating the base of the reactor building as a rigid foundation can be justified in most cases because of its massive basemat and heavy shear wall and stiff superstructure. Also, it is interesting to point out that the 30% differences in spectral response obtained is of the same order as those reported by Luco and Hadjian (1974). This suggested that the major contribution to the difference may arise from the two-dimensional characteristics of the FLUSH analysis. Overall, however, the agreement between the FLUSH and CLASSI response spectra is quite good.

SUMMARY AND CONCLUSIONS

The current state-of-the-art of the impedance approach for seismic soil-structure interaction analysis is presented. The impedance approach must properly consider important parameters such as frequency dependent impedance function, input motion, embedment, and soil layering etc., in the analysis in order to obtain realistic results. Both the impedance approach and the direct finite element method can be used to assess the effect of soil-structure interaction on the structural response. Each method is affected by limitations due to its inherent assumption.

The overall reliability of both analytical approaches is best judged through comparison with field measurements. Benchmark analysis and test problems should be used to verify the accuracy of seismic soil-structure interaction analysis procedure. The direct finite element method has been used to predict the seismic response recorded at the Humbolt Bay Nuclear Power Plant. Benchmark analysis using impedance approach to predict the actual structural response measured during earthquake is urgently needed.

APPENDIX I - REFERENCE

1. Agrawal, P. K., Chu, S. L., and Shah, H. H., "Comparative Study of Soil Spring and Finite Element Models for Seismic Soil-Structure Interaction Analysis of Nuclear Power Plants," presented at the December 17-18, 1973, Specialty Conference on Structural Design of Nuclear Plant Facilities, Held at Chicago, Ill. (Sargent and Lundy Engineers, Report No. SAD-130).

2. Apsel, R. J., "Dynamic Green's Functions for Layered Media and Application to Boundary-Value Problems," Ph.D. Dissertation, University of California, San Diego, California, 1979.

3. Chen, W. H., and Chatterjee, M., "Impedance Approach and Finite Element Method for Seismic Response Analysis of Soil-Structure System," Proceeding of the 3rd Canadian Conference on Earthquake Engineering, McGill University, Montreal, Canada, June 4-6, 1979a.

4. Chen, W. H., Chatterjee, M., and Day, S. M., "Seismic Response Analysis for a Deeply Embedded Nuclear Power Plant," Paper K7/6, 5th International Conference on Structural Mechanics in Reactor Technology, Berlin, Germany, August 13-19, 1979b.

5. Chen, W. H., Chatterjee, M., and Unemori, A., "Comparison of Analysis Method for Seismic Soil-Structure Interaction," Seventh World Conference on Earthquake Engineering, Istanbul, Turkey, September 8-13, 1980.

6. Day, S. M., "Finite Element Analysis of Seismic Scattering Problems," Ph.D. Dissertation, Institute of Geophysics and Planetary Physics, University of California, San Diego, California, 1977.

7. Dasgupta, G., Sackman, J. L., and Kelly, J. M., "Substructure Deletion in Finite Element Method," Proceedings of the 4th International Conference on Structural Mechanics in Reactor Technology, Vol. M, p-4, San Francisco, California, August, 1977.

8. Dominquez, J., "Dynamic Stiffness of Rectangular Foundations," Publication No. R78-20, Department of Civil Engineering, Massachusetts Institute of Technology, August, 1978a.

9. Dominguez, J., "Response of Embedded Foundations to Travelling Waves," Publication No. R78-24, Department of Civil Engineering, Massachusetts Institute of Technology, August, 1978.

10. Kobori, T., Mina, R. and Suzuki, T., "Dynamic Ground Compliance of Rectangular Foundation On An Elastic Stratum Over A Semi-Infinite Rigid Medium," Annual Report, Disaster Prevention Research Institute of Kyoto University, No. 10A, 1967, pp. 315-341.

11. Lee, T. H., Wesley, D. A., "Soil-Foundation Interaction of Reactor Structures Subjected to Seismic Excitation," Proceedings of First Conference on Structural Mechanics in Reactor Technology, Paper K3/5, Berlin, Germany, 1971, pp. 211-233.

12. Luco, J. E., and Hadjian, A. H., "Two-Dimensional Approximations to the Three-dimensional Soil-Structure Interaction Problem," Nuclear Engineering and Design, Vol. 31, 1974, pp. 195-203.

13. Lysmer, J., Udaka, T., Tsai, C. F., and Seed, H. B., "FLUSH - A Computer Program for Approximate 3-D Analysis of Soil-Structure Interaction Problems", Earthquake Engineering Research Center, EERC 75-30, University of California, Berkeley, California, Nov. 1975.

14. NUREG/CR-0693, "Seismic Input and Soil-Structure Interaction," Final Report, Prepared by D'Appolonia Consulting Engineers for Division of System Safety, Office of Nuclear Reactor Regulation, U.S. Nuclear Regulatory Commission, Washington, D.C., February, 1979.

15. Report EPRI NP-1091, "Applications in Soil-Structure Interaction," Vol. 1-3, Prepared by URS/John A. Blume and Associates, Engineers, San Francisco, California, June, 1979.

16. Seed, H. B., and Idriss, I. M., "Soil Modulus and Damping Factors for Dynamic Response Analyses," EERC 70-10, Earthquake Engineering Research Center, University of California, Berkeley, December, 1970.

17. Seed, H. B., Lysmer, J., and Hwang, R., "Soil-Structure Interaction Analysis for Seismic Response," Journal of the Geotechnical Engineering Division, ASCE, Vol. 101, No. GT5, May, 1975, pp. 439-457.

18. Veletsos, A. S., and Verbic, B., "Vibration of Viscoelastic Foundations," Report 18, Department of Civil Engineering, Rice University, Houston, Texas, 1973.

19. Wei, Y., "Steady State Response of Certain Foundation Systems," Ph.D Thesis, Rice University, Houston, Texas, 1971.

20. Wong, H. L., "Dynamic Soil-Structure Interaction," Earthquake Engineering Research Laboratory, Report No. EERL-75-01, California Institute of Technology, Pasadena, California, 1975.

21. Wong, H. L., and Luco, J. E., "Dynamic Response of Rectangular Foundations to Obliquely Incident Seismic Waves," Earthquake Engineering and Structural Dynamics, Vol. 6, 1978, pp. 3-16.

22. Wong, H. L., and Luco, J. E., "Soil-Structure Interaction - A Linear Continuum Mechanics Approach (CLASSI)," Report CE-79-3, Department of Civil Engineering, University of Southern California, Los Angeles, California, 1979.

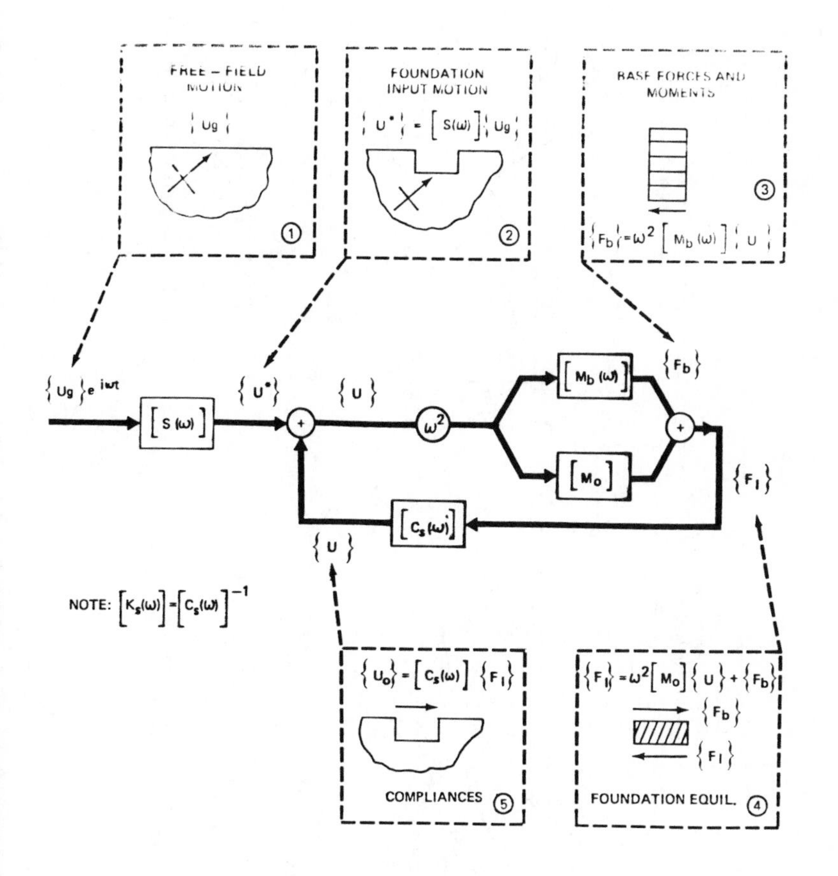

Fig. 1 Impedance Approach For Seismic Soil-Structure Interaction Analysis (After J. E. Luco)

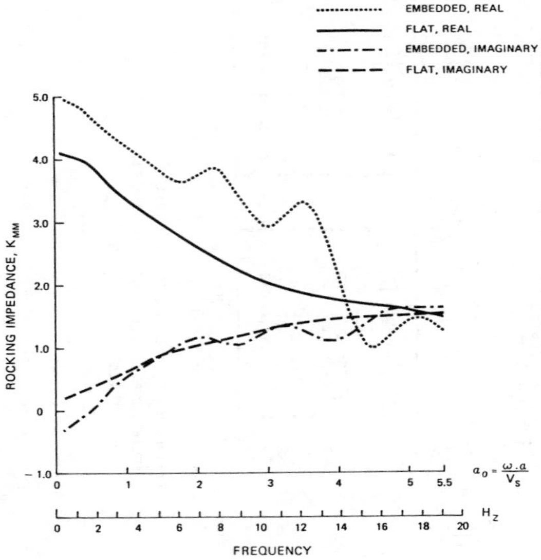

EMBEDDED, REAL
FLAT, REAL
EMBEDDED, IMAGINARY
FLAT, IMAGINARY
ROCKING IMPEDANCE, K_{MM}
5.0
4.0
3.0
2.0
1.0
0
− 1.0
0
1
2
3
4
5
5.5
$a_o = \frac{\omega . a}{V_s}$
0
2
4
6
8
10
12
14
16
18
20
H_Z
FREQUENCY

Fig. 2 Rocking Impedance

REAL
IMAGINARY
MODULUS
FREE FIELD
HORIZONTAL INPUT MOTION, $\Delta U/U_o$
1.4
1.2
1.0
0.8
0.6
0.4
0.2
0
− 0.2
− 0.4
− 0.6
− 0.8
− 1.0
0
1
2
3
4
5
5.5
$a_o = \frac{\omega . a}{V_s}$
0
2
4
6
8
10
12
14
16
18
20
H_Z
FREQUENCY

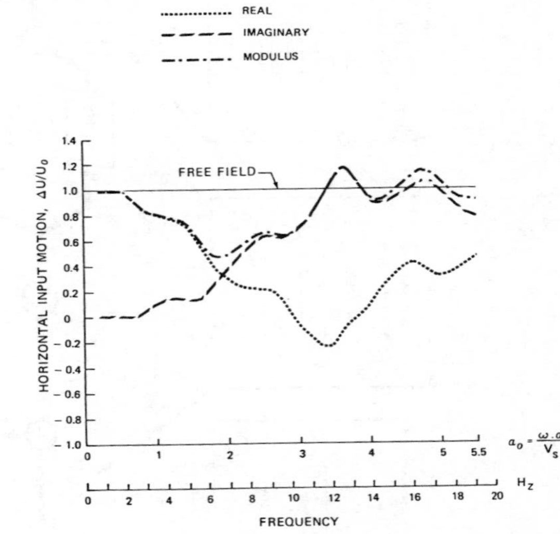

Fig. 3 Horizontal Input Motion

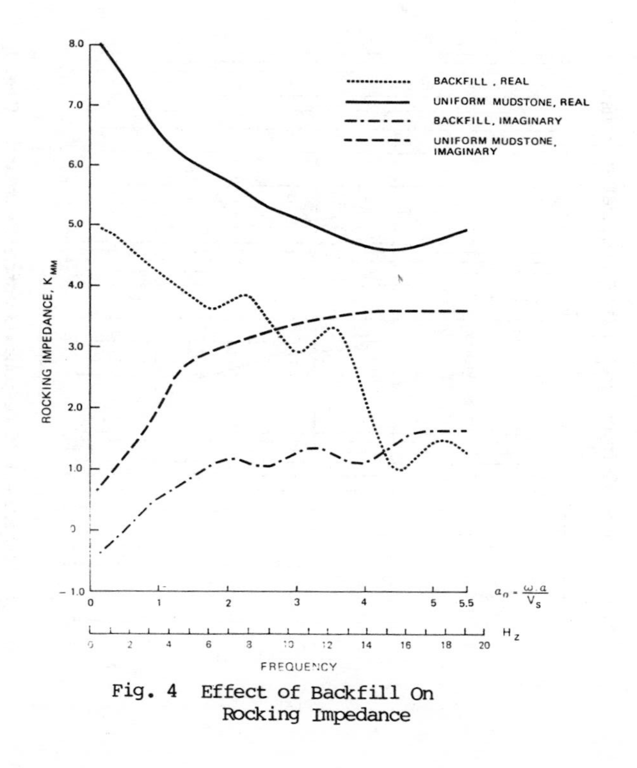

Fig. 4 Effect of Backfill On Rocking Impedance

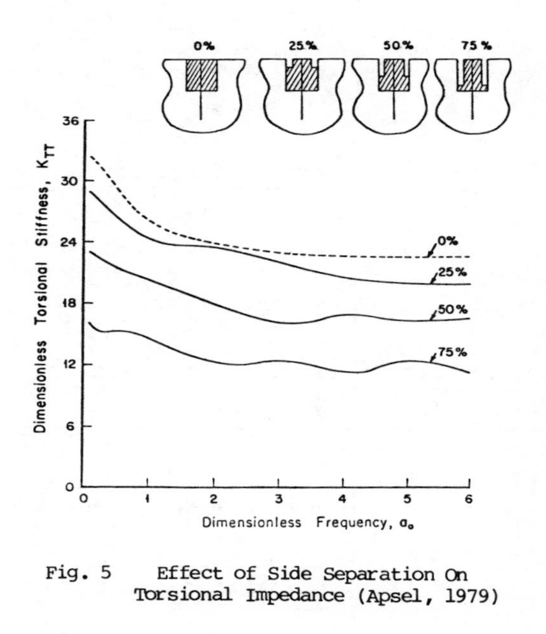

Fig. 5 Effect of Side Separation On Torsional Impedance (Apsel, 1979)

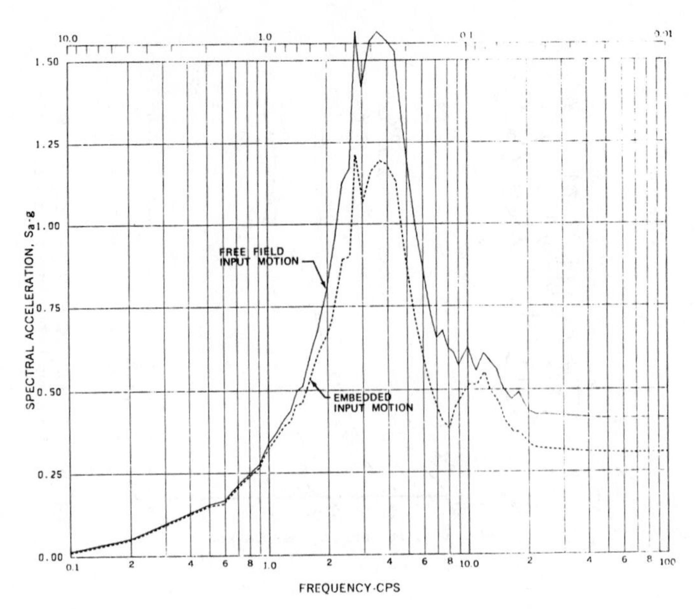

Fig. 6 Effect of Input Motion On Structural Response

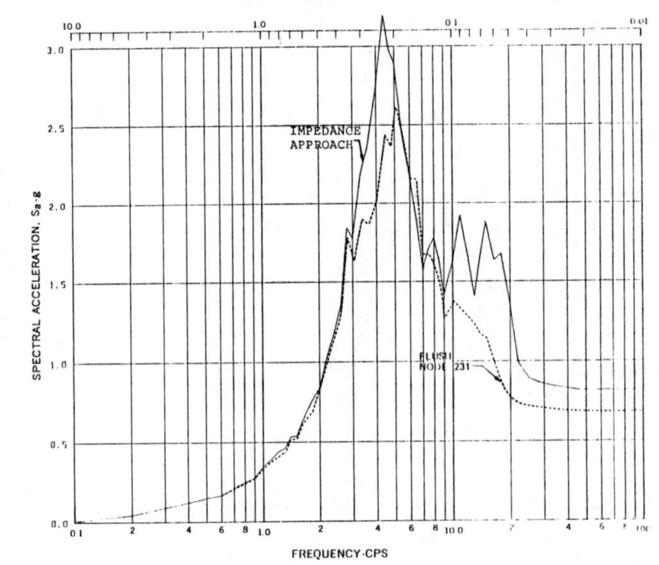

Fig. 7 Comparison of Response Spectra-Horizontal, Case I

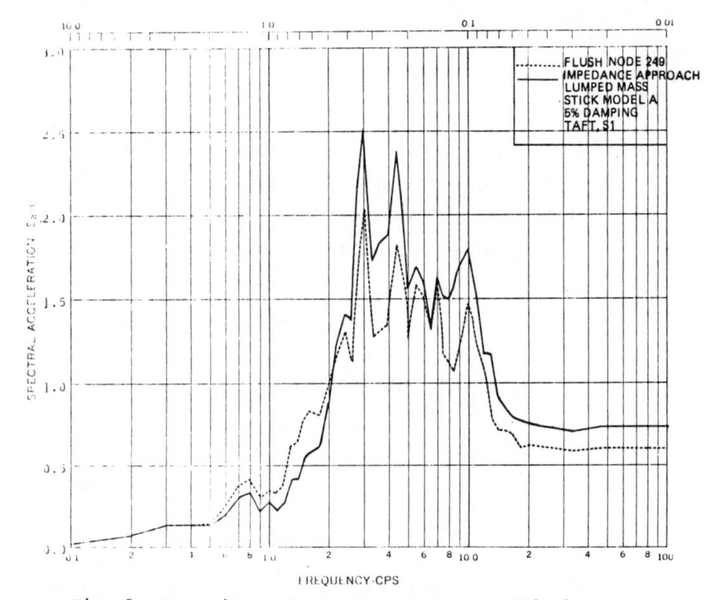

Fig. 8 Comparison of Response Spectra-Vertical, Case I

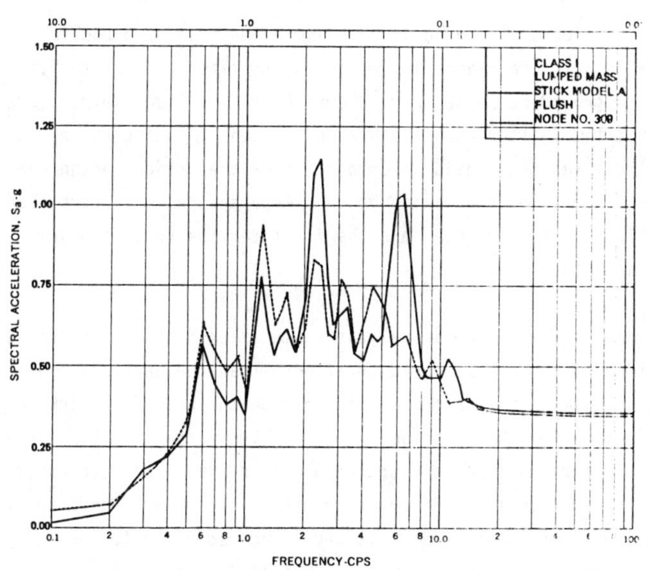

Fig. 9 Comparison of Response Spectra-Horizontal, Case II

TIME HISTORY CONSOLIDATION TO COMBINE MOTIONS DEVELOPED FROM THE SOIL STRUCTURE INTERACTION ANALYSIS USING THREE SETS OF SOIL PARAMETERS

C. -W. LIN
WESTINGHOUSE ELECTRIC CORPORATION
P. O. Box 355
Pittsburgh, Pennsylvania 15230

INTRODUCTION

Current techniques of conducting a seismic analysis of a nuclear power plant begin with the generation of the synthesized time histories for the design ground response spectra defined at the free field. Three orthogonal time-histories are generated. Each time-history represents one of the three translational components of the design earthquake motion which envelop the design ground response spectra at the free field. These time-histories are the input for the soil structure interaction study.

The original finite-element analysis formulation for soil structure interaction (SSI) is based upon the theory that the earthquake energy and its primary motion is propagated through the bedrock, which forms the foundation for the upper soil strata. The motion, in turn, reaches the earth's surface (and the nuclear plant) by vertically propagating waves. In order that the interaction effect of a heavy structure (such as a nuclear power plant) with its underlying soil strata can be evaluated, the bedrock motion without the influence of the surface structure is determined by a one-dimensional soil model.

With the motion established at bedrock, a second set of models is developed to comprise of at least two two-dimensional, vertically perpendicular slices through the nuclear plant, including simplified building representations and extending down to the bedrock, where the primary motion has been established.

The procedure is to excite these two models with the previously established bedrock motion. Each slice contributes a vertical and horizontal translation and a rocking motion in the plane of the slice. Only one of the two translational motions, however, is predominant. For instance, when the vertical input is applied at the bedrock, the vibration response at the structure base is primarily along the vertical direction with some amount of rocking, while the horizontal translational motion is negligible. The time-history motions obtained at the structure's base, subsequently, are used as input for the detailed building and equipment analyses.

The soil parameters used in the soil-structure interaction study are obtained from in-situ testing and laboratory tests. The test results are, generally, dependent on the time and location of the in-situ testing and the soil samples taken. It is customary to conduct the soil-structure interaction study for three different sets of soil parameters. Each one represents either the upper-bound, medium, or the lower-bound result.

Consequently, for each slice of the finite-element soil-structure interaction model, three sets of time-history motions will be obtained at the structural base. This necessitates that the sub-system analysis, where the structure base motions are used as the input, be conducted three times. For a coupled reactor coolant system and building non-linear time-history analysis, this would increase the total qualification cost significantly.

In this paper, a method is developed to combine into one set the three sets of time-history motions obtained from the three sets of soil parameters used. A numerical example showing the actual application of the method is also presented.

The time-history motions finally combined will maintain a fixed time-phase relationship between the translational and the rotational motions. They are, therefore, realistic; as a result, only the one set of time-history motions needs to be used as the input at the structure base for the analysis of the sub-system instead of the three sets otherwise required.

TIME-HISTORY CONSOLIDATION DESCRIPTION

A method developed in this paper to combine the three sets of time-history motions at the structure base from the SSI study is based on the knowledge that in each slice of the finite-element model used, one translational motion and one rocking motion are the predominate motions at the base of each input applied at the bedrock. Essentially, the model used in the SSI studies can be idealized as a two degree-of-freedom model (translation and rocking), with the soil act- ing as springs. In the latter part of this paper the equation of motion for this idealized soil-structure model consisting of four unknown quantities is given. The complete procedure of time-history consolidation is shown in Figure 1.

The first step in the time-history consolidation process is to determine the unknown parameters of the idealized model. A more detailed discussion on determining these parameters is given in the next section.

After the idealized model has been established, the second step is to determine the envelope time-history motion for the base translation. This is accomplished by constructing an envelope response spectrum for the three sets of base translational motions resulting from the three soil conditions included in the soil-structure interaction study. A synthesized time-history can then be generated matching this envelope response spectrum. The vertical time-history used to generate this synthesized time-history can be any one of the three translational time-history motions. From the efficiency point of view, however, the translation time-history corresponding to the median soil condition

(T_2) would be best suited for the purpose. This becomes the reference time-history motion. The envelope time-history has a response spectrum T shown in figure 1. If T is assumed to be the proper envelope translational input the corresponding rocking envelope will not be, in general, the envelope R. The following steps will adjust these differences.

After the synthesized time-history is obtained, the difference between the envelope translational time-history ($\overline{T}$) and the reference timetory (T_2) can be determined at each time step (step 3).

Step 4 will find the necessary changes or additions to the reference rocking motion (which is from the same set of the time-history motions that includes the reference translational motion) to obtain agreement between translational and rocking envelopes. This can be done by applying the difference of the translational motions (ΔT) to the two degree-of-freedom model and computing the corresponding rocking motion (ΔR).

The computed rocking motion is added to the reference rocking motion, (step 5) which in time is compared with the envelope of the three rocking motions on the response spectrum basis (step 6). If the new rocking motion does compare favorably with the envelope, ($\overline{R} \geq R$), then the envelope translation time-history ($\overline{T}$) and the final rocking time-history ($\overline{R}$) are the consolidated time-history set to be used for the sub-system analysis.

STEP 1 - DETERMINE THE PARAMETERS FOR THE IDEALIZED MODEL

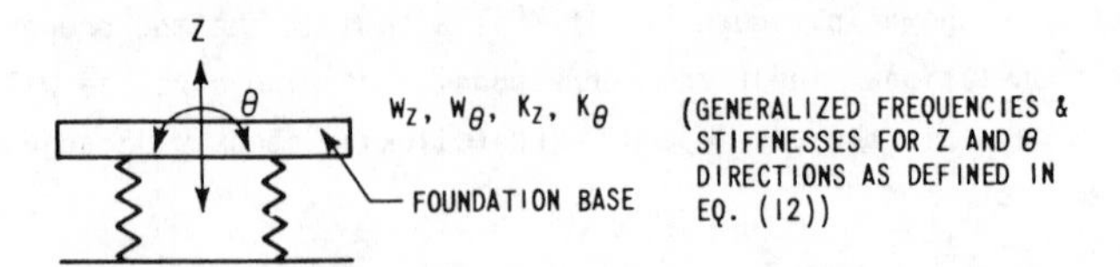

STEP 2 - OBTAIN ENVELOPE SPECTRA FOR EACH SOIL-STRUCTURE SLICE AND GENERATE SYNTHESIZED TIME HISTORY FOR TRANSLATION

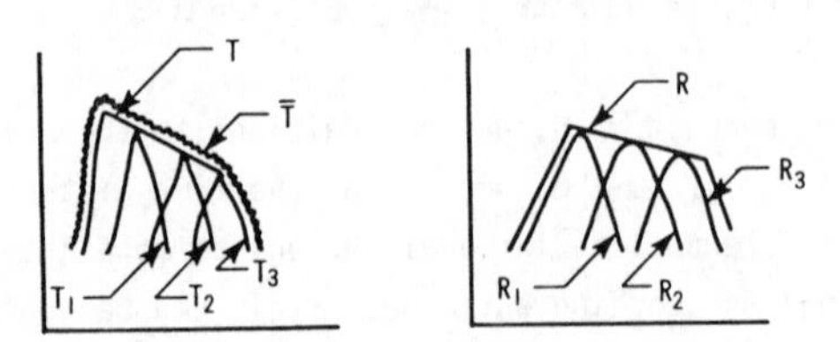

T_i, R_i : TRANSLATION AND ROTATION SPECTRA FOR i's SOIL CONDITION

T : ENVELOPE OF T_i

R : ENVELOPE OF R_i

$\bar{T}$: RESPONSE SPECTRA GENERATED TO ENVELOPE T.

STEP 3 - COMPUTE THE DIFFERENCE BETWEEN T AND THE REFERENCE TRANSLATIONAL MOTION (T_2)

$$\Delta T = \bar{T} - T_2$$

STEP 4 - COMPUTE DIFFERENCE IN ROTATION MOTION DUE TO ΔT

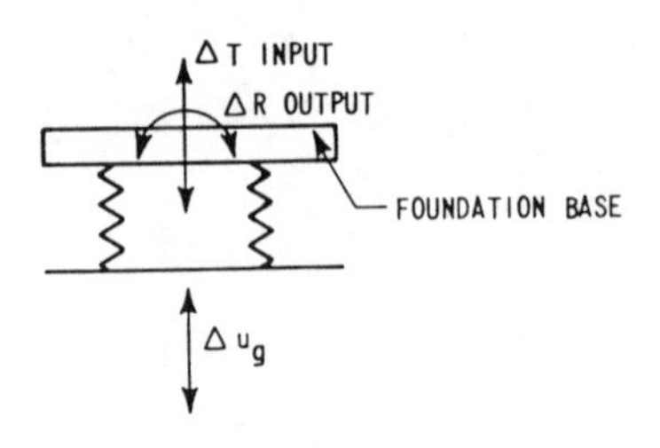

Figure 1. Procedure for Time History Consolidation (sheet 1 of 2)

STEP 5 - COMPUTE THE ADJUSTED ROTATIONAL MOTION

$\overline{R} = \Delta R + R_2$ FOR EACH TIME STEP

STEP 6 - COMPARE $\overline{R}$ WITH R ON A RESPONSE SPECTRA BASIS

IF $R \geq \overline{R}$ - PROCEDURE COMPLETE

IF $R < \overline{R}$ - CONTINUE WITH STEP 7

STEP 7 - GENERATE $\overline{\overline{R}}$ SO THAT $\overline{\overline{R}} \geq R$

STEP 8 - COMPUTE $\Delta R = \overline{\overline{R}} - \overline{R}$

STEP 9 - COMPUTE NEW DIFFERENCE IN TRANSLATIONAL MOTION DUE TO ΔR

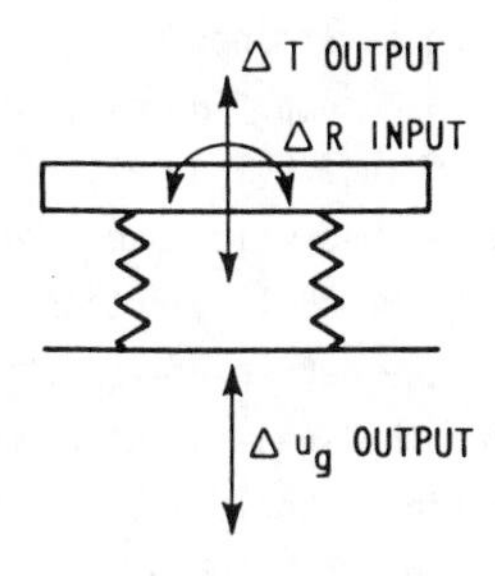

STEP 10 - CALCULATE THE ADJUSTED TRANSLATION MOTION

$\overline{\overline{T}} = \Delta T + \overline{T}$

IF $\overline{\overline{T}} > T$, PROCEDURE COMPLETE, OTHERWISE REPEAT STEPS FROM STEP 2 USING $\overline{\overline{T}}$ TO GENERATE ENVELOPE TRANSLATIONAL MOTION.

Figure 1. Procedure for Time History Consolidation (sheet 2 of 2)

If the new rocking motion does not compare favorably with the envelope, however, it can be adjusted using the DEBLIN2 method. A time-history ($\overline{R}$) is generated starting with $\overline{R}$ and enveloping R (step 7). A new difference between this adjusted rocking motion ($\overline{R}$) and the rocking motion R is calculated (step 8) and used as input to the two degree-of-freedom model to determine the adjustment required for the translational motion (ΔT). This is then added to the envelope translational motions ($\overline{T}$) and compared again with the envelope to determine if further changes are necessary.

The process is continued until satisfactory results are obtained. Consequently, this process generates the new consolidated translational and rocking motions with a realistic time-phase representative of the idealized soil-structure motion.

The method to compute the four unknown quantities in the equations of motions of the two degree-of-freedom soil-structure model has not been described. These quantities can be determined using the formulae for idealized soil spring calculation and computing directly the mass and mass moment of inertia for the structure. The latter computation is tedious. Instead, a method is developed which uses only the three sets of translational and rocking time-history motions to determine these unknown quantities. If the two degree-of-freedom model is suitable for the purpose of idealizing the soil-structure motion, any of the three sets of time-history motions should satisfy its equations of motion. Likewise, the difference motion between any of the two sets should also satisfy the equations of motion. Using the median soil time-history motions (set 2) as the reference time-history motions, two different motions can be computed. That is, set 2 minus set 1, and set 2 minus set 3. Substitution of these different motions into the equations of motion results in a total of four equations which can be used to determine the four unknowns.

There are several reasons for using the difference of motions. One reason is to be consistent with the process of using the difference of motions between the envelope and the reference motion when consolidating the time-histories. Second, using a linear model ignores the fact that the soil is really nonlinear. Working with the difference of motions will minimize the error. Third, the original equations of motion contain bedrock input (u_g). By using the difference motions, this input term is eliminated. This is based on the fact that the bedrock control motion is the same for all three soil conditions.

Since the equations of motion are non-linear, the four unknowns cannot be solved in the linear sense. This can be overcome by solving the equations of motion at a specific time; for instance, at the time where the reference translational motion reaches its peak value, or at the time where the response spectrum of the reference translation motion reaches its peak value. The latter approach insures that the adjustments calculated for the envelope time-history motions will add to the motions, thus, increasing the convergence speed of the procedure.

DERIVATION OF EQUATIONS

A. General Formulation

Presented in this section are equations used to determine the change in the rotation (or translation) time-history necessary, given a change in the translation (or rotation) time-history (from DEBLIN2 modifications), to maintain the original phase relationship between the two components of motion. (Analytical treatment to perform step 4).

The method used approximates the foundation soil with the two degree-of-freedom mass-spring model shown in figure 2.

The equations of motion are:

$$M\ddot{Z}_r + C(k_1 + k_2)\, Z_r - C(k_1 l_1 - k_2 l_2)\theta = -M\ddot{u}_g \tag{1}$$

$$J_C\, \ddot{\theta} + C(k_2 l_2{}^2 + k_1 l_1{}^2)\, \theta - C(k_1 l_1 - k_2 l_2)\, Z_r = 0 \tag{2}$$

where:

θ = absolute rotation displacement

Z_r = relative translation displacement

$\ddot{u}_g$ = control input motion

C = the amount of variation assumed for the soil parameters

M = foundation mass

J_C = rotatory inertia of the foundation

k_1, k_2 = spring constants of the soil

l_1, l_2 = distance of the c.g. to the soil springs k_1 and k_2, respectively and over-dot means differentiation with respect to time.

Following the procedures established previously, assuming now that a new time-history has been determined which envelopes the three translational time-histories for the three sets of soil parameters, one can calculate a new rotational (rocking) time-history and a revised input motion from equations (1) and (2). That is, let the response time-history motion and the input motion be changed to:

$$Z_r + \Delta Z_r \tag{3}$$

$$\theta + \Delta\, \theta \tag{4}$$

$$\ddot{u}_g + \Delta\, \ddot{u}_g \tag{5}$$

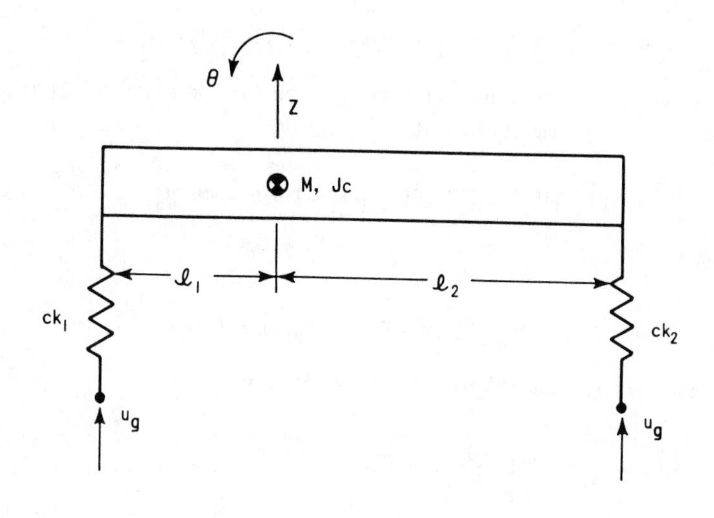

Figure 2. A Schematic Representation of the Two Degree-of-Freedom Foundation Soil Model

where ΔZ_r, $\Delta\theta$ and Δu_g are the additions to the translational time-history which resulted in the new envelope time-history, the additional rotational time-history to be added to the original time-history, and the additional input motion as a result of the change in the translational time-history, respectively. As a result of these changes, equations (1) and (2) become:

$$M(\ddot{Z}_r+\Delta\ddot{Z}_r) + C(k_1+k_2)\,(Z_r+\Delta Z_r) - C(k_1 l_1-k_2 l_2)\,(\theta+\Delta\theta) = -M(\ddot{u}_g+\Delta\ddot{u}_g) \quad (6)$$

and

$$J_C(\ddot{\theta}+\Delta\ddot{\theta}) + C(k_2 l_2^2+K_1 l_1^2)\,(\theta+\Delta\theta) - C(k_1 l_1-k_2 l_2)\,(Z_r+\Delta Z_r) = 0 \quad (7)$$

By subtracting equation (1) from equation (6) and equation (2) from equation (7), one arrives at

$$M\Delta\,\ddot{Z}_r + C(k_1+k_2)\,\Delta\,Z_r - C(k_1 l_1-k_2 l_2)\,\Delta\,\theta = -M\Delta\,\ddot{u}_g \quad (8)$$

and

$$J_C\Delta\ddot{\theta} + C(k_1 l_1^2+k_2 l_2^2)\,\Delta\theta - C(k_1 l_1-k_2 l_2)\,\Delta\,Z_r = 0 \quad (9)$$

equations (8) and (9) may also be written as

$$\frac{1}{\omega_Z^2}\,\Delta\,\ddot{Z}_a + \Delta\,Z_r - \frac{1}{k_Z}\,\Delta\,\theta = 0 \quad (10)$$

and

$$\frac{1}{\omega_\theta^2}\,\Delta\,\ddot{\theta} + \Delta\,\theta - \frac{1}{k_\theta}\,\Delta\,Z_r = 0 \quad (11)$$

where

$$\omega_Z^2 = C\,\frac{k_1 + k_2}{m}$$

$$k_Z = \frac{k_1 + k_2}{k_1 l_1 - k_2 l_2}$$

$$\omega_\theta^2 = C\,\frac{k_1 l_1^2 + k_2 l_2^2}{J_C}$$

$$k_\theta = \frac{k_1 l_1^2 + k_2 l_2^2}{k_1 l_1 - k_2 l_2} \tag{12}$$

$$\Delta\,\ddot{Z}_a = \Delta\,\ddot{Z}_r + \Delta\,\ddot{u}_g \tag{13}$$

Equations (1) and (11) indicate that when ΔZ_a is known, ΔZ_r and $\Delta\theta$ can be readily determined. This is done by eliminating between equations (10 and 11) ΔZ_r, which yields

$$\Delta\,\ddot{\theta} + \omega_\theta^2\,(1 - \frac{1}{k_\theta k_Z})\,\Delta\,\theta = -\frac{\omega_\theta^2}{k_\theta k_Z^2}\,\Delta\,\ddot{Z}_a \tag{14}$$

or

$$\Delta\,\ddot{\theta} + \omega_t^2\,\Delta\,\theta = -\frac{\omega_\theta^2}{k_\theta \omega_Z^2}\,\Delta\,\ddot{Z}_a \tag{15}$$

which has the following solution

$$\Delta\,\theta = \Delta\,\theta_0 \cos\,\omega_t t + \frac{\Delta\,\theta_0}{\omega_t}\sin\,\omega_t t + \frac{-\omega_\theta^2}{\omega_t k_\theta \omega_Z^2}\int_0^t \Delta\,\ddot{Z}_a\,(\gamma)\,\sin\,\omega_t\,(t-\gamma)\,d\,\gamma \tag{16}$$

and by differentiating equation (16) twice with respect to t, one has

$$\Delta\,\ddot{\theta} = -\omega_t[\Delta\,\theta_0 \cos\,\omega_t t + \frac{\Delta\,\theta_0}{\omega_t}\sin\,\omega_t t + \frac{-\omega_\theta^2}{\omega_t k_\theta \omega_Z^2}\int_0^t \Delta\,\ddot{Z}_a\,(\gamma)\,\sin\,\omega_t\,(t-\gamma)\,d\,\gamma] - \frac{\omega_\theta^2}{k_\theta \omega_Z^2}\,\Delta\,\ddot{Z}_a \tag{17}$$

A new rotational time-history therefore, is obtained by adding to the original time-history, $\ddot{\theta}$ the $\Delta\ddot{\theta}$ calculated from equation (17). It is to be noted that although the new translational time-history envelopes the three translational time-histories for the three sets of soil parameters, there is no guarantee that the new rotational time-history calculated will also envelope the three rotational time-histories for the three sets of soil parameters. However, a comparison between the newly calculated rotational time-history and the envelope rotational time-history can be made which would result in a new additional rotational time-history that can be used to further adjust the translational time-history. This iterative process can be carried out until both translational and rotational time-histories are satisfactory to the need.

The process of calculating a new addition to the translational time-history can be accomplished by using equation (11) written in the following form:

$$\Delta Z_r = k_\theta \left[\frac{\Delta \ddot{\theta}}{\omega_\theta^2} + \Delta \theta \right] \tag{18}$$

By knowing ΔZ_r, equation (10), when rearranged as the following:

$$\Delta \ddot{Z}_a = \omega_Z^2 \left[\frac{\Delta\theta}{k_Z} - \Delta Z_r \right] \tag{19}$$

can then be used to calculate $\Delta\ddot{Z}_a$.

B. Numerical Solution

The actual computation of equations (16), (17), (18) and (19) is done by using an iterative numerical procedure with parabolic representation of the time-histories between digitized points. For instance, let:

$$\Delta \ddot{Z}_a(t) = \Delta \ddot{Z}_{an} + S_n \left(\frac{t}{h}\right) + S_{n-1}^2 \left(\frac{t^2}{2h^2} - \frac{t}{2h}\right) \tag{20}$$

where

$$S_n = \Delta\ddot{Z}_{a_{n+1}} - \Delta\ddot{Z}_{a_n}$$

$$S^2_{n-1} = S_n - S_{n-1}$$

$$\Delta\ddot{Z}_a = \Delta\ddot{Z}_a\,(t) \text{ at } t = t_n$$

$$h = \Delta t = t_{n+1} = t_n \tag{21}$$

Equation (16) becomes

$$\Delta\,\theta_{n+1} = \Delta\,\theta_n \cos\omega_t h + \frac{\Delta\,\dot{\theta}_n}{\omega_t}\sin\omega_t h$$

$$+ \frac{\omega_\theta{}^2\Delta\ddot{Z}_{a_n}}{k_\theta\omega_Z{}^2} - \left\{\frac{1}{\omega_t{}^2} + \frac{1}{\omega_t{}^2}\cos\omega_t h\right\}$$

$$+ \frac{\omega_\theta{}^2}{k_\theta\omega_Z{}^2}\; S_n \left\{\frac{1}{\omega_t{}^2} + \frac{1}{\omega_t{}^3 h}\sin\omega_t h\right\}$$

$$+ \frac{\omega_\theta{}^2}{k_\theta\omega_Z{}^2\omega_t{}^2}\; S^2_{n-1}\left\{\frac{1}{(\omega_t h^2)}\,[1 - \cos\omega_t h] - \frac{1}{2\omega_t h}\sin\omega_t h\right\} \tag{22}$$

Similarly, one has

$$\Delta\,\dot{\theta}_{n+1} = -\omega_t\Delta\theta_n\sin\omega_t h + \Delta\,\theta_n\cos\omega_t h$$

$$+ \Delta\,\ddot{Z}_{a_n}\left\{-\frac{1}{\omega_t}\sin\omega_t h\right\}\frac{\omega_\theta{}^2}{k_\theta\omega_Z{}^2} + S_n\left\{\frac{\cos\omega_t h-1}{h\omega_t{}^2}\right\}\frac{\omega_\theta{}^2}{k_\theta\omega_Z{}^2}$$

$$+ S^2_{n-1}\left\{-\frac{1}{\omega_t{}^2 h} + \frac{1}{2\,\omega_t{}^2 h}\,(1 - \cos\omega_t h)\right.$$

$$\left. + \frac{\sin\omega_t h}{\omega_t} - \frac{1}{(\omega_t h)^2}\right\}\frac{\omega_\theta{}^2}{k_\theta\omega_Z{}^2} \tag{23}$$

and $\Delta\theta_{n+1}$ can be calculated directly from equation (14) for each time-step.

C. Model Parameters Determination

The mathematical models adopted in section 2 require that the parameters such as C, M, J_C, k_1, k_2, ℓ_1, ℓ_2 be known. Except for C, which is the variation of the soil parameters adopted in the soil-structure interaction and is a known quantity, the other constants can be reorganized as done in equation (12) and be determined by the procedure discussed in this section. This eliminates the need for determining the parameters from the physical model.

Equations (1) and (2) can be written as the following equations for the three sets of soil parameters:

$$M\ddot{Z}_1 + (k_1 + k_2)\, C_1 Z_1 + C_1(k_2 l_2 - k_1 l_1)\theta_1 = -M\ddot{u}_g \qquad (24)$$

$$M\ddot{Z}_2 + (k_1 + k_2)\, C_2 Z_2 + (k_2 l_2 - k_1 l_1)\, C_2 \theta_2 = -M\ddot{u}_g \qquad (25)$$

$$M\ddot{Z}_3 + (k_1 + k_2)\, C_3 Z_3 + (k_2 l_2 - k_1 l_1)\, C_3 \theta_3 = M\ddot{u}_g \qquad (26)$$

$$J_C\ddot{\theta}_1 + C_1\, (k_2 l_2^{\,2} + k_1 l_1^{\,2})\, \theta_1 + C_1\, (k_2 l_2 - k_1 l_1)\, Z_1 = 0 \qquad (27)$$

$$J_C\ddot{\theta}_2 + (k_2 l_2^{\,2} + k_1 l_1^{\,2})\, C_2 \theta_2 + (k_2 l_2 - k_1 l_1)\, C_2 Z_2 = 0 \qquad (28)$$

$$J_C\ddot{\theta}_3 + (k_2 l_2^{\,2} + k_1 l_1^{\,2})\, C_3 \theta_3 + (k_2 l_2 - k_1 l_1)\, C_3 Z_3 = 0 \qquad (29)$$

The subscripts 1, 2 and 3 for the variables Z, Z, θ and θ indicate whether the motion is of the first, second, or third set of the soil parameters, and C_1, C_2, C_3 represent the ratio of the soil parameters variations, for each of the three sets. For instance, when the soil parameters are assumed to vary by plus or minus fifty percent (as is generally acceptable) C_1, C_2, and C_3 may have the value of 1.5, 1.0, and 0.5, respectively.

Equations (24) and (29) can be rearranged to yield:

$$\frac{1}{\omega_Z^2}(\ddot{Z}_1 - \ddot{Z}_2) + (C_1Z_1 - C_2Z_2) - \frac{1}{k_Z}(C_1\theta_1 - C_2\theta_2) = 0 \tag{30}$$

$$\frac{1}{\omega_Z^2}(\ddot{Z}_2 - \ddot{Z}_3) + C_2Z_2 - C_3Z_3) - \frac{1}{k_Z}(C_2\theta_2 - C_3\theta_3) = 0 \tag{31}$$

$$\frac{1}{\omega_\theta^2}(\ddot{\theta}_1 - \ddot{\theta}_2) + (C_1\theta_1 - C_2\theta_2) - \frac{1}{k_\theta}(C_1Z_1 - C_2Z_2) = 0 \tag{32}$$

$$\frac{1}{\omega_\theta^2}(\ddot{\theta}_2 - \ddot{\theta}_3) + (C_2\theta_2 - C_3\theta_3) - \frac{1}{k_\theta}(C_2Z_2 - C_3Z_3) = 0 \tag{33}$$

where ω_Z^2, ω_θ^2, k_Z, and k_θ are defined in equation (12) and can be solved from equation (30) and (31) to yield

$$\omega_Z^2 = \frac{(C_2\theta_2 - C_3\theta_3)(\ddot{Z}_1 - \ddot{Z}_2) - (C_1\theta_1 - C_2\theta_2)(\ddot{Z}_2 - \ddot{Z}_3)}{(C_2Z_2 - C_3Z_3)(C_1\theta_1 - C_2\theta_2) - (C_2\theta_2 - C_3\theta_3)(C_1Z_1 - C_2Z_2)} \tag{34}$$

$$\omega_\theta^2 = \frac{(C_2Z_2 - C_3Z_3)(\ddot{\theta}_1 - \ddot{\theta}_2) - (C_1Z_1 - C_2Z_2)(\ddot{\theta}_2 - \ddot{\theta}_3)}{(C_2\theta_2 - C_3\theta_3)(C_1Z_1 - C_2Z_2) - (C_2Z_2 - C_3Z_3)(C_1\theta_1 - C_2\theta_2)} \tag{35}$$

$$k_Z = \frac{(C_2\theta_2 - C_3\theta_3)(\ddot{Z}_1 - \ddot{Z}_2) - (C_1\theta_1 - C_2\theta_2)(\ddot{Z}_2 - \ddot{Z}_3)}{(C_2Z_2 - C_3Z_3)(\ddot{Z}_1 - \ddot{Z}_2) - (C_1Z_1 - C_2Z_2)(\ddot{Z}_2 - \ddot{Z}_3)} \tag{36}$$

$$k_\theta = \frac{(C_2Z_2 - C_3Z_3)(\ddot{\theta}_1 - \ddot{\theta}_2) - (C_1Z_1 - C_2Z_2)(\ddot{\theta}_2 - \ddot{\theta}_3)}{(C_2\theta_2 - C_3\theta_3)(\ddot{\theta}_1 - \ddot{\theta}_2) - (C_1\theta_1 - C_2\theta_2)(\ddot{\theta}_2 - \ddot{\theta}_3)} \tag{37}$$

When ω_Z, ω_θ, k_Z and k_θ are certain to be constants, the relationships in equations (34) to (37) exist for any given time within the duration of the time-histories. The present soil structure analysis adopts finite-element models taking into account the non-linear properties of both the soil damping and the strain-stress relationship. As a result, both the translational and the rotational motions are no longer linear and their relationship is at least mildly non-linear in nature.

Nevertheless, equations (34) and (37) can be used to determine ω_Z, ω_θ, k_Z and k_θ at a given time which is selected to have the additional translational time-history. This is done to insure that the new time-history will envelop the three sets of time-histories. Also, the fact that the coefficients ω_Z, ω_θ, k_Z and k_θ are determined from equations (34) to (37) assures that the additional translational and rotational motions follow an idealized foundation soil motion. The time-phase relationship between these two motions is, therefore, properly synchronized. It is to be noted also, that, although a non-linear formulation of this problem is possible, it is mathematically complex and technically not feasible at the present time.

DEMONSTRATION PROBLEM AND CONCLUSIONS

The following demonstration problem uses time-histories that were developed from the soil structure analysis of a nuclear power plant. There were three kinds of soil conditions: soft, medium and hard; and three input directions: north-south, east-west, and vertical. The time-histories of the median soil condition in the east-west direction were used in this demonstration problem to show the procedure for consolidating a set of real time-histories.

Several modifications were necessary to arrive at a satisfactory envelope of the rotation response spectrum. The rotation and translation acceleration time-history were run through the program a final time to assure that the phase relationship was maintained. Figures 3 and 4 illustrate that the corresponding spectra have enveloped the design response spectra for the three soil conditions. These results show that the computer program yields excellent results and is able to satisfactorily consolidate three sets of time-histories taken from real soil structure analyses into one set.

CONCLUDING REMARKS

A method has been developed to combine the three sets of time-history motions at the foundation as a result of the soil-structure interaction study conducted for three soil parameter variations. The

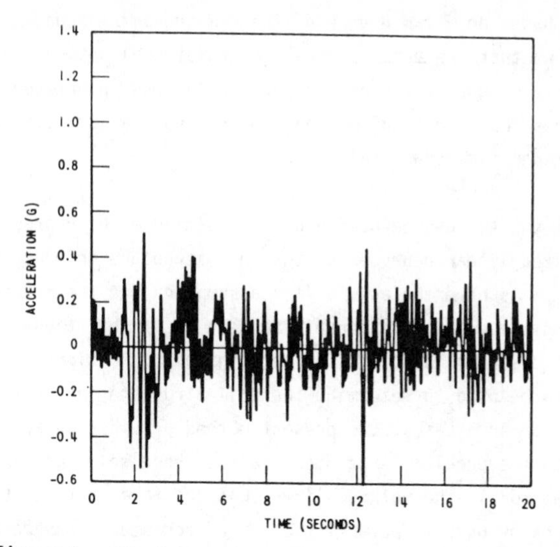

Figure 3. Final Translation Acceleration Time History

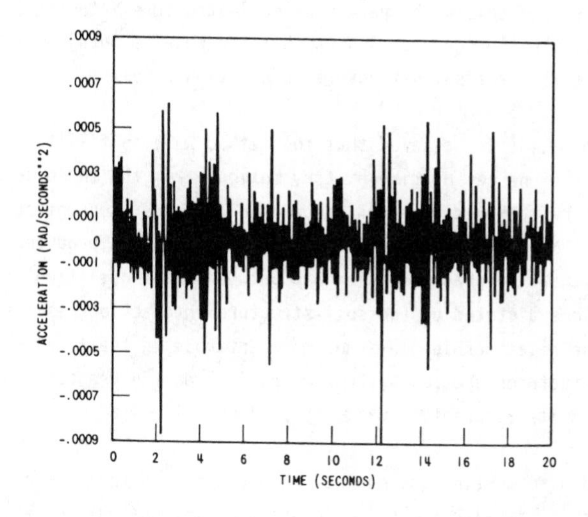

Figure 4. Final Rotation Acceleration Time History

method is based on a two degree-of-freedom idealized foundation model which allows that the translational and rotational motions be combined while maintaining a predetermined phase relationship between the two. This insures that the final motions determined realistically account for the proper time phase relationships.

The use of the two degree-of-freedom foundation model represents only the idealized linear behavior of the soil-structure interaction studies. It is nevertheless, a fair assumption for the purpose of maintaining the phase relationship between the translational and rotational motions. In addition, for certain soil conditions, soil springs can be used to determine the soil-structure interaction effects. For such cases, the present method provides an accurate solution to the problem. For other cases, where soil springs are not used in the model, the method allows that the three sets of time-history motions be used to determine a set of acceptable parameters for the model. This eliminates the need of determining the soil spring based on the formulae reported in the literature, but may otherwise have no relationship with the actual soil-structure interaction study conducted. This procedure assures that the results obtained will be consistent with the soil-structure interaction results.

Consequently, it is believed that the method will fulfill the intent of producing one set of time-history motions from the three sets, which otherwise would have had to be used in the subsequent sub-system analysis. More importantly, the time-history motions produced will maintain the time-phase relationship between the translation and rotation motions dictated by the soil-structure interaction results. This time-phase relationship plays an important role in the final analysis of the structures and sub-systems which include the reactor coolant system and its associated auxiliary pipings.

In addition to meeting the goal of maintaining a fixed time-phase relationship, it also produces time-history motions which envelope the required envelope response spectra closely. Therefore, the results are conservative, but, simultaneously, as optimized as possible. This should help to prevent the final sub-system analysis from being overly conservative.

C.-W. Lin

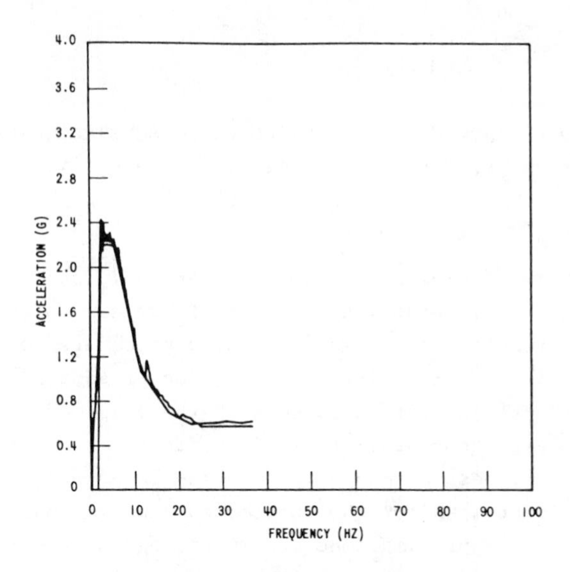

Figure 5. Final Translation Response Spectrum

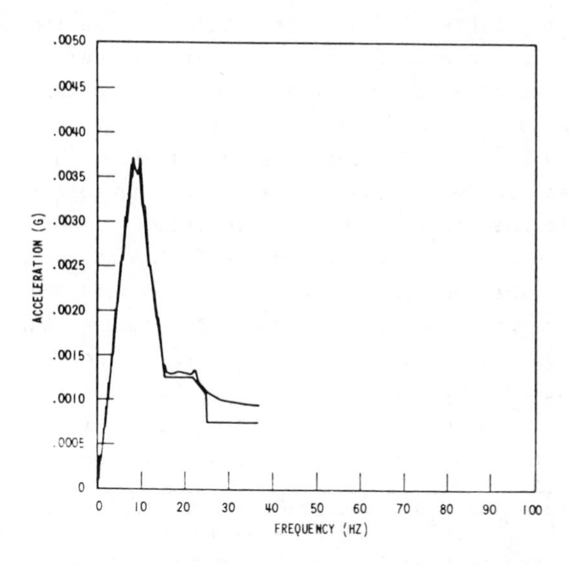

Figure 6. Final Rotation Response Spectrum

SOIL-STRUCTURE INTERACTION USING SUBSTRUCTURES

by Anand K. Singh,[1] M. ASCE; Tzu-I. Hsu,[2] M. ASCE; and Nancy A. Holmes[2]

INTRODUCTION

The finite element method is extensively used in the soil-structure interaction analysis of nuclear power plants. The present method requires a two-step procedure to complete the analysis. In the first step, a simplified two-dimensional structural model is used with the soil model to determine the soil-structure interaction effects. In the second step, the response of a detailed three-dimensional structural model is determined for the interacted base motions completed in step one. Approximations are introduced by the two-dimensional simplification of the structural model required by the first step.

This paper presents a substructure formulation for soil-structure interaction analysis which calculates the structural response in one step. The soil and the structure are treated as separate substructures. The structure is represented by its modal characteristics, i.e., mode shapes, frequencies, and participation factors. The soil is represented by plane strain elements with viscous boundaries and transmitting boundaries. No modal solution of the soil is performed. The coupled system is analyzed using the complex frequency response analysis method. The substructure formulation results in significant cost savings since the modal representation of the structure reduces the total degrees of freedom in the coupled soil-structure interaction model and eliminates the second step of the two-step procedure.

---

1. Associate and Assistant Head, Structural Analytical Division, Sargent & Lundy, Chicago, IL.

2. Senior Engineering Analyst, Structural Analytical Division, Sargent & Lundy, Chicago, IL.

Based on the above formulation, results for a typical power plant structure are presented and compared to those obtained by the two-step procedure. The mode shapes and frequencies of the simplified two-dimensional structural model and the detailed three-dimensional model are also compared.

BACKGROUND

The substructure method for computing the vibration characteristics of structural systems has been the subject of increasing interest over the past several years. Hurty [5,6] presented a procedure for analyzing structural systems in which the displacement behavior of each substructure is described by the rigid body modes, constrained modes, and normal modes. Gladwell [4] presented a method in which the substructure modes are computed by assuming that all other substructures are connected to this substructure but are rigid and entail no internal deformation. These substructure modes are then used to obtain the natural frequencies and modes of vibration of the entire structure. Craig and Bampton [3] presented a method where the final structural vibration modes are based on substructure normal modes, degrees of freedom at common boundary nodes, and the substructure constrained modes. Benfield and Hruda [1] developed a method in which only the substructure vibration modes are retained as generalized coordinates when all of the structure modes are obtained. This method is particularly suitable for structures with a large number of interface nodes, such as finite element shell models. Singh [9] extended the substructure method to compute the seismic response of nuclear power plant structures and piping systems and concluded that the approach is both accurate and economical.

METHOD OF ANALYSIS

For the soil-structure interaction problem where the soil medium is one substructure and the structure represents the other substructure, the approach suggested by Benfield and Hruda [1] is very efficient. In this paper, the above approach is extended to compute the soil-structure interaction response. The structure is represented by its fixed base modal characteristics, whereas the soil medium is represented by mass, stiffness, and damping matrices. The two substructures are then coupled to obtain the structural response, including the effects of soil-structure interaction. The response is obtained using the frequency response method [7].

Figure 1 shows the schematic details of a soil-structure system for a nuclear power plant. The structure is modeled by a lumped mass-spring system and the soil is modeled by plain strain elements with viscous and transmitting boundaries [8] to account for the three-dimensional nature of the soil media. The common boundary interface node is labeled 1. For the soil-structure system shown in Figure 1, the equations of motion for a given base rock motion are:

$$\begin{bmatrix} m_{ii} & & \\ & m_{bb} & \\ & & m_{ss} \end{bmatrix} \begin{Bmatrix} \ddot{x}_i \\ \ddot{x}_b \\ \ddot{x}_s \end{Bmatrix} + \begin{bmatrix} c_{ii} & c_{ib} & 0 \\ c_{bi} & c_{bb} & c_{bs} \\ 0 & c_{sb} & c_{ss} \end{bmatrix} \begin{Bmatrix} \dot{x}_i \\ \dot{x}_b \\ \dot{x}_s \end{Bmatrix} + \begin{bmatrix} k_{ii} & k_{ib} & 0 \\ k_{bi} & k_{bb} & k_{bs} \\ 0 & k_{sb} & k_{ss} \end{bmatrix} \begin{Bmatrix} x_i \\ x_b \\ x_s \end{Bmatrix}$$

$$= -\begin{bmatrix} m_{ii} & & \\ & m_{bb} & \\ & & m_{ss} \end{bmatrix} \begin{Bmatrix} \phi_{ri} \\ \phi_{rb} \\ \phi_{rs} \end{Bmatrix} \ddot{y}(t) - \begin{Bmatrix} 0 \\ V \end{Bmatrix} + \begin{Bmatrix} 0 \\ F \end{Bmatrix} - \begin{Bmatrix} 0 \\ T \end{Bmatrix} \qquad (1)$$

where the subscripts i, b, and s refer to the structure interior, common boundary interface, and the soil degrees of freedom, respectively; m, c, and k are the mass, the damping, and the stiffness matrix components, respectively, of a unit thickness; and $\ddot{x}$, $\dot{x}$, and x are the acceleration, velocity, and displacement vectors relative to the base motion. ϕ_{ri}, ϕ_{rb}, and ϕ_{rs} are the rigid body vectors with components equal to unity in the direction parallel to the support motion, and zero otherwise. $\ddot{y}(t)$ is the prescribed base acceleration motion. The forces $\{V\}$, $\{F\}$, and $\{T\}$

originate from the viscous boundaries and transmitting boundaries and are defined in Reference 8 as follows:

$$\{V\} = \frac{1}{B}[C](\{\dot{x}\} - \{\dot{x}\}_f) \tag{2a}$$

$$\{F\} = [G]\{x\}_f \tag{2b}$$

$$\{T\} = ([R]+[L])(\{x\}-\{x\}_f) \tag{2c}$$

where B is the width of the structure normal to the excitation direction and the subscript f refers to free field parameters. [C] is a diagonal matrix which depends on the properties of the free field. [G] is a frequency-independent stiffness matrix formed from the complex moduli in the free field. [R] and [L] are the frequency-dependent boundary stiffness matrices.

The equations of motion for free vibration of the substructure representing the structure are given by:

$$\begin{bmatrix} m_{ii} & 0 \\ 0 & m_{bb} \end{bmatrix} \begin{Bmatrix} \ddot{x}_i \\ \ddot{x}_b \end{Bmatrix} + \begin{bmatrix} k_{ii} & k_{ib} \\ k_{bi} & k_{bb} \end{bmatrix} \begin{Bmatrix} x_i \\ x_b \end{Bmatrix} = 0 \tag{3}$$

For linear problems, the displacement of an internal degree of freedom can be expressed as:

$$\{x_i\} = \{x_i\}_c + [\phi_c]\{x_b\} \tag{4}$$

where $\{x_i\}_c$ is the displacement of internal degrees of freedom when common boundary nodes are assumed fixed and $[\phi_c]$ are the constrained modes for the substructure representing the displacement of internal degrees of freedom induced by a unit displacement of the common boundary degrees of freedom with all other boundary freedoms fixed. The constrained modes are computed by solving for each substructure a set of linear equations given by

$$[k_{ii}][\phi_c] = -[k_{ib}] \tag{5}$$

Thus, for free vibrations,

$$[m_{ii}]\{\ddot{x}_i\}_c + [k_{ii}]\{x_i\}_c = 0 \tag{6}$$

Expressing $\{x_i\}_c$ in terms of normalized mode shapes of equation 6,

$$\{x_i\}_c = [\phi]\{p\} \tag{7a}$$

where

$$[\phi]^T[k_{ii}][\phi] = [\omega_n^2] \tag{7b}$$

and

$$[\phi]^T[m_{ii}][\phi] = [I] \tag{7c}$$

$\{p\}$ represents the generalized normal coordinates and ω_n is the frequency of the constrained substructure in mode n. Based on equations 4 and 7, the displacement in the soil-structure system can be expressed as

$$\begin{Bmatrix} x_i \\ x_b \\ x_s \end{Bmatrix} = \begin{bmatrix} \phi & \phi_c & 0 \\ 0 & I & 0 \\ 0 & 0 & I \end{bmatrix} \begin{Bmatrix} p \\ x_b \\ x_s \end{Bmatrix} \tag{8}$$

Transformation of equation 1 by equation 8 yields the equations of motion in terms of the generalized coordinates, as follows:

$$\begin{bmatrix} I & \overline{m}_{ib} & 0 \\ \overline{m}_{bi} & \overline{m}_{bb} & 0 \\ 0 & 0 & m_{ss} \end{bmatrix} \begin{Bmatrix} \ddot{p} \\ \ddot{x}_b \\ \ddot{x}_s \end{Bmatrix} + \begin{bmatrix} \overline{c}_{ii} & \overline{c}_{ib} & 0 \\ \overline{c}_{bi} & \overline{c}_{bb} & c_{bs} \\ 0 & c_{sb} & c_{ss} \end{bmatrix} \begin{Bmatrix} \dot{p} \\ \dot{x}_b \\ \dot{x}_s \end{Bmatrix} + \begin{bmatrix} \omega^2 & 0 & 0 \\ 0 & \overline{k}_{bb} & k_{bs} \\ 0 & k_{sb} & k_{ss} \end{bmatrix} \begin{Bmatrix} p \\ x_b \\ x_s \end{Bmatrix}$$

$$= - \begin{Bmatrix} \gamma_i \\ \gamma_b \\ \gamma_s \end{Bmatrix} \ddot{y}(t) - \begin{Bmatrix} 0 \\ V \end{Bmatrix} + \begin{Bmatrix} 0 \\ F \end{Bmatrix} - \begin{Bmatrix} 0 \\ T \end{Bmatrix} \tag{9}$$

where

$$\bar{m}_{ib} = \bar{m}_{bi}^T = \phi^T m_{ii}\,\phi_c \quad ; \quad \bar{m}_{bb} = m_{bb} + \phi_c^T m_{ii}\,\phi_c ;$$

$$\bar{k}_{bb} = k_{bb} - \phi_c^T k_{ii}\,\phi_c \quad ; \quad \bar{c}_{ib} = \bar{c}_{bi}^T = \phi^T c_{ii}\,\phi_c + \phi^T c_{ib} ;$$

$$\bar{c}_{ii} = \phi^T c_{ii}\,\phi \quad ; \quad \bar{c}_{bb} = c_{bb} - \phi_c^T c_{ii}\,\phi_c ;$$

$$\gamma_i = \phi^T m_{ii}\,\phi_{ri} \quad ; \quad \gamma_b = \phi_c^T\, m_{ii}\,\phi_{ri} + m_{bb}\,\phi_{rb} ;$$

and $\gamma_s = m_{ss}\,\phi_{rs}$

The first few modes are generally adequate to represent the structural behavior.

The forces $\{T\}$, related to the transmitting boundary, are frequency-dependent. The damping in the soil may vary from element to element but is frequency-independent. The structural damping is specified in terms of modal damping. Under these conditions, equation 9 is best solved using the frequency response method. The following assumptions for the damping matrix greatly reduce the computations:

$$\bar{c}_{ii} = \phi^T c_{ii}\,\phi = \left[2\beta_n \frac{\omega_n^2}{\Omega}\right] \tag{10a}$$

$$\bar{c}_{ib} = \bar{c}_{bi}^T = 0 \tag{10b}$$

$$c_s = \frac{2\beta_s}{\Omega}\,k_s \tag{10c}$$

where β_n is the modal damping ratio for mode n of the structure, Ω is the frequency for which the response is being computed, c_s is the damping matrix of a soil element which has a modal damping ratio of β_s and stiffness matrix k_s. Expressing the input motions and response in terms of a truncated Fourier Series yields

$$\{\ddot{y}(t)\} = \mathrm{Re} \sum_{j=0}^{N/2} \{\ddot{y}(\Omega_j)\}\exp(i\Omega_j t) \tag{11a}$$

$$\{x(t)\} = \mathrm{Re} \sum_{j=0}^{N/2} \{x(\Omega_j)\} \exp(i\Omega_j t) \tag{11b}$$

where N is the number of digitized points in the input motion. The amplitudes $\ddot{y}(\Omega_j)$ and $x(\Omega_j)$ can be found by the Fast Fourier Transform algorithm (9). Substitution of equations 10 and 11 into 9 yields

$$\begin{aligned}(-\Omega_j^2[\overline{M}]+[K^*]+[R]_j+[L]_j+\frac{i\Omega_j[C]}{B})\ x(\Omega_j) &= \\ -\ \{Q\}\ddot{y}(\Omega_j) + ([G]+[R]_j+[L]_j &+ \frac{i\Omega_j}{B}[C])\{x(\Omega_j)\}_f\end{aligned} \tag{12}$$

where

$$[\overline{M}] = \begin{bmatrix} I & \overline{m}_{ib} & 0 \\ \overline{m}_{bi} & \overline{m}_{bb} & 0 \\ 0 & 0 & m_{ss} \end{bmatrix} \qquad \{Q\} = \begin{Bmatrix} \gamma_i \\ \gamma_b \\ \gamma_s \end{Bmatrix}$$

$$[\overline{K}] = \begin{bmatrix} [\omega_n^2(1+i2\beta_n)] & 0 & 0 \\ 0 & \overline{k}^*_{bb}+k^* & \overline{k}^{*T}_{sb} \\ 0 & \overline{k}^*_{sb} & \overline{k}^*_{ss} \end{bmatrix}$$

in which

$$k^* = -\phi_c^T(1+2i\beta_n)k_{ii}\phi_c$$

$\overline{k}^*_{bb}$, $\overline{k}^*_{sb}$, and $\overline{k}^*_{ss}$ are the stiffness matrices obtained using complex moduli. The linear set of equations represented by equation 12 determines the displacement amplitudes $\{x(\Omega_j)\}$

at each frequency Ω_j, j=0,1...N/2. The equations can be solved by Gaussian elimination. The displacements in the time domain follow from equation 11 by the inverse Fast Fourier Transform.

NUMERICAL EXAMPLE

In order to demonstrate the advantages of the proposed substructure method, a soil-structure interaction analysis was performed for a typical nuclear power plant using the two-step and the proposed substructure approaches. Figure 2 shows the schematic details of the three-dimensional structural model for horizontal excitation. Each floor slab is modeled as a rigid horizontal diaphragm having three in-plane degrees of freedom, two horizontal translations, and one rotation about the vertical axis. For rocking, each slab can move in two additional degrees of freedom (rotation about two horizontal axes). However, since the bending stiffness of the shear wall is large, all slabs are assumed to rock by the same angle.

The mass of the slab and the mass of the equipment and structural components are lumped at the centroid of the included masses. The mass of shear walls is distributed to the connecting floor slabs. Each shear wall and braced vertical frame is modeled as a set of elastic shear springs between floors. To simulate the correct torsional stiffness, these springs in the model are located at the physical location of the shear wall or braced frame in the plant. The containment, the reactor pedestal, and the reactor pressure vessel are modeled by lumped masses connected by flexural members. The flexural and shear stiffnesses of these members correctly simulate the stiffnesses of the components modeled.

The two-step approach requires a two-dimensional structural model, generally in terms of a plane strain finite element model. Such a model was obtained by matching the shear stiffness and mass distribution of the detailed three-dimensional (3-D) and equivalent two-dimensional (2-D) structural models. The masses and stiffness of the 2-D plane strain structural model were divided by the structural width in the third direction to keep the soil-structure mass and stiffness ratios in the correct proportion. Tables 1 and 2 show the modal periods and modal participation factors obtained from the 3-D detailed model and the 2-D plane strain model. Figures 3 and 4 show the comparison of the normalized mode shapes for the two models. The closeness of the results verifies the equivalence of the 2-D and the 3-D models. These two models can be further verified by comparing the responses obtained using the NRC Regulatory Guide 1.60 consistent time history applied at the base of the two models. Table 3 shows a comparison of maximum displacements at selected slabs. Again, the responses are very close for the two models being considered.

The 2-D plane strain model was used in the first step of a two-step soil-structure interaction analysis. The FLUSH program [8] was used to obtain the interaction time history at the base of the structure. In the second step, this time history was then applied to the base of the detailed 3-D structural model. The floor response spectra at selected locations were generated. For the same soil-structure system described above, the soil-structure interaction responses were also generated using the substructure approach and the 3-D structural model. Figures 5 through 7 show the comparison of response spectra for the one-step and two-step results. It can be observed that the two responses are comparable, although the two-step approach gives higher responses in certain frequency ranges. This difference may be due to a "frequency mismatch" between the 2-D model used in the soil-structure interaction analysis and the 3-D model used to generate the floor response spectra. Part of the difference can also be attributed to the absence of rocking input motion in the second step of the two-step approach. The substructure approach automatically includes the rocking effects.

Table 4 presents the relative computer costs of the two-step and the substructure approaches: the proposed substructure approach costs 40% less. The lower cost is a direct result of the reduction in the total degrees of freedom in the coupled soil-structure system by the modal representation of the structure and the elimination of the second step of the two-step approach. Note that the above figures do not include the savings in man-time and computer costs resulting from the elimination of the two-dimensional plane strain finite element model in the substructure approach.

CONCLUSIONS

The paper presents a substructure method for soil-structure interaction analysis in which a detailed three-dimensional structural model is used. The proposed method eliminates the need for a two-dimensional plane strain model for the structure. The proposed method is shown to be accurate and considerably less expensive than the two-step approach.

APPENDIX - REFERENCES

1. Benfield, W. A. and Hruda, R. F., "Vibration Analysis of Structures by Component Mode Substitution," AIAA Journal, Vol. 9, No. 7, July 1971, pp. 1255-1261.

2. Cooley, J. W. and Tukey, J. W., "An Algorithm for the Machine Calculation of Complex Fourier Series," Mathematics of Computation, Vol. 19, April 1965.

3. Craig, R. R. and Bampton, M. C. C., "Coupling of Substructures for Dynamic Analysis," AIAA Journal, Vol. 6, No. 7, July 1968, pp. 1313-1319.

4. Gladwell, G. M. L., "Branch Mode Analysis of Vibrating Systems," Journal of Sound and Vibration, Vol. 1, No. 1, January 1964, pp. 41-59.

5. Hurty, W. C., "Vibration of Structural Systems by Component Mode Synthesis," Journal of the Engineering Mechanics Division, ASCE, Vol. 86, No. EM4, Proc. Paper 2572, August 1960, pp. 51-69.

6. Hurty, W. C., "Dynamic Analysis of Structural Systems Using Component Modes," AIAA Journal, Vol. 3, No. 4, April 1965, pp. 678-685.

7. Hurty, W. C., and Rubinstein, M. F., Dynamics of Structures, Prentice-Hall, Inc., Englewood Cliffs, New Jersey, 1964.

8. Lysmer, J., Udaka, T., Tsai, C-F., and Seed, B. H., "FLUSH, A Computer Program for Approximate 3-D Analysis of Soil-Structure Interaction Problems," EERC Report No. 75-30, University of California, Berkeley, 1975.

9. Singh, A. K., "Dynamic Analysis Using Modal Synthesis," Journal of the Power Division, ASCE, Vol. 104, No. PO2, Proc. Paper No. 13652, April 1978, pp. 131-140.

TABLE 1

PREDOMINANT E-W MODAL PERIODS AND MODAL PARTICIPATION FACTORS FOR EQUIVALENT AND DETAILED MODELS

DETAILED MODEL			EQUIVALENT MODEL		
Mode	Period (sec.)	Participation Factor	Mode	Period (sec.)	Participation Factor
2	.200	22.1	1	.200	26.0
4	.190	43.2	2	.190	45.9
6	.170	73.4	3	.160	73.9
15	.073	28.9	4	.073	44.9

TABLE 2

PREDOMINANT N-S MODAL PERIODS AND MODAL PARTICIPATION FACTORS FOR EQUIVALENT AND DETAILED MODELS

DETAILED MODEL			EQUIVALENT MODEL		
Mode	Period (sec.)	Participation Factor	Mode	Period (sec.)	Participation Factor
1	.240	9.7	1	.240	8.9
5	.180	33.1	2	.170	35.8
8	.140	52.1	3	.140	63.0
18	.071	38.8	5	.066	41.1

TABLE 3

COMPARISON OF MAXIMUM DISPLACEMENTS (FT)
FOR A 3-D and A 2-D FIXED BASE STRUCTURE MODEL

	3-D		2-D	
Slab No.	E-W	N-S	E-W	N-S
1	.0010	.0003	.0012	.0004
4	.0060	.0032	.0056	.0034
6	.0084	.0046	.0078	.0063
8	.0457	.0271	.0540	.0250

TABLE 4

COMPARISON OF RELATIVE COMPUTER COSTS

	ONE-STEP ANALYSIS	TWO-STEP ANALYSIS
COMPUTER COST	60	100

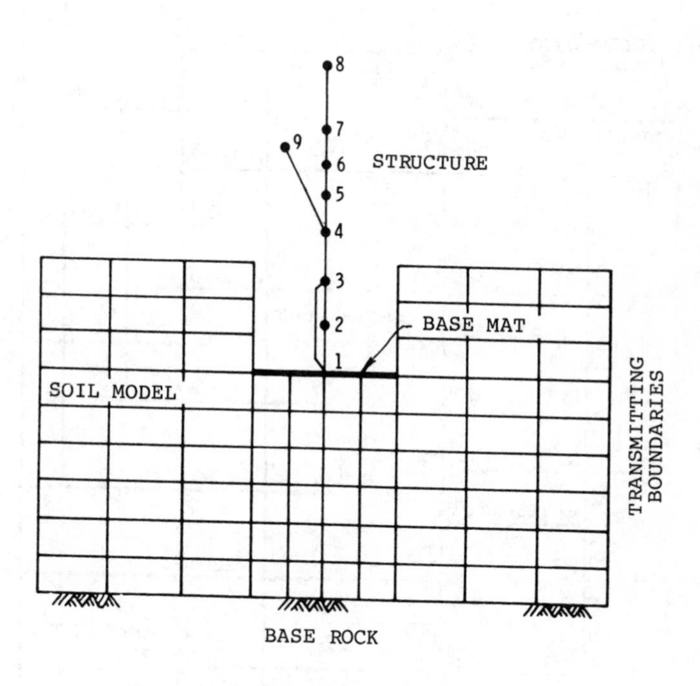

Figure 1 Schematic Detail Of A Soil-Structure Subsystem

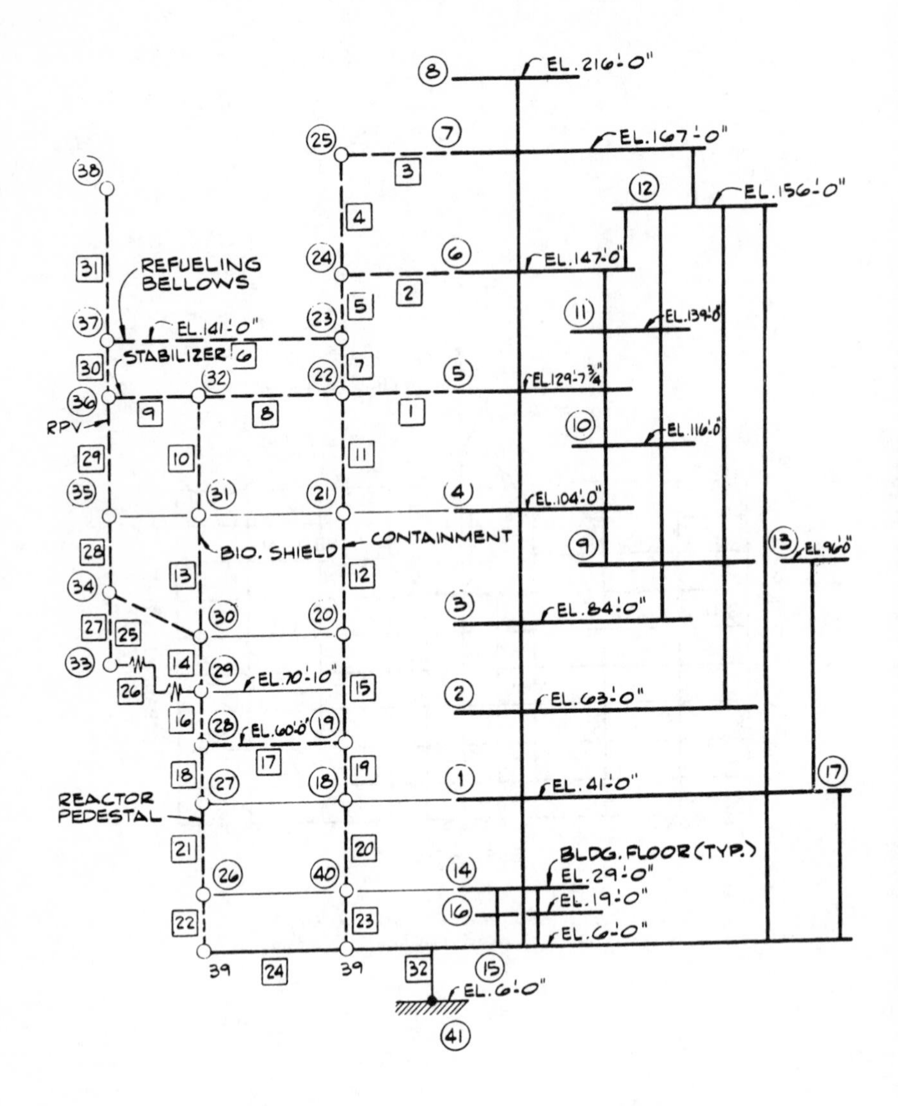

Figure 2 Horizontal Structural Model

Singh

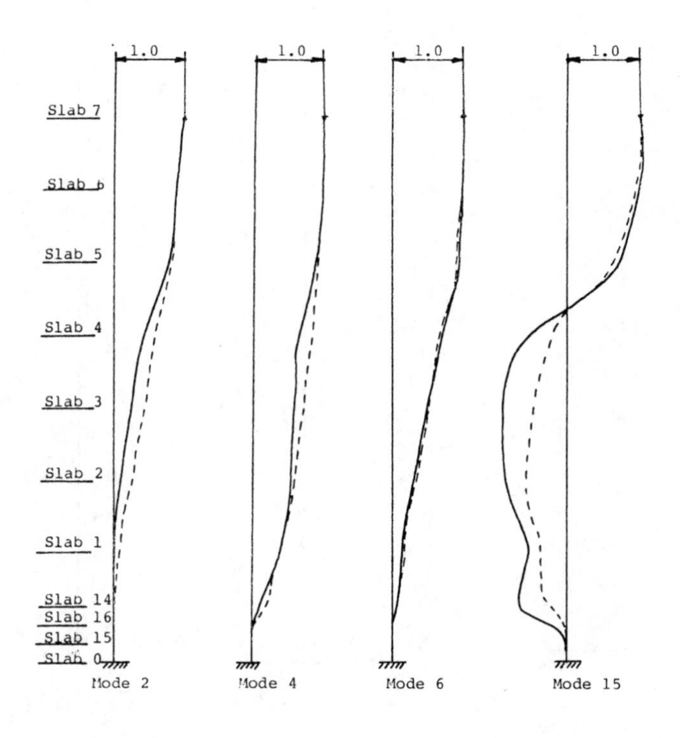

Figure 3 Normalized Mode Shape Comparison for the Detailed and Equivalent model (E-W Model)

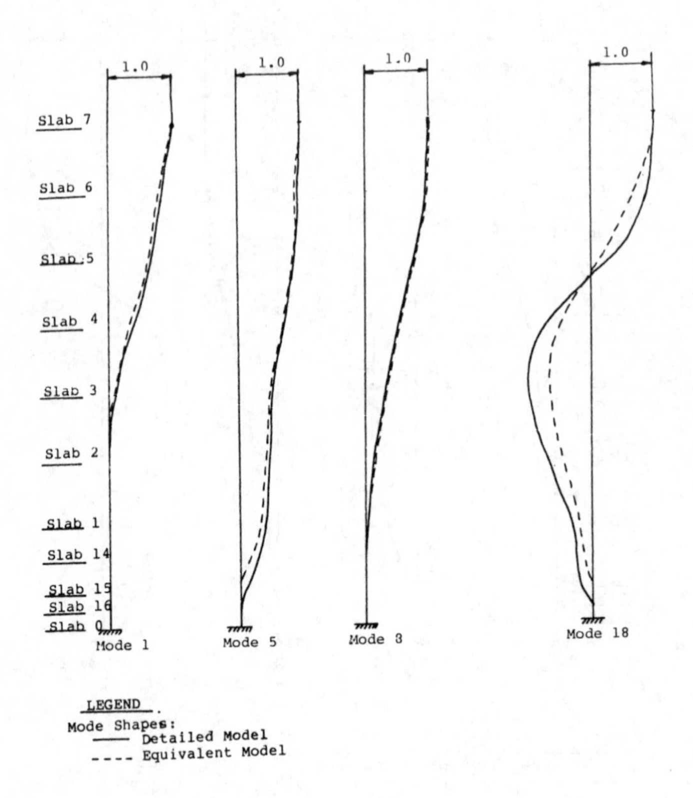

Figure 4 Normalized Mode Shape Comparison for the Detailed and Equivalent Model (N-S Model)

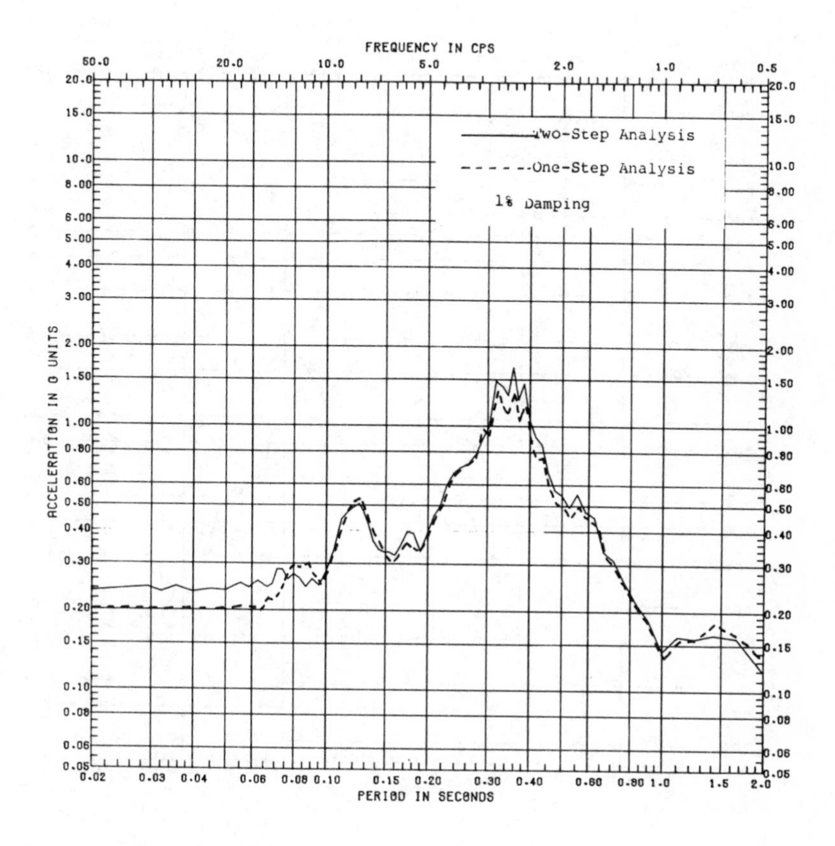

Figure 5 Comparison of Response Spectra (El. 6'-0")

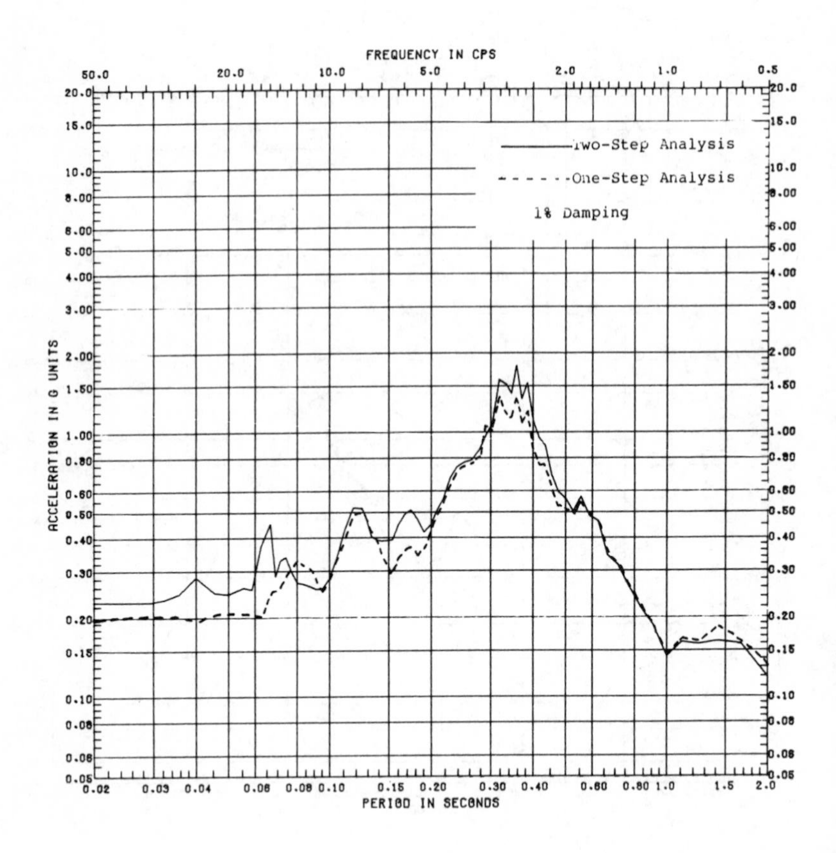

Figure 6 Comparison of Response Spectra (El. 41'-0")

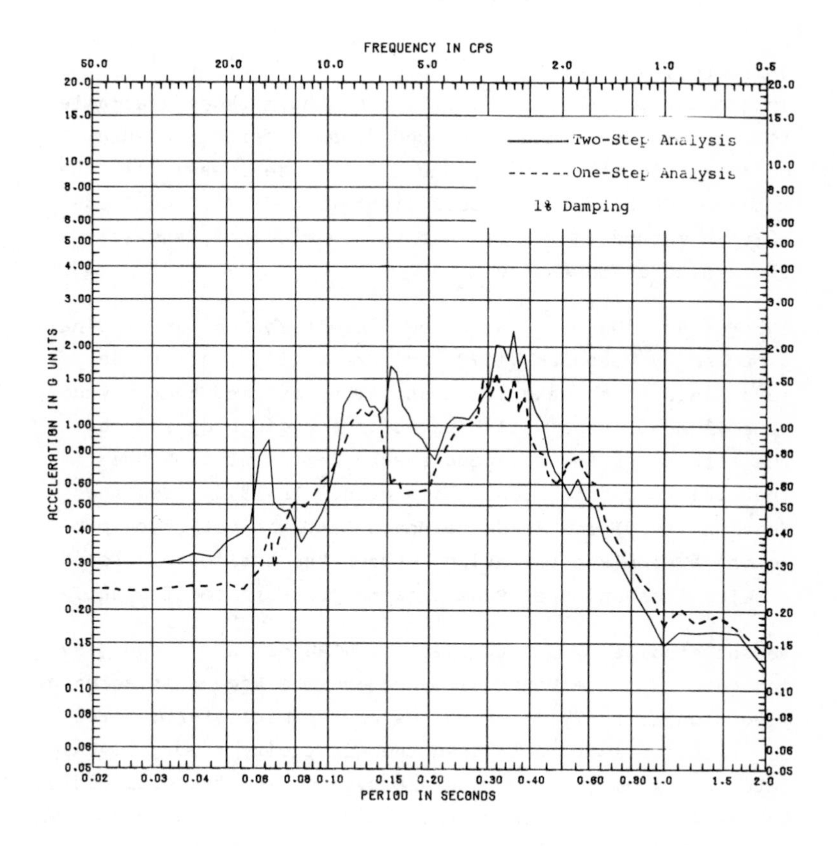

Figure 7 Comparison of Response Spectra (El. 84'-0")

EARTHQUAKE INDUCED LATERAL EARTH PRESSURE

W. Huang,[1] M.ASCE; Y. N. Chen,[2] A.M.ASCE, & S. Singh,[3] M.ASCE

INTRODUCTION

This paper presents a comparison of the methods available to determine earthquake-induced dynamic earth pressure on retaining walls. Four methods are reviewed: (1) the Mononobe-Okabe seismic coefficient method; (2) the Seed-Whitman method; (3) Scott's shear beam method; and (4) the finite element method.

The Mononobe-Okabe seismic coefficient method was originally proposed by Mononobe (1929) and Okabe (1926) after the 1923 Kanto earthquake in Japan. They assumed that a wedge of soil behind the wall behaved as a rigid body and that the effects of the earthquake were represented by the inertial forces throughout its mass. In 1970, Seed and Whitman (1970) modified the Mononobe-Okabe method by proposing a simple correlation between the horizontal ground acceleration and the dynamic earth pressure coefficient.

Recent studies by Scott (1974) and Aggour and Brown (1974) indicate that the Mononobe-Okabe and the Seed-Whitman methods give significantly lower values of dynamic lateral force for rigid walls. Scott used the elasticity theory to calculate the dynamic earth pressure by modeling the soil behind the wall as a one-dimensional shear beam. Aggour and Brown also employed the elasticity theory, but used finite element techniques to solve the problem.

---

1. Supervisor, Structural Analytical Division, Sargent & Lundy, Chicago, Illinois 60603.
2. Senior Engineering Analyst, Structural Analytical Division, Sargent & Lundy, Chicago, Illinois 60603.
3. Assistant Head, Structural Analytical Division, Sargent & Lundy, Chicago, Illinois 60603.

A comparison of the four methods shows that both the shear beam method and the finite element method yield considerably higher dynamic earth pressures than those obtained using the Mononobe-Okabe or the Seed-Whitman method for rigid walls. However, if the flexibility of the retaining wall is considered, the dynamic earth pressure is lower than that obtained by the rigid wall assumption. As demonstrated by the finite element results, which will be presented later, the dynamic earth pressure for a 2-ft thick, 40-ft high wall comes very close to the values calculated by the Mononobe-Okabe or Seed-Whitman method.

METHODS OF ANALYSIS

Mononobe-Okabe Method.- The Mononobe-Okabe method has been used widely by soil engineers to determine the dynamic earth pressure on a retaining wall. It assumes that a soil wedge behind the wall is formed at the point of incipient failure when the wall moves away from the backfill. The effect of earthquake motion is represented by the horizontal and vertical components of the inertial force acting on the soil wedge. The total active force during an earthquake is computed by Coulomb's theory, wherein the inertial forces are applied on the wedge in addition to the body force. The calculated total force is applied at one-third the wall height from the base.

For a wall having a rough and vertical back face and horizontal backfill slope, the coefficients of the total active earth pressure are plotted against different horizontal seismic coefficients in Figure 1.

Seed-Whitman Method.- The value of K_{AE} in Figure 1 represents the total active lateral coefficient. For design purposes, Seed and Whitman (1970) separated this total pressure coefficient into two components: the coefficient of static pressure K_A, and the coefficient of dynamic pressure increment ΔK_{AE}; that is,

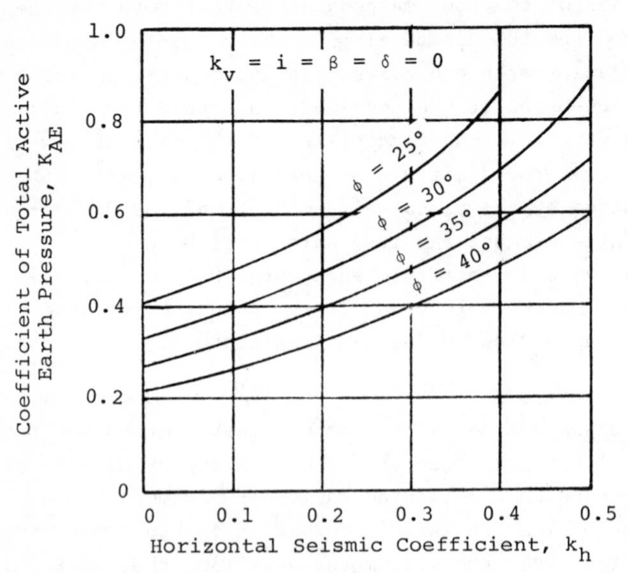

Figure 1 Coefficient of Total Active Earth Pressure By Mononobe-Okabe Method

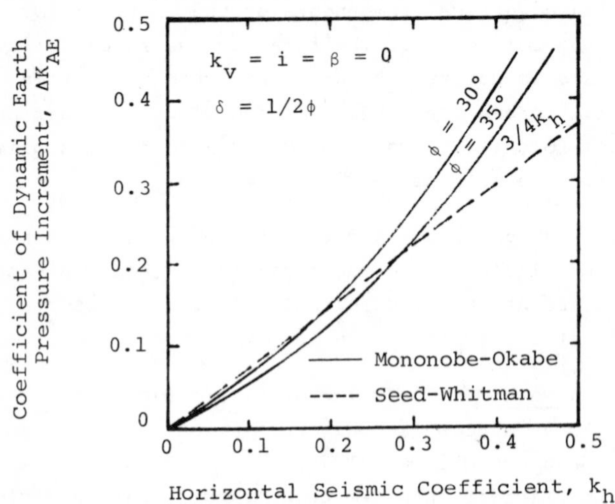

Figure 2 Comparison of Dynamic Earth Pressure Coefficients by Seed-Whitman Method and Mononobe-Okabe Method

$$K_{AE} = K_A + \Delta K_{AE} \qquad (1)$$

For different horizontal seismic coefficients k_h, the corresponding values of ΔK_{AE} calculated by the Mononobe-Okabe method for a wall with rough and vertical back face and horizontal backfill are plotted in Figure 2. For the same type of wall, Seed and Whitman (1970) suggested the following equation for the dynamic lateral force increment:

$$K_{AE} = 3/4k_h \qquad (2)$$

As shown in Figure 2, Eq. 2 correlates well with the Mononobe-Okabe method when the horizontal seismic coefficient is less than 0.3 and the angle of friction of the backfill soil ranges from 30 to 35 degrees. Seed and Whitman also pointed out that the dynamic lateral force increment would act at about 0.6 times the wall height from the base.

Shear Beam Method.- This method was applied by Scott (1974) to analyze the soil-wall interaction during an earthquake. The soil is treated as a one-dimensional shear beam attached to the retaining wall by a Winkler foundation of spring constant k. Soil pressure is developed at a depth x from the top of the wall proportional to the instantaneous lateral displacement of shear beam u at that depth. Two cases were considered in Scott's analysis: shear modulus constant with depth for cohesive backfills and shear modulus increasing parabolically with depth for cohesionless backfills. The wall was assumed to be rigid in the analysis.

Scott used the following equation of motion for the soil shear beam with uniform soil density and spring constant:

$$\frac{\partial^2 u}{\partial t^2} = \frac{1\partial}{\rho\partial x}\left(G\frac{\partial u}{\partial x}\right) - \frac{2ku}{\rho L} \qquad (3)$$

where t is the time; g, the soil shear modulus; ρ, the density of the soil; and L, the length of the shear beam.

For constant shear modulus, Eq. 3 was solved by the separation of variables, with the lateral deflection u as a harmonic function of x and t. Scott found that the first mode motion was principally responsible for the pressure distribution on the wall during an earthquake. Therefore, the dynamic pressure distribution p and the fundamental frequency ω_1 were found to be as follows:

$$p = \frac{4kS_{a1}}{\omega_1^2} \cos\left(\frac{\pi x}{2H}\right) \tag{4}$$

$$\omega_1 = \left(\frac{\pi^2 G}{4H^2 \rho} + \frac{2k}{\rho L}\right)^{1/2} \tag{5}$$

where H is the height of the wall and S_{a1} is the value of the absolute acceleration response at the funamental frequency ω_1 in the design earthquake response spectrum.

For simplicity, Scott correlated the spring constant k with the soil shear modulus G and the Poisson's ratio ν by the expression

$$k = \frac{8G(1-\nu)}{L(1-2\nu)} \tag{6}$$

Eq. 6 is derived by comparing the fundamental frequency obtained from the one-dimensional shear beam model and the two-dimensional plane strain solution without permitting vertical displacments.

By substituting for k in Eq. 4 from Eq. 6, the dynamic earth pressure distribution becomes:

$$p = \frac{32S_{a1}G(1-\nu)}{\pi\omega_1^2 L(1-2\nu)} \cos\left(\frac{\pi x}{2H}\right) \tag{7}$$

The area under the cosine pressure distribution curve gives the total force acting on the wall per unit length of the wall,

$$P = \frac{64S_{a1}HG(1-\nu)}{\pi^2\omega_1^2 L(1-2\nu)} \tag{8}$$

The point of action of this curve is located at $2H/\pi$ or about 0.64H above the base of the wall.

The above solutions are obtained based on the assumption that the soil shear modulus is independent of the shear strain. However, it is well known that the soil shear modulus and damping ratio do change with the induced shear strain during earthquake excitation. If the strain-dependent shear modulus is considered, Eq. 3 becomes nonlinear and a closed-form solution is not readily available. It is suggested that an iteration procedure be used to determine the strain-compatible shear modulus as an improvement to the Scott method. This iteration procedure is the same as the one used in the finite element method, FLUSH (Lysmer et al, 1975).

An initial soil shear modulus G is tried and the fundamental frequency ω_1 is determined from Eqs. 5 and 6. The S_{a1} value is obtained from the response spectrum of input earthquake corresponding to the damping ratio selected from the damping versus shear strain curve. The ω_1 and S_{a1} values are thus used in the following equation to compute the shear strain.

$$\gamma = \frac{4S_{a1}}{\pi H \omega_1^2} \tag{9}$$

Eq. 9 is derived from the shear beam deflection equation of the case of uniform soil shear modulus along the depth. From the shear modulus versus shear strain curve, a new shear modulus G can be obtained corresponding to the newly computed γ. If the new and old shear moduli do not match in value, another cycle of iteration is needed and the procedure continues until the shear moduli converge. This strain-compatible shear modulus is then used in the Scott method to compute the dynamic earth pressure.

For a special case where the soil shear modulus G was assumed to vary with the depth x by the following parabolic equation

$$G = G_o \frac{x}{H} (2 - \frac{x}{H}) \qquad (10)$$

where G_o is the shear modulus at the base, the deflection of the shear beam is a Legendre polynomial function of depth x. It was again found that the first mode was primarily responsible for the lateral pressure distribution. Therefore, the dynamic earth pressure distribution becomes

$$P = \frac{3kS_{a1}}{2\omega_1^2} (1 - \frac{x}{H}) \qquad (11)$$

which is a triangular distribution with the maximum value at the ground surface. The fundamental frequency ω_1 in this case is defined as

$$\omega_1 = (\frac{2G_o}{H^2 \rho} + \frac{2k}{\rho L})^{1/2} \qquad (12)$$

and the total force acting on the wall as the area under the pressure diagram is defined as

$$P = \frac{3kS_{a1}H}{4\,\omega_1^2} \qquad (13)$$

Since the pressure distribution is triangular, the maximum force acts at a height 2H/3 from the base of the wall.

By modifying the spring constant k, the Scott method also considers the case of a rigid wall connected to its base by a torsional spring. For practical purposes, it was suggested that the modification factor be kept around 0.5. In other words, for the type of wall described above, k/2 should be used in stead of k in Eqs. 4 and 5 to determine the dynamic earth pressure increment for the constant shear modulus case, and in Eqs. 12 and 13 for the parabolically increasing shear modulus case.

Finite Element Method.- The finite element analysis was performed using strain-dependent shear modulus and damping values of soil for each element in a finite element representation of the soil-wall system. Because of the strain-dependent shear modulus and damping of soil, an iteration procedure was employed in FLUSH. This iteration procedure is similar to those described in the last paragraph under the Scott method, except that the shear strain is not computed by Eq. 9. The FLUSH output gives the maximum earth pressure acting on the wall. Because the finite element model also includes the retaining wall, the flexibility of the wall can be incorporated into the analysis.

A comparison of the four methods of modeling the soil-wall system is presented in Table 1.

COMPARISON OF RESULTS

Dynamic Earth Force on Rigid Wall.- An example problem consisting of a 40-ft high concrete retaining wall with rigid base was analyzed using the four methods. Both cohesive and cohesionless soils were used in the study as backfill material. For the cohesive soil, the unit weight was assumed to be 127 pcf, the angle of internal friction 28°, and the Poisson's ratio 0.4. The strain-dependent shear modulus and damping ratio for the cohesive soil are shown in Figure 3. For the cohesionless backfill, the unit weight, angle of internal friction, and the Poisson's ratio were 123 pcf, 34°, and 0.4, respectively. The coefficient of dynamic shear modulus K_2 and the damping ratio are shown in Figure 4.

For the earthquake motion, a horizontal acceleration coefficient k_h = 0.3 was used in the Mononobe-Okabe and the Seed-Whitman methods. The earthquake response spectrum used for the Scott method was the wide-band spectrum specified in the Nuclear Regulatory Commission (NRC) Regulatory Guide 1.60, with 0.3 g zero period acceleration. The earthquake motion used in the finite element analysis was a ten-second synthetic time history consistent with the NRC Regulatory Guide 1.60 response spectrum of 0.3 g. In the analysis by Scott and by the finite element method, the earthquake

Table 1 Comparison of Methods of Dynamic Earth Pressure Analysis

ITEM	MONONOBE-OKABE	SEED-WHITMAN	SCOTT	FINITE ELEMENT
Flexibility of Wall	Rigid	Rigid	Rigid wall with rigid base; Rigid wall with hinged base	Variable
Required Soil Properties	Unit weight, angle of internal friction	Unit weight	Unit weight, Poisson's ratio, strain-dependent shear modulus, and damping	Unit weight, Poisson's ratio, strain-dependent shear modulus, and damping
Shape of Soil Behind Wall	Wedge	Wedge	Horizontal layer	Any shape
Earthquake Motion	Seismic coefficient	Seismic coefficient	Response spectrum	Time history

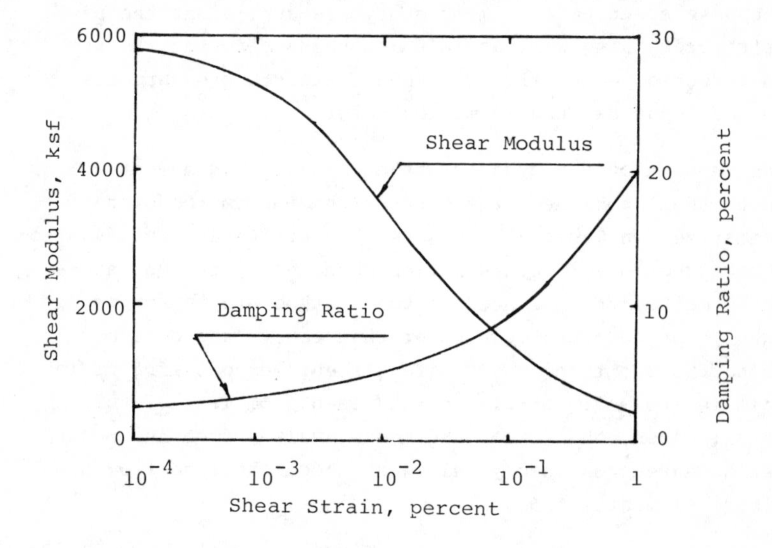

Figure 3 Shear Modulus and Damping Ratio for Cohesive Backfill

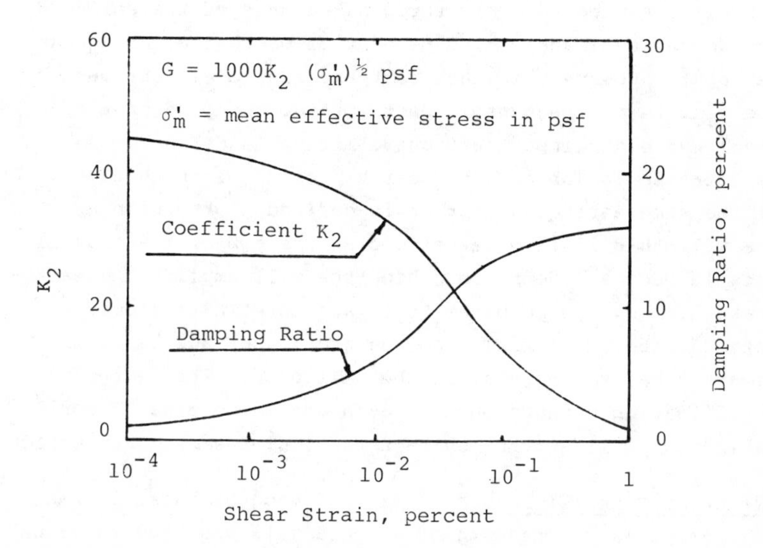

Figure 4 Shear Modulus and Damping Ratio for Cohesionless Backfill

response spectrum or time history was applied at the base of the retaining wall and also at the surface of the backfill to investigate the effect on dynamic earth pressure due to the input earthquake motion level.

The results of the dynamic earth pressure and moment acting on the wall obtained from each of the four methods are summarized in Table 2. As shown in this table, the dynamic forces due to earthquake motion obtained by the Mononobe-Okabe method and the Seed-Whitman method are approximately equal, but both are much lower than the values obtained using the Scott and the finite element methods if a rigid wall is assumed. It also can be seen from Table 2 that in all cases, the forces and moment values from the Scott method agree reasonably well with those obtained from the finite element method.

Effect of Level of Earthquake Motion Definition.- For the Mononobe-Okabe and Seed-Whitman methods, the earthquake force is assumed to apply through the mass of the soil wedge behind the wall and is independent of where the earthquake is input. However, in the Scott and finite element methods, the results of the dynamic earth pressure depend on the level where the input earthquake motion is defined. As illustrated in Table 2, the earth pressure is much higher if the same earthquake motion is defined at the base of the wall than if it were defined at the top of the backfill. This is because the soils behind the wall amplify the earthquake motion. In both the Scott and the finite element methods, the acceleration at the ground surface level is always greater than that at the wall base. The degree of amplification depends on the depth and properties of soil behind the wall and the characteristics of earthquake motion.

Effect of Backfill Length.- The influence of backfill length on dynamic earth pressure of a rigid wall was studied using both the Scott method and the finite element method. The same example problem of cohesive backfill was used, except

Table 2 Summary of Dynamic Force and Moment on Retaining Wall of 40-ft Height

Type of Backfill		Mononobe-Okabe	Seed-Whitman	Scott		Finite Element		
				Earthquake at Base	Earthquake at Top	Earthquake at Base		Earthquake at top
						Rigid Wall	2-ft Thick	
Cohesive (G constant with depth)	F	26.6	22.7	94.6	72.4	92.3	26.5	62.0
	M	354.9	548.6	2408.0	1843.6	2159.4	297.5	1430.3
	$\overline{y}$	0.33H	0.60H	0.65H	0.64H	0.59H	0.28H	0.58H
Cohesionless (G varies with effective overburden pressure)	F	22.4	22.1	–	–	63.9	–	38.3
	M	299.1	531.3	–	–	1296.9	–	863.5
	$\overline{y}$	0.33H	0.60H	–	–	0.51H	–	0.56H
Cohesionless (G varies parabolically with depth)	F	22.4	22.1	54.0	27.5	61.4	–	33.0
	M	299.1	531.3	1080.0	733.3	1188.4	–	642.1
	$\overline{y}$	0.33H	0.60H	0.67H	0.67H	0.48H	–	0.49H

Notes: 1. F = Force (kips); M = Base Moment (kip-ft); $\overline{y}$ = Moment arm about base.

2. L/H = 10 used in Scott method; L/H = 3 and transmitting boundary used in finite element method; L = length of backfill; H = wall height (40 ft).

that the length of the backfill was allowed to vary. The results are shown in Figure 5.

According to the Scott method, the backfill length causing maximum dynamic earth pressure on a rigid wall depends on the Poisson's ratio of the soil. When the shear modulus is constant with depth, the value of this critical backfill length is

$$L = \frac{8H}{\pi} \left(\frac{(1-\nu)}{(1-2\nu)}\right)^{1/2} \tag{14}$$

For Poisson's ratio $\nu = 0.4$, the maximum dynamic earth pressure occurs when the backfill length is about 4.41 times the height of the wall. This value agrees well with that determined by the finite element method, as shown in Figure 5b.

Effect of Wall Flexibility.- Wall thicknesses of two, four, and ten feet were used in the finite element analysis in addition to the infinitely rigid wall case. For all cases, the height of the wall was kept at 40 ft. The results shown in Figure 6 indicate that the dynamic earth pressure decreases as the wall becomes more flexible, especially the earth pressure near the upper part of the wall. The values of dynamic force and moment from the finite element analysis of a 2-ft thick wall with cohesive backfill are also presented in Table 2. The force and moment are greatly reduced from those based on the rigid wall assumption and agree well with those obtained from the Mononbe-Okabe or Seed-Whitman method.

Figure 5 also compares the two cases of the Scott solution for a wall completely fixed at the base and one partially fixed at the base, i.e., the one which allows the wall to tilt with a torsional spring at the base. It is seen that the dynamic force obtained in the former case is about 40% greater than that obtained in the latter case.

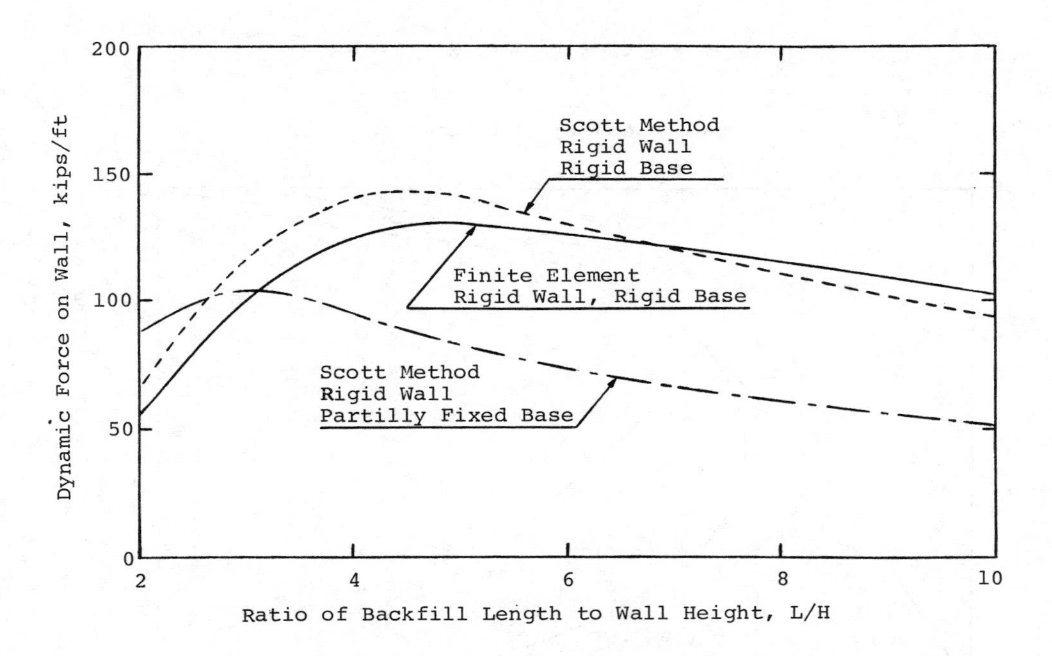

Figure 5. Effect of Backfill Length

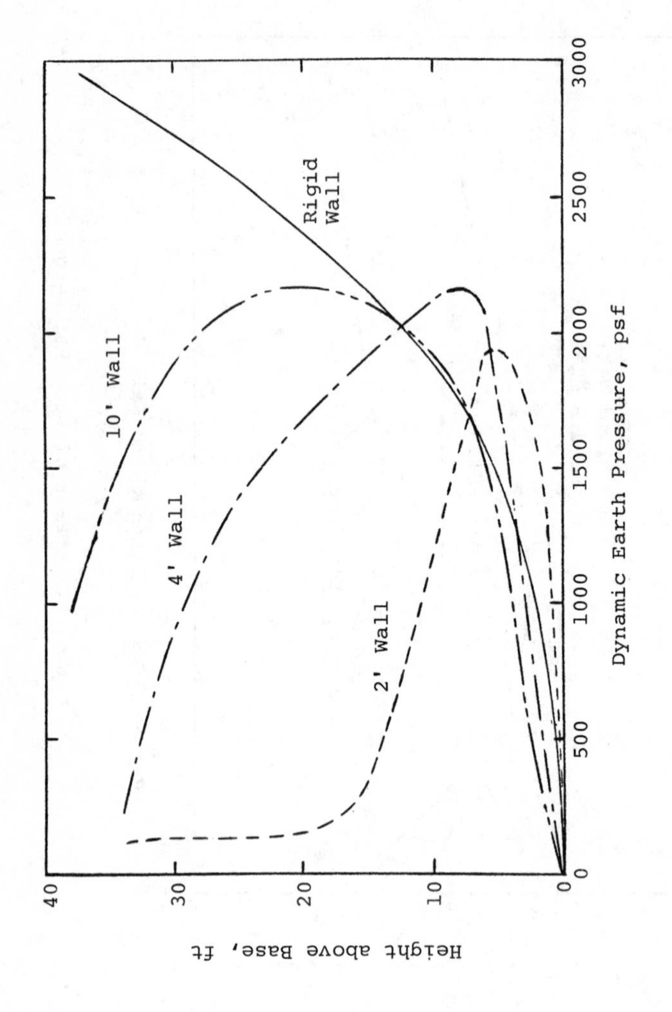

Figure 6. Effect of Wall Flexibility

CONCLUSIONS

Based on the comparison of the four methods, the following conclusions can be drawn:

1. The maximum dynamic earth pressure on a rigid wall with rigid base obtained from the Scott method or the finite element method is greater than that computed by the Mononobe-Okabe method or the Seed-Whitman method. The results from the Scott method, an analytical method based on the wave propagation theory, agree well with those from the finite element solution.

2. The finite element method indicates a substantial reduction in total force and moment if the flexibility of the wall is included in the analysis. For the given example the forces obtained from the finite element method for a 2-ft thick wall compared well with the Mononobe-Okabe method and the Seed-Whitman method.

3. The dynamic earth pressure obtained from the Scott method or the finite element method depends on the definition of earthquake input level. The earth pressure is much larger for the same earthquake motion if it is defined to input at the base of the wall rather than at the top of the backfill.

4. The Scott method and the finite element method both indicate that for a rigid wall with rigid base, the dynamic earth pressure due to an earthquake depends on the length of backfill modeled in the analysis. From the Scott method, the critical backfill length, which is defined as the length of backfill causing the maximum dynamic earth pressure on the wall, is about 4.41 times the height of the wall for a Poisson's ratio equal to 0.4. The finite element approach results verify this critical backfill length.

REFERENCES

Aggour, M. S. and Brown, C. B., "Retaining Walls in Seismic Areas," Proceedings, 5th World Conference on Earthquake Engineering, Rome, 1974, Vol. 2, pp. 2624-2627.

Lysmer, J., Udaka, T., Tsai, C. F., and Seed, H. B., "FLUSH, A Computer Program for Approximate 3-D Analysis of Structure-Soil-Structure Interaction," EERC Report No. 75-30, Earthquake Engineering Research Center, University of California, Berkeley, November 1975.

Mononobe, N., "Earthquake-Proof Construction of Masonry Dams," Proceedings, World Engineering Conference, 1929, Vol. 9, p. 275.

Okabe, S., "General Theory of Earth Pressure," Journal of the Japanese Society of Civil Engineers, Vol. 12, No. 1, 1926.

Scott, R. F., "Earthquake-Induced Earth Pressures on Retaining Walls," Proceedings, 5th World Conference on Earthquake Engineering, Rome, 1974, Vol. 2, pp. 1611-1620.

Seed, H. B. and Whitman, R. V., "Design of Earth Retaining Sructures for Dynamic Loads," Proceedings, Conference on Lateral Stresses and Earth-Retaining Structures, ASCE, 1970, pp. 103-147.

THE SIGNIFICANCE OF SITE RESPONSE IN SOIL-STRUCTURE INTERACTION ANALYSES FOR NUCLEAR FACILITIES

by

H. Bolton Seed[1] and John Lysmer[2]

GENERAL CONSIDERATIONS

The basic problem of soil-structure interaction is illustrated in Fig. 1. It involves the determination of the motions of one or more structures at a given site from a knowledge of a given motion (the control motion) at a specified point (the control point) of the site prior to construction (the free field).

A complete soil-structure interaction analysis for any structure must necessarily consist of two distinct parts; a site response analysis and an interaction analysis. Unless the nature of the seismic wave field into which the structure is being placed is known with a reasonable degree of accuracy, there is no way in which the resulting interaction of the structure with the soil deposit and the wave field can be determined.

The site response analysis involves the determination of the temporal and spatial variations of the free-field motions. The interaction analysis involves the determination of the motions of a structure placed in the above seismic environment. These are different types of problems but each needs to be addressed to determine a solution to the soil-structure interaction problem.

Each of the above problem types can in principle be formulated in terms of continuum models or discretized models, and it is not possible here to describe all of the possible forms of equations of motion which have been proposed. It is, however, useful for a better understanding of the nature of and the connection between the two problem types to consider the equations of motion for the three linear models shown in Fig. 2. The models are identical in the sense that all are of the

[1,2]Professor of Civil Engineering, University of California, Berkeley, CA.

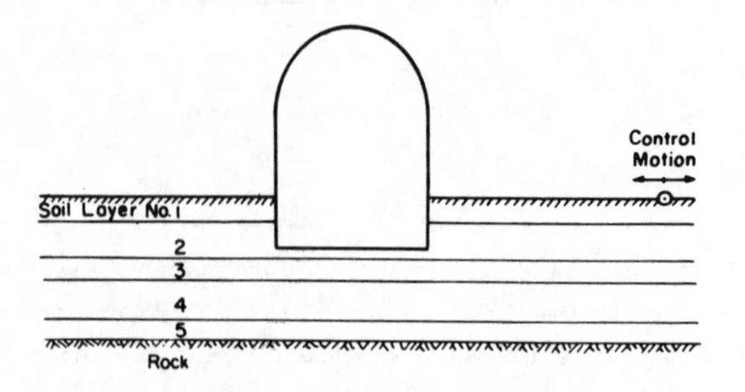

Fig. 1 SOIL STRUCTURE INTERACTION PROBLEM

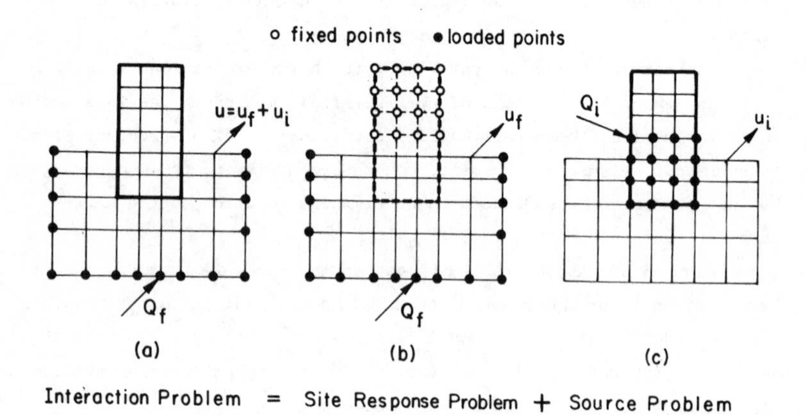

Fig. 2 SUPERPOSITION THEOREM FOR INTERACTION PROBLEMS

finite element type and all are spanned by the same finite element mesh. Also, all masses and stiffnesses are the same, except that the structural part of the model shown in Fig. 2(b) has no stiffness and mass, and that for this model the structural nodes above ground level are assumed to be fixed in space (actually these points can be given any specified motion without loss of generality).

Since the fixed nodes have no influence on the motion of the ground, Fig. 2(b) represents a free-field site response problem. It has the equation of motion

$$[M_f]\{\ddot{u}_f\} + [C_f]\{\dot{u}_f\} + [K_f]\{u_f\} = \{Q_f\} \tag{1}$$

where $[M_f]$, $[C_f]$, $[K_f]$ are the mass, damping, and stiffness matrices, respectively, for the free field, and $\{u_f\}$ is a vector containing the nodal point displacements. Since the source of excitation is outside the model the load vector $\{Q_f\}$ has non-zero elements on the external boundary only. Solutions to the equation of motion, Eq. (1), can be obtained by standard methods, see Desai and Christian (1976). It will here be assumed that a free-field solution is available. Thus $\{u_f\}$ and $\{Q_f\}$ are known.

Figure 2(a) represents the corresponding interaction problem. The total displacements can be written

$$\{u\} = \{u_f\} + \{u_i\} \tag{2}$$

where $\{u_f\}$ are the known free-field displacements and $\{u_i\}$ are the interaction displacements. Assuming that the external boundary is very far away from the structure the equation of motion for the interaction problem is

$$[M]\{\ddot{u}\} + [C]\{\dot{u}\} + [K]\{u\} = \{Q_f\} \tag{3}$$

where $\{Q_f\}$ is the same load vector as in Eq. (1) and $[M]$, $[C]$, and $[K]$ are the total mass, damping, and stiffness matrices, respectively. Substitution of Eqs. (1) and (2) into Eq. (3) yields

$$[M]\{\ddot{u}_i\} + [C]\{\dot{u}_i\} + [K]\{u_i\} = \{Q_i\} \tag{4}$$

where

$$\{Q_i\} = ([M_f] - [M])\{\ddot{u}_f\} + ([C_f] - [C])\{\dot{u}_f\} + ([K_f] - [K])\{u_f\} \tag{5}$$

The load vector, $\{Q_i\}$, in Eq. (5), can be computed from the known free-field displacements. It depends only on the difference in properties between the structure and the excavated soil. Thus Q_i has non-zero elements only at the structure and Eq. (4) is the equation of motion for the source problem illustrated by Fig. 2(c). This problem is well-posed and can be solved for the interaction displacements, $\{u_i\}$. The total displacements for the soil-structure interaction problem can be found by superposition as indicated by Eq. (2).

Equations (2) and (4) remain valid even as the distance to the boundary goes to infinity and the mesh size shrinks to infinitesimal dimensions. Hence, the above formulation can be extended to continuum mechanics and three dimensions.

The above formulations reveal three important characteristics of the soil-structure interaction phenomenon:

1. The only free field ground motions which are of importance for the interaction phenomena are those within the volume to be excavated for the embedded part of the structure.
2. For an embedded structure the amount of interaction depends on the difference in mass and stiffness between the structure and the volume of excavated soil, see Eqs. (4) and (5).
3. Soil-structure interaction analysis implies in many cases the use of superposition, see Eq. (2). Thus true nonlinear analyses may, for many types of motion specification, not be possible.

The first observation has far reaching consequences; especially for embedded structures on relatively soft sites since, for such sites, both theory and observation indicate that the free-field motions vary significantly with depth. This implies that the site response analysis is an important, and in the opinion of the writers perhaps the most important, part of a soil-structure interaction analysis. The significance of site response with regard to the design of nuclear facilities is discussed in the following pages.

THE SITE RESPONSE PROBLEM

Site response problems involve the determination of the temporal and spatial variation of motions within a site. In principle, these motions can be determined from a large model which includes the source of the earthquake. However, in practice the source parameters and the regional geology cannot be determined in sufficient detail to solve this problem with a high degree of accuracy in the frequency range of interest for design. Thus, current methods of site response analysis normally attempt to predict the above variation of motions from a single specified control motion at some control point within the site. This problem is mathematically ill-posed and unique solutions can be obtained only by the introduction of restrictive assumptions regarding the geometry of the site and the nature of the wave field causing the control motion. In practice, consistent solutions can be obtained only for horizontally layered sites. Possible wave patterns include: vertically propagating or inclined plane body waves, and horizontally propagating surface waves. Only the case of vertically propagating waves can currently be solved by truly nonlinear methods.

CONTROL MOTION AND CONTROL POINT

The inherent problem in site response analysis is the choice of wave field to be used in the analysis and it is therefore natural to classify and discuss the different available methods according to the type of wave field assumed. However, before doing so it should be mentioned that the choice of an appropriate control motion and control point is just as, and in many cases, more important than the choice of wave field.

The control motion should be chosen with due respect to observed relations between earthquake magnitude, epicentral distance, maximum acceleration, duration, frequency content, see Idriss (1978) and Ref. 1, and the site-dependent characteristics established by Hayashi et al. (1971), Seed et al. (1976a, 1976b), Faccioli (1978), and Ref. 1.

Except for the obvious case in which the control motion is an observed record at the control point, the preferable control point is a point either at the ground surface or, as discussed below, at an assumed rock outcrop at the depth of which the motion is specified. This is so because most of our data base of strong motion earthquake records from which the control motion has to be estimated was obtained at surface stations and, even more important, because the frequency content of motions at points below the ground surface is strongly influenced by reflections at the free surface. Thus the specification that the control motion at depth should be a broad-band spectrum or a motion recorded at another site or depth may result in completely unrealistic computed motions for the site.

At the present time, the control motion for the design of nuclear power plants is usually specified as a broad-band spectrum at the ground surface in the free field as shown in Fig. 1.

With the control motion, the control point and the site properties fixed, the solution to the site response problem depends entirely on the nature of the wave field producing the ground surface motions. This wave field may consist of many components including:

(1) some Rayleigh waves

(2) some Love waves

(3) some plane vertically propagating waves

(4) some plane body waves inclined at an angle to the vertical

and (5) some other wave types such as spherical and cylindrical waves which are usually not considered.

At the present time seismologists cannot advise engineers in sufficient detail on the relative contents of the different possible wave forms which make up the surface motion. Thus, even though soil-structure interaction analyses can be performed for an arbitrary wave field, in practice, such analyses cannot be made due to lack of data on the characteristics of the wave field involved.

Under these conditions, a typical engineering approach is to make analyses for extreme cases of possible wave fields, i.e., for a motion represented by all Rayleigh waves or for a motion represented entirely by a system of body waves, and to determine the influence of the motion specification on the results of the analysis. If the differences are small, then precise specification of the components of the wave field is considered unnecessary. If the differences are large, then increasing efforts must be made to determine even a crude assessment of the relative components of different wave types, or alternatively, conservative choices of wave components may have to be made for different parts of the analysis. It is important therefore to examine the characteristics of the different wave forms which might contribute to the surface control motion.

HORIZONTALLY PROPAGATING WAVES

For horizontally layered sites it is relatively easy to set up linear methods of analysis for horizontally propagating waves. However, many possible choices of wave patterns exist (inclined body waves at different angles of incidence, different modes of surface waves, etc.) and it is currently impossible to determine from available seismological data the exact contributions of each wave type to earthquake motions near the surface in the frequency range of interest to earthquake engineers. However, some estimates have been made (Trifunac and Brune, 1970; Randall, 1971; Chandra, 1972; Nair and Emery, 1975; Liang and Duke, 1977; and Toki, 1977). These estimates involve considerable uncertainties however; in view of these uncertainties analyses of site response for motions represented only by horizontal propagating waves are mainly of scientific interest and represent an extreme bound of the probable motions. Nevertheless, as will be discussed below, some practical conclusions can be drawn from such analyses.

The free-field motions caused by horizontally propagating waves will be discussed in three parts: Surface waves are discussed immediately below, inclined body waves are discussed after the section on vertically propagating waves and, finally, the three wave types are discussed together in a section on motions at shallow depths.

Surface Waves

Rayleigh waves in a perfect elastic half-space are well-known and the theory for these are given in standard textbooks, e.g., Richart et al., (1970). However, for obvious reasons soil dynamics analysts are much more interested in surface waves in multi-layered systems, Thomson (1950), Haskell (1953), Ewing et al. (1957). Two types of waves may occur in such systems: Love waves, in which the motions are horizontal and perpendicular to the direction, x, of wave propagation, and Rayleigh waves which involve both vertical and horizontal motions in the vertical xz-plane. For plane harmonic waves the displacement fields are of the form:

Love waves: $$u_y = \sum_{s=1}^{\infty} L_s \cdot h_s(z) \cdot e^{i(\omega t - c_s x)} \qquad (6)$$

$$\text{Rayleigh waves: } \left\{ \begin{array}{l} u_x = \sum_{s=1}^{\infty} R_s \cdot f_s(z) \cdot e^{i(\omega t - k_s x)} \\ \\ u_z = \sum_{s=1}^{\infty} R_s \cdot g_s(z) \cdot e^{i(\omega t - k_s x)} \end{array} \right\} \qquad (7)$$

where ω and t are the frequency and time, respectively, and L_s and R_s are unknown mode participation factors. The infinite sets of wave numbers, c_s and k_s, and mode shapes, $h_s(z)$, $f_s(z)$, and $g_s(z)$, may in principle be determined by methods developed by Thompson (1950) and Haskell (1953). The wave numbers are directly related to the phase velocities of the different wave modes through $V_L = \omega/c$ and $V_R = \omega/k$. Thus the fundamental problem of site response analysis with surface waves is to determine the infinite set of factors L_s and R_s from a single given amplitude at the control point. This is clearly an ill-posed problem and solutions can only be obtained by further assumptions, the most common of which is to assume that only the fundamental Rayleigh or Love mode, corresponding to s = 1, exists. For undamped systems the frequency-dependent phase velocity and mode shape can be found by the Thomson-Haskell method and the amplitude L_1 or R_1 may be determined from the control motion.

Continuum analyses are possible for the case of viscoelastic layers over an undamped half-space (Ewing et al. (1957), Boncheva (1977)). However, for this case it is more practical to first discretize the semi-finite system by the use of finite elements as proposed by Lysmer (1970) and Waas (1972) for Rayleigh waves and Waas (1972), Lysmer and Waas (1972) for Love waves. Only Rayleigh waves will be discussed here. The theoretical model is shown in Fig. 3. It involves the assumption of a linear variation of displacements between layer interfaces and the existence of a stationary rigid base at some finite depth. If this depth is chosen to be considerably larger than the wave length of the Rayleigh waves of interest, a half-space is simulated by this model.

For an N-layer system these assumptions reduce the equation of motion for the layered system to a quadratic eigenvalue problem:

$$([A]k^2 + [B]k + [C] - \omega^2[M])\{v\} = \{0\} \qquad (8)$$

where [A], [B], [C] and [M] are simple 2N x 2N matrices which can be formed from the stiffnesses, damping ratios and mass densities of the

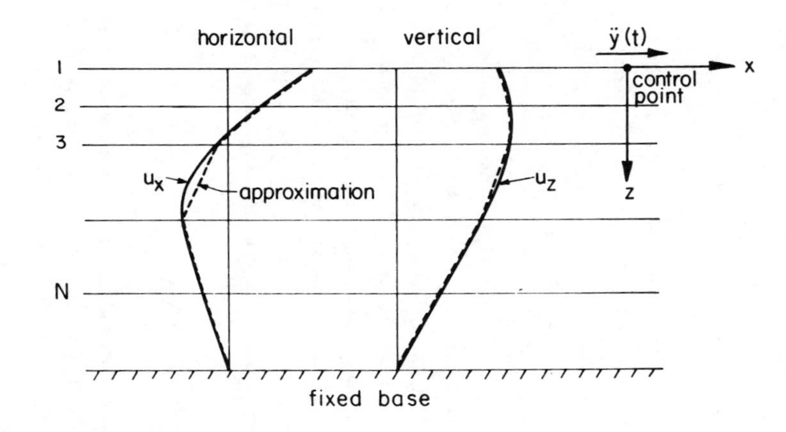

Fig. 3 TYPICAL SOIL PROFILE AND RAYLEIGH WAVE MODE SHAPE

layered system, and $\{v\}$ is an eigenvector (mode shape) which contains the 2N displacement amplitudes of the layer interfaces. The mode shape represents the functions $f_s(z)$ and $g_s(z)$ in Eq. (7). For a given frequency, ω, Eq. (8) can be solved by methods developed by Waas (1972). The solution consists of 2N possible wave numbers, k_s, and associated mode shapes $\{v\}_s$ and, in analogy with Eq. (7), the general solution to the equation of motion may be expressed in the form:

$$\{u\} = \sum_{s=1}^{2N} R_s \{v\}_s \cdot e^{i(\omega t - k_s x)} \tag{9}$$

For a damped system all the wave numbers will be complex with negative imaginary parts. Hence, Eq. (9) can also be written

$$\{u\} = \sum_{s=1}^{2N} e^{im(k_s)} \cdot R_s \{v\}_s \cdot e^{i(\omega t - Re(k_s)x)} \tag{10}$$

which represents a system of generalized Rayleigh waves (modes) which propagate in the positive x-direction, each with its own mode shape, $\{v\}_s$, phase velocity, $\omega/Re(k_s)$, and decay factor, $\exp\,[2\pi Im(k_s)/Re(k_s)]$, per wave length, $\lambda = 2\pi/Re(k_s)$. Experience with the method has shown that most of the Rayleigh modes decay extremely rapidly in the frequency range of interest to earthquake engineers and only a few terms of Eq. (9) need therefore be considered. If it is assumed that only the fundamental mode (defined as the mode with the largest value of $Re(k_s)$) is present, Eq. (9) reduces to

$$\{u\} = R_1 \cdot \{v\}_1 e^{i(\omega t - k_1 x)} \tag{11}$$

and the mode participation factor, R_1, can be determined at each frequency from the amplitude of the control motion at say the surface at $x = 0$. This method has been used by Chen and Lysmer (1979) to determine possible Rayleigh motion fields for several sites.

High-order Surface Waves

The fundamental Rayleigh mode defined above is, as shown by Lysmer (1970), identical to the fundamental mode considered by seismologists in layered systems overlying a deformable half-space. The rest of the terms of Eq. (9) represent higher-order Rayleigh modes (in the terminology of seismologists) and body waves. These modes will have longer wavelengths and will propagate faster than the fundamental mode which by definition has the shortest wavelength and thus the lowest phase velocity. While most of these higher modes can be neglected, since they decay rapidly in the direction of wave propagation, others may decay less rapidly than the fundamental mode. This phenomenon occurs only at relatively high frequencies on sites with a marked increase in stiffness with depth; say a sand profile over rock. These low-decay modes could conceivably contribute significantly to the motion at a surface control point. However, studies by Chen and Lysmer (1979) have shown that such modes, when they occur, are associated with energy propagation in deeper high-velocity layers and that they cause near surface motions which are similar to those caused by vertical or slightly inclined body waves. That this is so is not surprising when one considers the propagation mechanism of these modes. The very facts that the waves travel at high velocities and decay slowly indicate that the major part of the energy propagation occurs in deeper layers with high body wave velocities and low damping. This immediately implies that insignificant amounts of energy are propagated horizontally in the softer surface layers or, in other words, that the higher frequency motions in the surface layers are maintained through nearly vertical energy propagation through a mechanism similar to that of slightly inclined body waves. The result is that the upper parts of the mode shapes, i.e. the variation of displacements with depth, are virtually identical to those found in analyzing vertically propagating or slightly inclined body waves.

Thus, in practical calculations the effects of higher-order surface wave components can be considered by assuming a certain content of slightly inclined or vertical body waves in the control motion. In view of this observation only Rayleigh wave fields consisting of fundamental modes will be considered in the following sections.

Effect of Layering

The importance of using layered system theory, rather than the simpler half-space theory, in dealing with structures on a layer of soil overlying rock is illustrated by the computed variations of accelerations and shear stresses with depth for the cases of (1) a homogeneous half-space and (2) a 128 ft layer of sand overlying a half-space shown in Figs. 4 and 5, respectively.

Figure 4 shows the computed results for a uniform half-space with properties similar to those of an average soil deposit. It may be seen that analyses for a typical control motion with a peak horizontal acceleration of 0.25g at the ground surface in such a half-space indicate that: (a) if the motion is assumed to consist entirely of Rayleigh waves, the vertical component of acceleration at any depth will be 1.5 to 4 times greater than the horizontal component of acceleration at the same depth; (b) if the motion is assumed to result solely from vertically propagating shear waves, the horizontal accelerations at any depth will be greater than those computed at the same depth for the Rayleigh wave assumption; and (c) the values of horizontal shear stress computed for a control motion represented entirely by Rayleigh waves will be about 50% greater than those computed on the assumption that the control motion is produced by vertically propagating shear waves.

A totally different picture is obtained if similar analyses are made for a 128 ft layer of sand overlying a rock formation, for which computed response values are shown in Fig. 5. In this analysis the stiffness and damping of the sand was allowed to vary with depth and with the cyclic strain level developed due to the motion. For this more realistic representation of the case of a soil deposit overlying rock it may be seen that (a) if the control motion is represented by a system of Rayleigh waves, the vertical components of motion are about two thirds of the horizontal components at any depth--a result more consistent with the observed ratio of horizontal to vertical accelerations in a large number of earthquakes; (b) down to depth of about 60 ft, there is no great difference in the values of peak horizontal accelerations whether the computations are based either on the assumption that the control motion results only from Rayleigh waves or only from vertically propagating shear waves; and (c) the values of maximum horizontal shear stress computed for a control motion represented entirely by Rayleigh waves are typically about 30% less than those computed on the assumption that the control motion is produced only by vertically propagating shear waves.

It is clear from the above examples, that for soil deposits, it is more conservative to compute maximum horizontal shear stresses based on the assumption that the motions result from vertically propagating waves

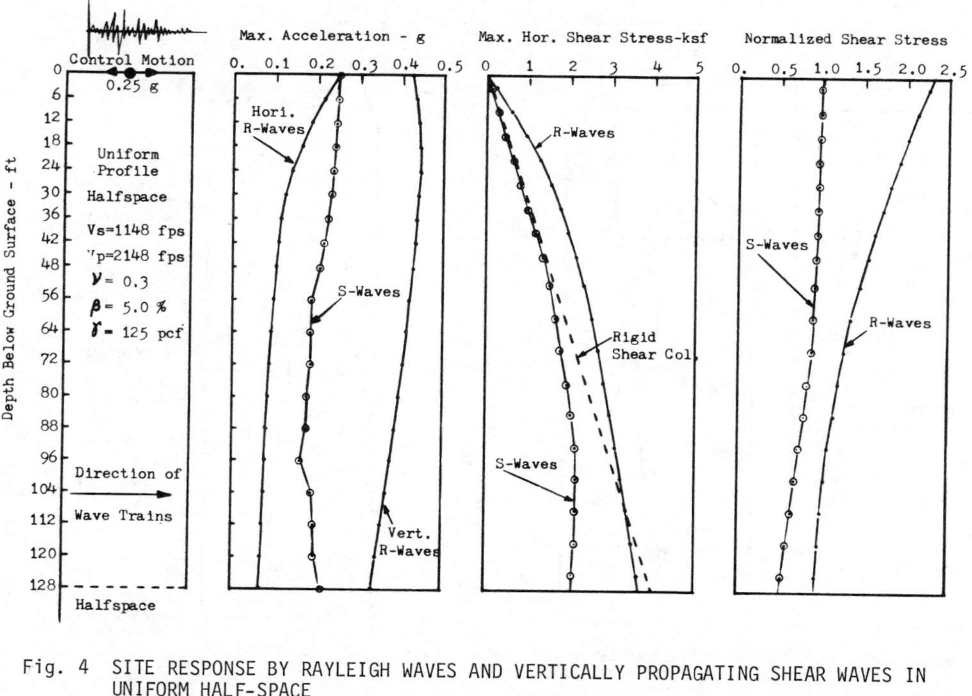

Fig. 4 SITE RESPONSE BY RAYLEIGH WAVES AND VERTICALLY PROPAGATING SHEAR WAVES IN UNIFORM HALF-SPACE

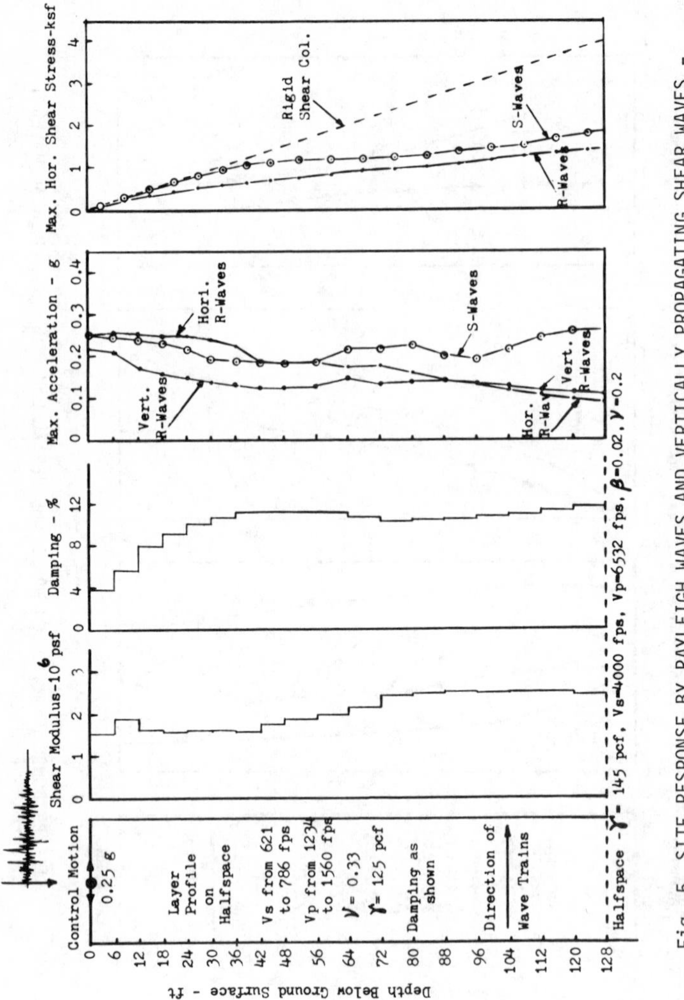

Fig. 5 SITE RESPONSE BY RAYLEIGH WAVES AND VERTICALLY PROPAGATING SHEAR WAVES - 128 FT LAYER OF SAND OVER UNIFORM HALF-SPACE

than it is to assume a field consisting of Rayleigh waves--yet the opposite result would be obtained if the layering of the system were not considered. Similarly, erroneous conclusions concerning vertical motions could result from the use of a homogeneous half-space to represent a layered soil deposit. Finally, it may be seen that if the system layering is correctly considered, horizontal accelerations will vary with depth but they will not differ significantly whether the motions are represented entirely by Rayleigh waves or entirely by vertically propagating waves. This topic will be discussed at length in a later section of this paper.

Effect of Distance of Propagation

It is important to note that Figs. 4 and 5 show only the variation of response with depth directly under the point on the ground surface where the control motion is specified. In dealing with horizontally propagating Rayleigh waves, it is also important to consider the manner in which the motions will vary with distance of propagation. The results of such a study are shown in Fig. 6. In this analysis a typical NRC control motion with a peak acceleration of 0.25g was represented by a system of fundamental Rayleigh waves which were then allowed to propagate horizontally from the control point location across a horizontal deposit of sand overlying rock. The figure shows the changes in surface motion characteristics with distance of propagation. It is readily apparent that in such a system, the high frequency components of the fundamental Rayleigh waves are rapidly damped out as a result of the relatively high damping characteristics of the soil and in fact, at a distance of a few hundred feet, virtually all motions with frequencies higher than 1 to 2 Hz have decayed to insignificant values. Since most soil deposits extend horizontally for thousands of feet, it is thus unrealistic to expect that the high frequency components of motions in such deposits could result from horizontal propagation of fundamental Rayleigh waves.

As to the higher-order surface waves discussed in a previous section, some of these modes may not decay as rapidly as the fundamental modes. However, as discussed, if such waves do occur, they will, at shallow depths, be similar to slightly inclined body waves and they can be treated as such. They will, because of their long wavelengths produce

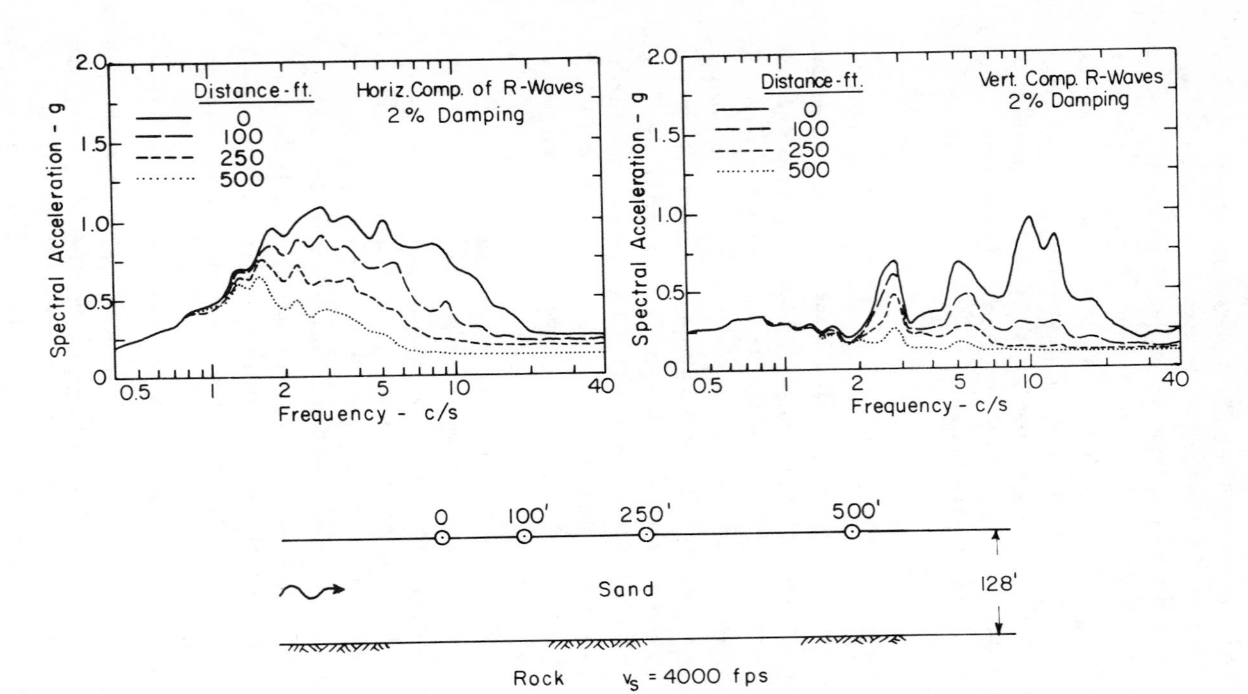

Fig. 6 ATTENUATION OF RAYLEIGH WAVE CHARACTERISTICS WITH DISTANCE TRAVELLED THROUGH SOIL DEPOSIT

motions which are essentially in phase at any shallow depth within reasonable horizontal distances. Thus they are unlikely to produce significant rocking and/or torsion in nuclear structures.

Thus, for structures on soil deposits and cases where the frequencies of concern are greater than 1 or 2 Hz, there seems to be no realistic basis for considering that Rayleigh waves make any significant contribution to the site response. The same logic would also apply to Love waves.

However there are two cases where this argument would not necessarily apply:

(1) for surface motions propagating in rock where the much lower damping and much higher wave velocities would lead to very low rates of horizontal attenuation of motions

and (2) for structures whose response is primarily dependent on the long period components of motion (say below 1 Hz) since such components are apparently not damped out readily even in soil deposits.

The above observations are apparently consistent with seismological observations of earthquake ground motions which do not indicate any significant contributions of Rayleigh waves in the high frequency range.

With regard to the design of nuclear power plants, the above discussion leads to two important conclusions:

(1) For structures located on soil deposits, there is no need to consider Rayleigh or Love waves as making any significant contribution to the ground motions for the purpose of evaluating soil-structure interaction effects.

and (2) For structures located on rock, surface waves may contribute to the ground motions but not at the highest frequencies required to be considered in design. Never-the-less consideration of the possible effects of some components of surface waves may be warranted in design.

Additional evidence supporting these statements has been presented in Gomez-Masso et al. (1979).

VERTICALLY PROPAGATING WAVES

The great majority of methods for site response analysis use Kanai's (1952) assumption of vertically propagating waves. This assumption leads to simple one-dimensional mathematical models for horizontally layered systems and has, partly because of the similarity between motions caused by different wave fields, led to remarkable success in predicting the major features of site response during earthquakes, especially since the analytical procedure was modified by introduction of the equivalent linear method by Seed and Idriss (1969).

Linear site response problems with vertically propagating waves can be solved by a multitude of numerical techniques which are described in texts on soil dynamics, e.g., Desai and Christian (1976). The most efficient method for computing free-field motions from a specified surface control motion appears to be the complex response method used in the program SHAKE, Schnabel et al. (1972). With these methods it is currently possible to analyze any layered viscoelastic soil system overlaying a viscoelastic half-space. The control motion can be specified at the ground surface, at any depth in the soil deposit or as an outcrop motion. Nonlinear effects can be approximated by the equivalent linear method.

Recent efforts have been directed towards the development of true nonlinear methods of analysis. Several methods have been proposed for performing nonlinear total stress analysis of site response problems with vertically propagating shear waves. The most important of these are: The method of characteristics, Streeter et al. (1974), Idriss et al. (1976), Taylor and Larkin (1978); the finite difference method, Joyner (1977); and implicit integration schemes, Martin (1975). In addition several methods of effective stress analysis have been proposed, Ghaboussi and Dikmen (1978), Zienkiewicz et al. (1978), Finn et al. (1977), Liou et al. (1977), and Martin and Seed (1979), which can predict the pore pressure build-up in saturated sands during seismic excitation.

Comparative studies of ground motion characteristics computed by the equivalent linear method and non-linear methods show relatively small differences except where motions are very strong and soils relatively weak--a situation not likely to occur at a nuclear plant site. Thus the development of non-linear analysis techniques has further confirmed the fact that equivalent linear methods are sufficiently accurate for virtually all practical purposes in evaluating the response of nuclear power plant sites.

A major problem of current methods of nonlinear site response analysis is a limitation on the location of the control point. It is not currently possible to specify the control motion at the surface, and specification at a deeper point within the profile is, for reasons to be discussed in the following section, not desirable. However, the motion may be specified at a deep outcrop. This problem has been solved by Joyner and Chen (1975) for the special case of a layered nonlinear system overlying a uniform linearly elastic half-space, see Fig. 7(a). Joyner and Chen specify the outcrop control motion, $y(t)$, at the surface of the elastic half-space shown in Fig. 7(b). The horizontal motions in the half-space, Fig. 7(b), are by simple wave theory

$$u_b(z,t) = \frac{1}{2}\, y(t + z/V_s) + \frac{1}{2}\, y(t - z/V_s) \tag{12}$$

where V_s is the shear wave velocity for the half-space. In the combined system, Fig. 7(a), an additional downward propagating wave occurs due to reflections from the upper soil layer. Hence the motions in that system are:

$$u_a(z,t) = \frac{1}{2}\, y(t + z/V_s) + u(t - z/V_s) \tag{13}$$

The function $u(t)$ must be such that: $u_a(o,t) = u_o(t)$, where $u_o(t)$ is the actual motion at $z = 0$. Hence, $u(t) = u_o(t) - \frac{1}{2}\, y(t)$ and

$$u_a(z,t) = \frac{1}{2} y(t + z/V_s) - \frac{1}{2} y(t - z/V_s) + u_o(t - z/V_s) \tag{14}$$

This leads to the following shear stress at $z = 0$

$$\tau_o(t) = \rho V_s\, \dot{y}(t) - \rho V_s\, \dot{u}_o(t) \tag{15}$$

where ρ is the mass density of the half-space. Joyner and Chen apply this stress boundary condition at the base of the upper soil column and thus achieve a system which can be analyzed by nonlinear methods and which correctly accounts for the effects of the underlying half-space.

The boundary condition expressed by Eq. (15) may be achieved by the physical model shown in Fig. 7(c). In this model the upper soil column is supported on a Lysmer-Kuhlemeyer (1969) dashpot and excited by a horizontal force at the base proportional to the known outcrop velocity time history.

The Joyner-Chen model is an important contribution to the art of nonlinear site response analysis and may, as suggested by Joyner (1975), be extended to approximately two-and three-dimensional soil-structure interaction analyses of nonlinear regions overlying a linear half-space for the special case of vertically propagating waves arriving from the half-space.

INCLINED BODY WAVES

Some energy may be arriving at the control point in the form of non-vertically propagating body waves. There is in fact evidence to suggest that most of the energy approaching the ground surface results from body waves inclined within about 30° of the vertical. This includes the effect of the high-order surface wave modes discussed in a previous section.

The response of horizontally layered sites to plane harmonic body waves arriving at a specified incident angle through an underlying elastic half-space has been investigated by several researchers. The fundamental work was done by Thomson (1950) and Haskell (1960,1962) who developed an efficient matrix method for computing the frequency-dependent transmission coefficients in a layered continuum for incident SH, SV and P-waves. Efficient computer codes for the Thomson-Haskell method were developed by Hannon (1964) and Teng (1967). Silva (1976) extended the Thomson-Haskell method to include damping in the soil layers. With this method it is possible to solve linear or equivalent linear site response problems with inclined body waves for systems consisting of viscoelastic soil layers overlying a uniform undamped half-space, provided the incident angle in the half-space is known. Since no damping is included in the half-space the resulting surface motions do not decay in the horizontal direction. More recently, Chen and Lysmer (1979), have developed a method which includes damping in the underlying half-space.

Analyses of surface response to inclined body waves have also been made by Joyner et al. (1976), who determined the transfer functions from bedrock to the soil surface for a soil deposit 186 m in depth for shear waves propagating at various angles to the vertical. The results of this study are shown in Fig. 8, and it is apparent that for angles of incidence up to 45°, there is a negligible difference between the motions computed

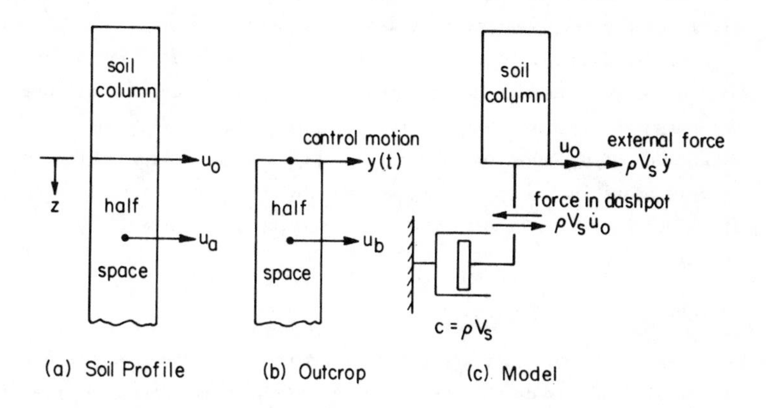

Fig. 7 CONTROL MOTION AT OUTCROP

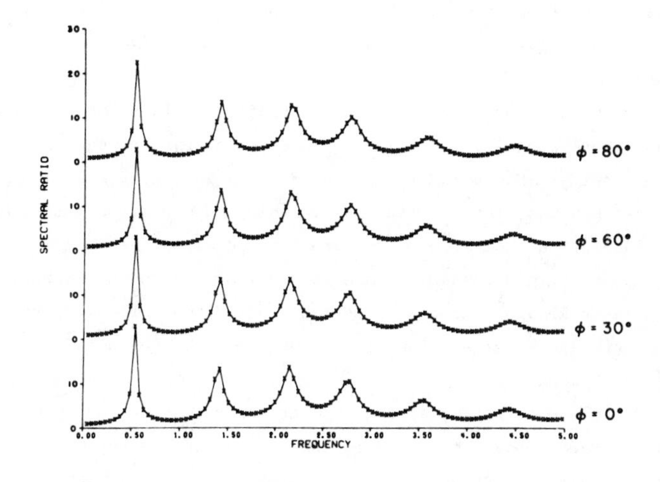

Fig. 8 INFLUENCE OF ANGLE OF SHEAR WAVE INCIDENCE ON COMPUTED SURFACE RESPONSE

(after Joyner et al, 1976)

for inclined waves and for vertically propagating waves. Similar conclusions have been reached by Udaka et al. (1978).

It is reasonable to conclude therefore that the variation of horizontal motions with depth within a soil deposit are for all practical purposes the same, whether they are computed for vertical or inclined directions of propagation within the depth range of interest to engineers. On this basis it is appropriate to use analyses for vertically propagating waves because of their greater simplicity and availability of solutions.

MOTIONS OF SHALLOW DEPTH

As discussed early in this paper, only the motions within a relatively shallow depth (the projected depth of embedment) of the free field will influence the motions of structures. The same discussion also indicates that both the spatial and temporal variation of the free-field motions within this depth are of importance in evaluating soil-structure interaction effects. It is therefore appropriate to discuss in more detail how the amplitude and frequency content of free-field motions vary with depth and, in particular, how they vary near the free ground surface.

If motions could not vary significantly within relatively shallow depths, there would be little point in pursuing this matter further. Not to consider this variation would be equivalent to assuming that the soil mass behaves like a rigid body in which case no interaction would take place and an analysis of the problem would be unnecessary. The belief that it is necessary to analyze soil-structure interaction therefore implies the tacit assumption that motions will vary with depth and appropriate consideration of this fact must be included in any analytical procedure.

The presence of the strong discontinuity represented by the free ground surface imposes predictable and observable limitations on how horizontal amplitudes and frequency contents of motions vary with depth near the ground surface and shows that these variations are significant. Thus the subject merits careful consideration in soil-structure interaction studies.

Theoretical Considerations

The potential effects of the free ground surface on the amplitude and frequency content of waves at various depths in a uniform deposit is shown in Figs. 9 and 10. Both figures show the variation of amplitude with the dimensionless depth z/λ_s in a perfect half-space, where $\lambda_s = V_s/f$ is the wavelength of shear waves at the frequency, f[Hz], considered.

Figure 9 corresponds to the case of vertically propagating shear waves for which the horizontal amplitude is

$$U = U_o \cos 2\pi \frac{z}{\lambda_s} \tag{16}$$

and Fig. 10 corresponds to the case of horizontally propagating Rayleigh waves.

The two types of wave fields are obviously quite different. It is remarkable, however, that both the shear wave field and the Rayleigh wave field produce monotonically decreasing horizontal displacements within the approximate depth

$$\begin{aligned} z &\simeq \frac{1}{4}\lambda_s \quad \text{to} \quad \frac{1}{5}\lambda_s \\ &\simeq \frac{V_s}{4f} \quad \text{to} \quad \frac{V_s}{5f} \end{aligned} \tag{17}$$

and that all horizontal displacements vanish at this depth. A similar phenomenon occurs for inclined shear waves and for layered soil systems where V_s in Eq. (17) can be replaced by the average shear wave velocity, $\overline{V}_s$ above the depth z. As can be seen from the dotted curve shown in Fig. 9 the existence of material damping does not change the substance of these observations.

Two important conclusions can be drawn from these analyses:

(1) Any horizontal motion computed (or observed) at the depth z must be deficient of components of the frequency

$$f \simeq \frac{\overline{V}_s}{4z} \quad \text{to} \quad \frac{\overline{V}_s}{5z} \tag{18}$$

i.e., its response spectrum will have a dip at the approximate frequency f, which incidentally is equal to the fixed base

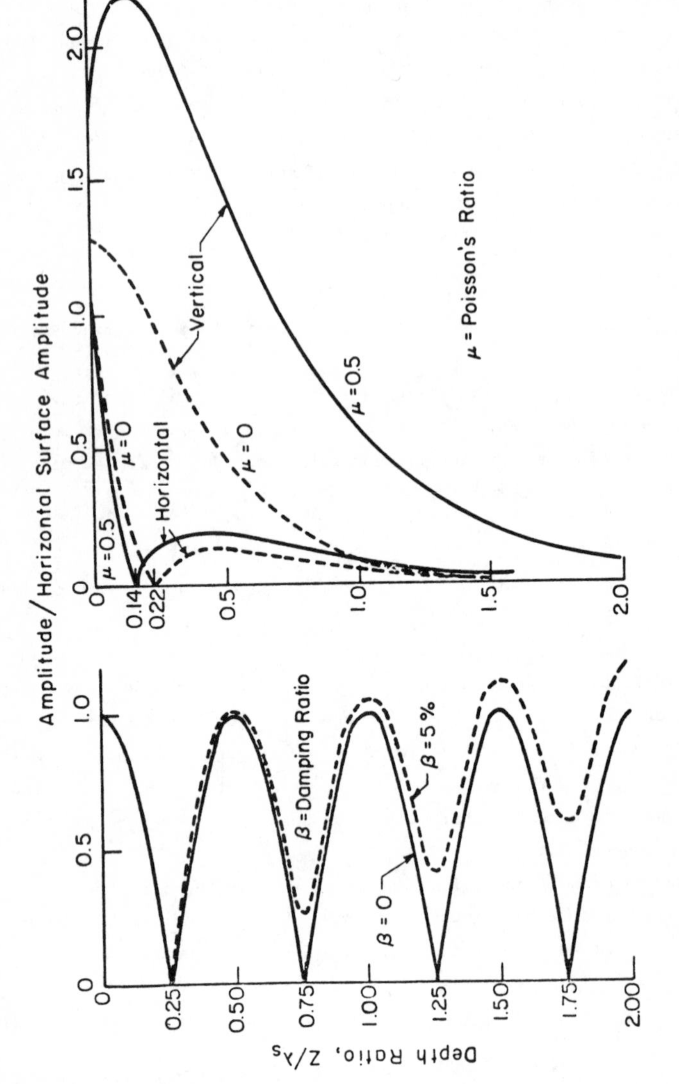

Fig. 9 SHEAR WAVES

Fig. 10 RAYLEIGH WAVES

natural frequency of the soil column above the depth z. Thus, the only level at which a smooth spectrum can exist is at a free surface, and specifying a control motion with a smooth spectrum at any other depth will, as experience has shown, lead to completely unrealistic results. This free surface can be the actual ground surface or a real or imaginary rock outcrop; however a smooth spectrum cannot exist within a soil deposit, whether the motions be due to near-vertically propagating body waves or to horizontally propagating Rayleigh waves.

(2) In a deposit with uniform properties, seismic motions will decrease with depth below the ground surface at least down to the depth

$$z \simeq \frac{\bar{V}_s}{4f_{max}} \text{ to } \frac{\bar{V}_s}{5f_{max}} \tag{19}$$

where f_{max} is the highest frequency present in the motion. This follows directly from Eq. (16) which shows that all components decrease in amplitude within the above depth. Because of variations in soil characteristics with depth this predicted reduction will often extend to depths greater than those indicated by Eq. (19). For a typical soil site, with say $\bar{V}_s$ = 1000 fps, and a seismic environment, with say f_{max} = 20 Hz, the above formula shows that a significant reduction in the free field motion may occur within the upper 10 ft (or deeper if the predominant frequency is lower) of the site. Thus in view of the discussion in connection with Eq. (4), even relatively shallow embedment may significantly influence the seismic response of structures on soft sites and both the embedment and the reduction in the amplitude of the seismic environment with depth should be considered in a rational interaction analysis.

Substitution of realistic values of V_s and f into Eq. (17) will show that z is typically larger than 20 ft for soil sites and 60 ft for rock sites. Thus typical structures experience only the upper part ($z/\lambda_s < 0.2$) of the motions shown in Figs. 9 and 10. In this "shallow" depth range horizontal motions produced by any seismic environment with the same horizontal surface control motion will be quite similar. It is therefore to be expected that the horizontal motions produced at points below the ground surface during earthquakes will be relatively independent of the type of wave field producing the motion.

The above observations were made for motions in a uniform half-space. For layered systems, the stiffness of which usually increases with depth, calculations have shown that the similarity between motions produced at shallow depth by different types of wave fields is even more pronounced.

An interesting example of these effects in a 600 feet deep soil deposit overlying a rigid half-space is shown by the analytical results presented in Figs. 11, 12, and 13. To study the response at different depths in this deposit, analyses were made using vertically propagating shear wave theory for 15 different excitation records. In eight of the analyses, existing records obtained on deep soil deposits were scaled to have a peak acceleration of 0.20g and considered to be developed at the ground surface. The distribution of acceleration with depth and the frequency characteristics of the motions developed at depths of 40 and 76 ft were then determined by deconvolution analyses.

For the same soil deposit, a second study was made in which seven records representative of rock motions were used as base excitation and the base motions were scaled in each case to produce a peak acceleration of 0.20g at the ground surface.

There was surprisingly little difference in the computed distribution of motions whether the excitation was applied at the ground surface or whether it was applied at the base of the soil deposit. The results of the two sets of studies were analyzed statistically to determine the mean acceleration distribution separately for the deconvolution analyses and for the base input analyses. The results of this analysis are shown in Fig. 11. On the whole the results are remarkably similar, all showing a marked drop in peak acceleration within the upper 100 ft.

The response spectra for the motions developed at depths of 40 and 76 ft were also computed and analyzed statistically for the two different groups. The 84 percentile spectra for surface motions, motions at 40 ft depth and motions at 76 ft depth for the deconvolution analyses are shown in Fig. 12. It may be seen that while the spectrum for the surface motions is of the broad band type, the spectrum for motions at a depth of 40 ft contains a marked suppression of frequencies corresponding

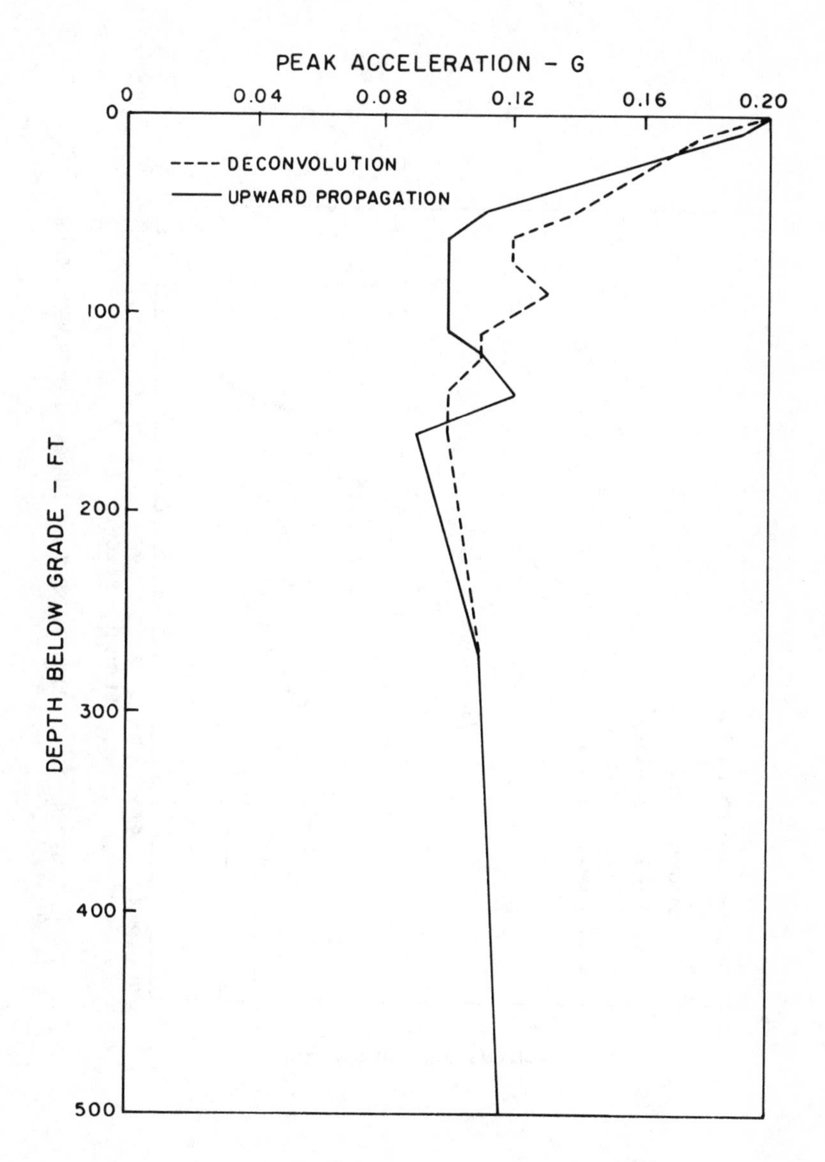

Fig. 11 VARIATION OF MEAN PEAK ACCELERATION WITH DEPTH OF DECONVOLUTION AND UPWARD PROPAGATION ANALYSES

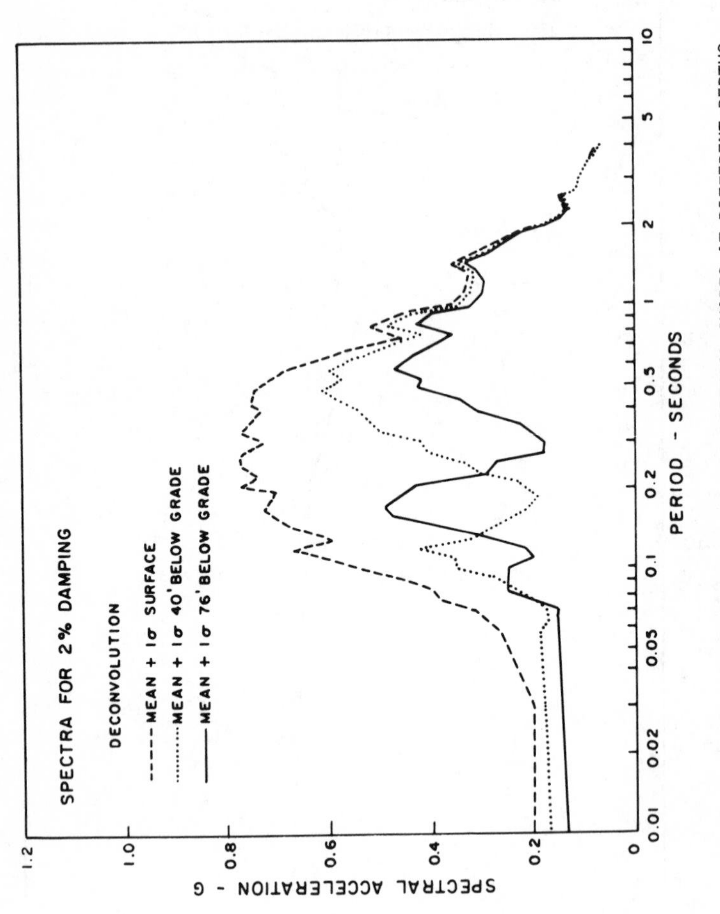

Fig. 12 STATISTICAL ANALYSIS OF COMPUTED SPECTRAL SHAPES AT DIFFERENT DEPTHS IN SOIL DEPOSIT

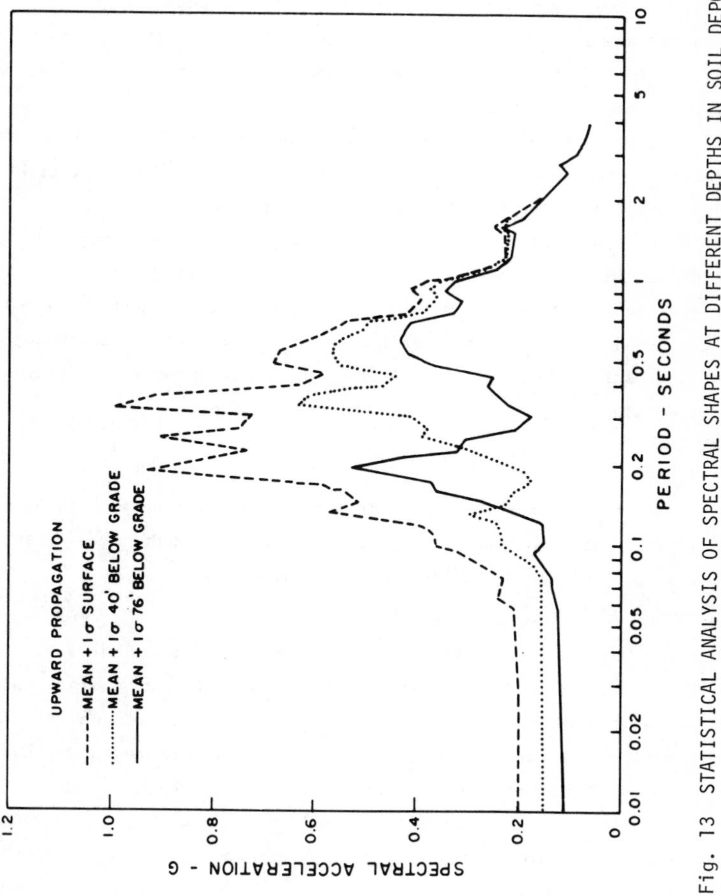

Fig. 13 STATISTICAL ANALYSIS OF SPECTRAL SHAPES AT DIFFERENT DEPTHS IN SOIL DEPOSIT

to a period of about 0.18 second while that for motions at a depth of 76 ft shows a marked suppression of frequencies corresponding to a period of about 0.3 second. The fixed base natural periods of this deposit for 40 ft of soil and 76 ft of soil were about 0.18 and 0.3 seconds respectively. Similar results are shown in Fig. 12 for the base excitation analyses. Thus it may be seen that the frequency suppression effect, as predicted by Eq. (18), is mainly a feature of the geometry and material characteristics of the deposit and depends only slightly on the assumed source of the wave motions involved.

In a deposit 600 ft deep extending to substantial distances in all directions there would not be expected to be any substantial contribution of surface waves to the motions in the frequency range of 1 to 20 Hz, the types of motions primarily investigated in this study, and thus the use of vertically propagating shear waves as the primary wave field is appropriate. Never-the-less the effect of the discontinuity provided by the ground surface on the amplitudes of motions and the frequency characteristics of motions at different depths is clearly illustrated by this example.

Field Evidence

Despite the fact that no concerted effort has been made to date to obtain field data to confirm the above theoretical predictions, a substantial body of field data does in fact exist.

Variation of peak acceleration with depth

The best data to show the variation of peak accelerations with depth is that obtained from vertical arrays of instruments, which only in recent years have been installed at a number of locations to record earthquake motions. Probably the most successful array has been that installed by the U. S. Geological Survey near Menlo Park, California, Joyner et al. (1976). Details of the instrument locations at depths of 0, 12 m, 40 m, and 186 m in relation to the soil profile are shown in Fig. 14 and the characteristics of the deposit in which they were installed are shown in Fig. 15. A number of records of small earthquakes were obtained from the instruments in this array during the period 1972 to 1977. Three typical records are shown in Figs. 16, 17 and 18. The marked decrease in amplitude of the recorded motions with depth is readily apparent.

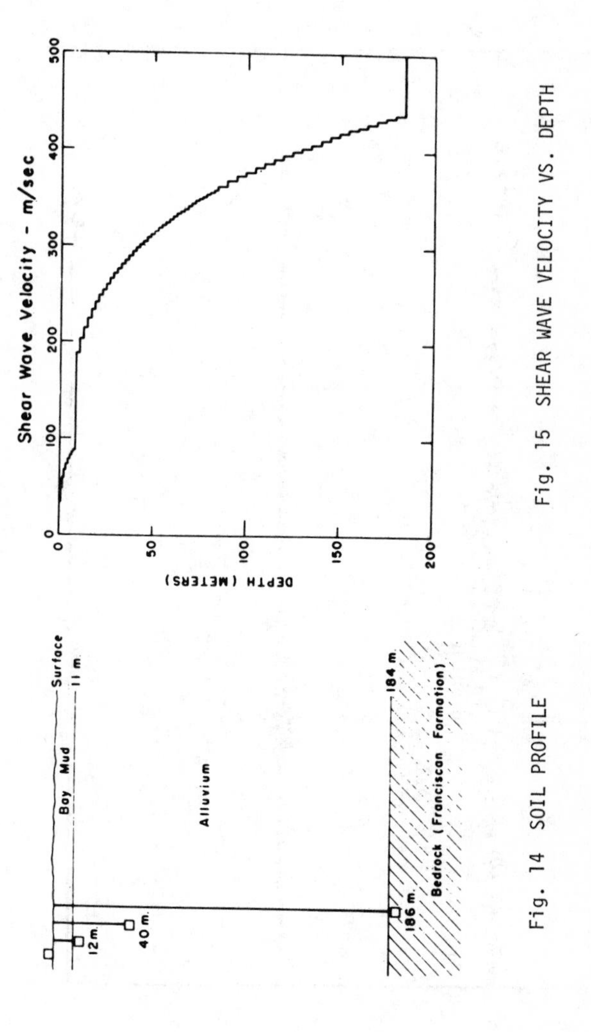

Fig. 14 SOIL PROFILE

Fig. 15 SHEAR WAVE VELOCITY VS. DEPTH

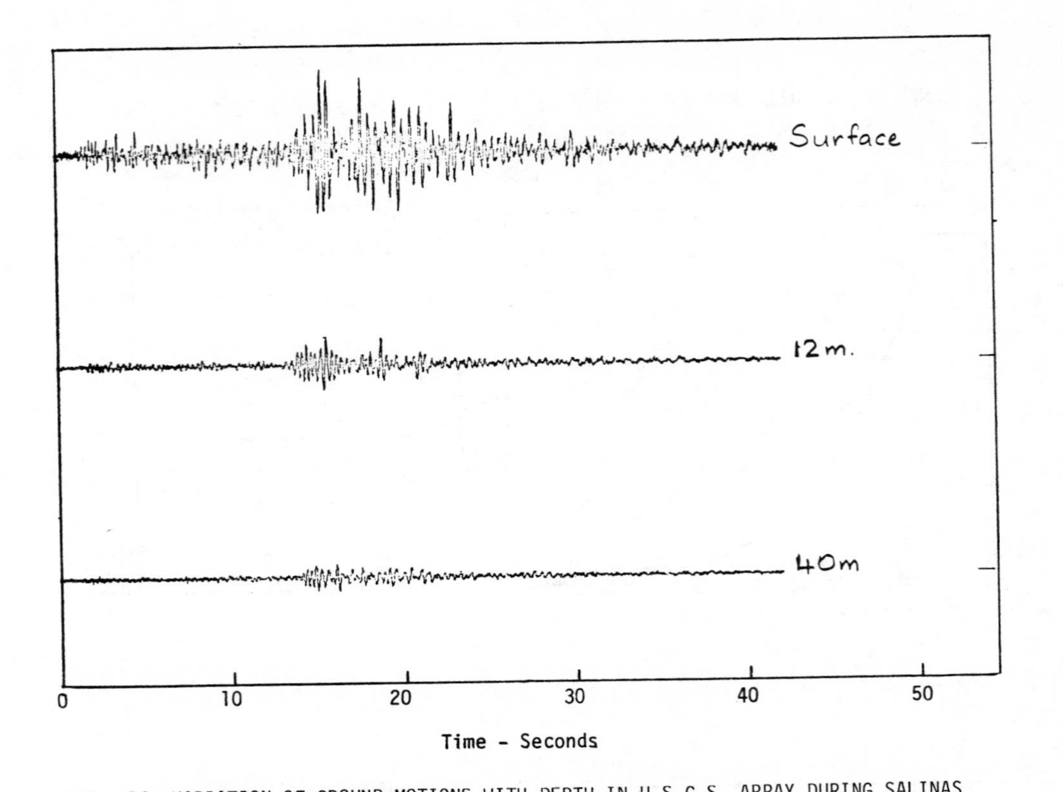

Fig. 16 VARIATION OF GROUND MOTIONS WITH DEPTH IN U.S.G.S. ARRAY DURING SALINAS (CALIFORNIA) EARTHQUAKE OF MARCH 10, 1972

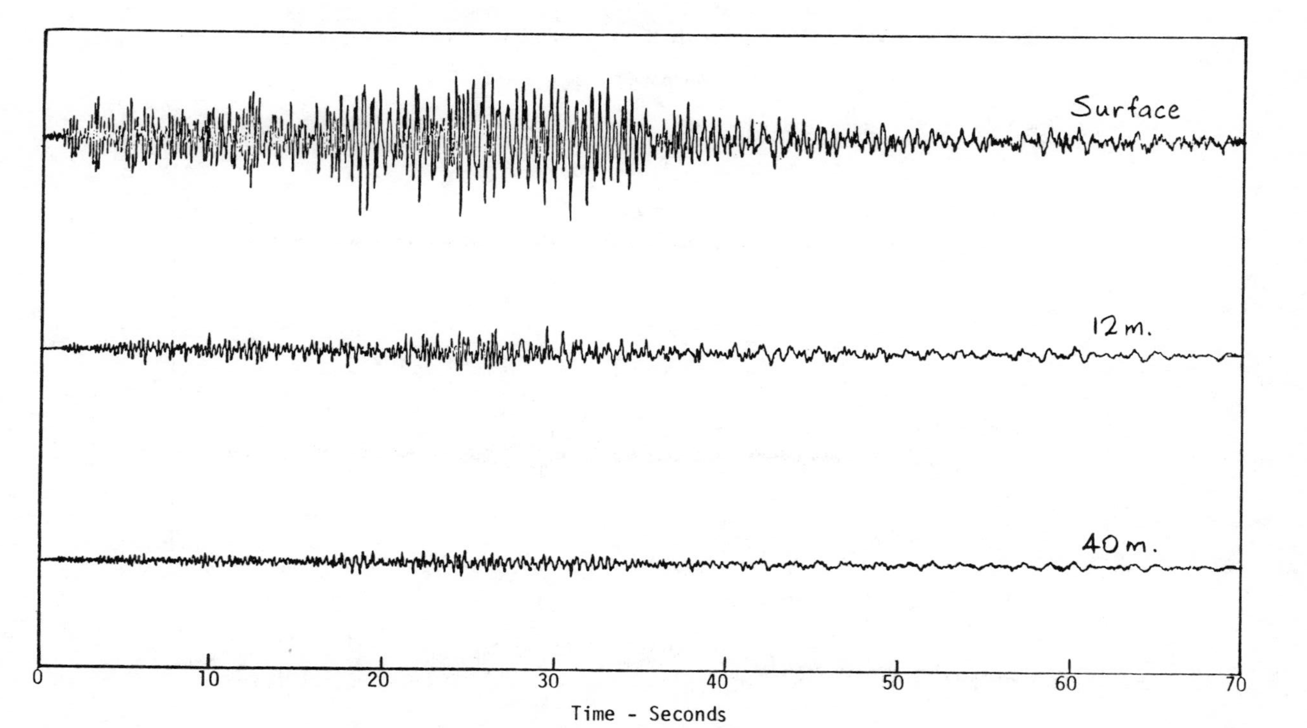

Fig. 17 VARIATION OF GROUND MOTIONS WITH DEPTH IN U.S.G.S. ARRAY DURING BEAR VALLEY (CALIFORNIA) EARTHQUAKE OF APRIL 9, 1972

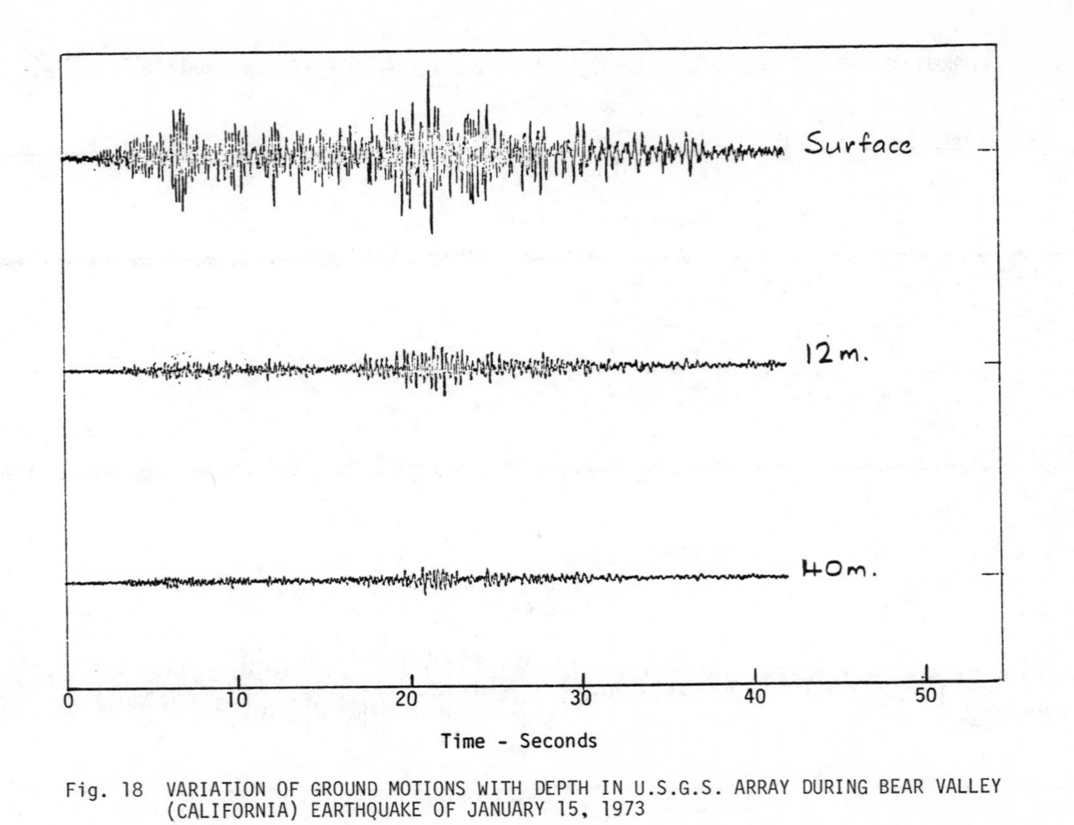

Fig. 18 VARIATION OF GROUND MOTIONS WITH DEPTH IN U.S.G.S. ARRAY DURING BEAR VALLEY (CALIFORNIA) EARTHQUAKE OF JANUARY 15, 1973

Similar decreases of motion amplitude with depth have been observed in four earthquakes recorded in a similar array at Richmond, California by the University of California Seismological Laboratory.

For somewhat stronger motions, an excellent set of data was obtained by records obtained in the basements of buildings in Tokyo in the Tokyo-Higashi-Matsuyama earthquake of July 1, 1968, Ohsaki and Higawara (1970). The recorded values of maximum acceleration for different basement depths are shown in Fig. 19. Although there is considerable scatter in the data, peak accelerations at a depth of 70 ft are typically only about 25% of those recorded at the ground surface. It may be argued that these results are influenced by soil-structure interaction effects, but such effects are likely to be small and in any case would tend to minimize the variation of peak acceleration with depth rather than amplify the effect.

Yet another source of data can be obtained from records obtained in nearby pairs of buildings, each pair involving one constructed at the ground surface and the other at a depth of about 15 ft below the ground surface, in the San Fernando, California earthquake of 1971. Such records for seven sets of buildings are listed in Table 1. It may be seen that in all seven cases, the peak acceleration recorded in the building with a basement was substantially less than that in the building constructed on the ground surface. While some variation of motions would be due to different spatial locations of these buildings, a statistical study of this data clearly shows that the substantial decrease in acceleration with depth is not a chance phenomenon but a pattern attributable to deterministic effects.

Finally, for very strong motions, an excellent set of records was obtained at the Humboldt Bay Power Station in the 1975 Ferndale earthquake. One of these records was obtained at a free-field ground surface location and another at the base of a caisson structure at a depth of 80 ft. The full set of records is shown in Fig. 20. The average maximum acceleration at the ground surface was 0.30g while the average at a depth of 80 ft was 0.13g. Clearly this difference needs to be taken into account if the effects of soil-structure interaction are to be analyzed in a meaningful way in this case.

Other data are available to show similar affects to those discussed above but it is believed that the cases presented provide sufficient

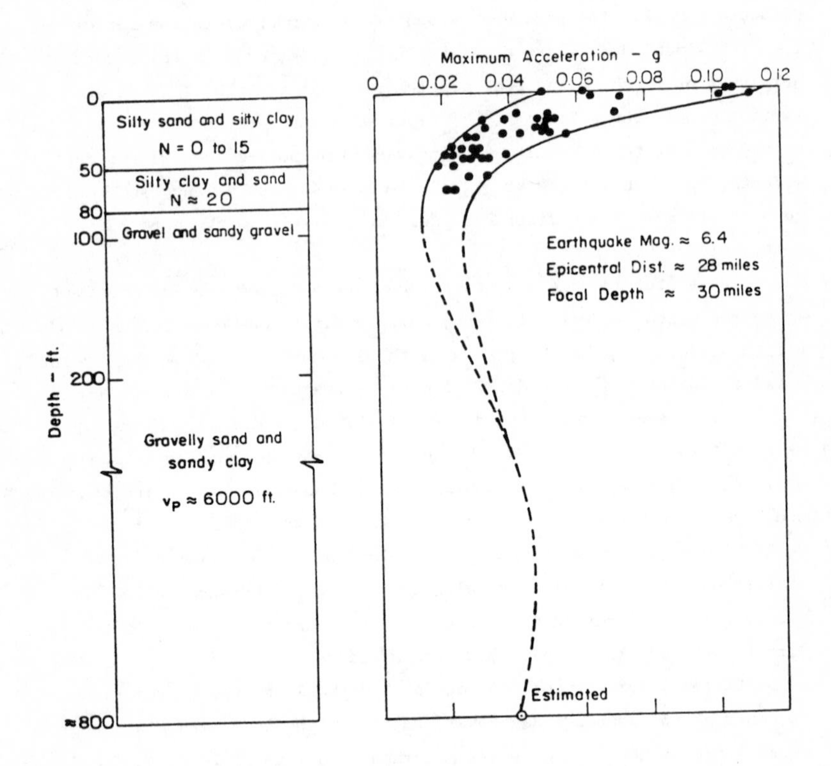

Fig. 19 VARIATION OF RECORDED MAXIMUM ACCELERATION WITH DEPTH FOR BUILDING IN TOKYO-HIGASHI-MATSUYAMA EARTHQUAKE, JULY 1, 1968

Table 1

Change in Maximum Acceleration Between Ground Level and Basement Level

Location	Maximum Acceleration		Percent Change in Ground Surface Accel. at Basement Level
	Ground Surface	Basement	
8244 Orion Blvd. 15107 Vanowen Blvd.	0.26g	0.12g	-54%
14724 Ventura Blvd. 15250 Ventura Blvd.	0.26g	0.23g	-12%
Hollywood Storage Bldg.	0.22g	0.15g	-32%
6430 Sunset Blvd. 6466 Sunset Blvd.	0.19	0.12g	-37%
1880 Century Park East 1800 Century Park East	0.13g	0.10g	-23%
222 South Figueroa 234 South Figueroa 445 South Figueroa	0.15g 0.20g	0.14g	-20%
3407 West Sixth 616 S. Normandie 3470 Wilshire 3550 Wilshire	0.18g	0.12g 0.14g 0.17g	-33% -22% - 6%

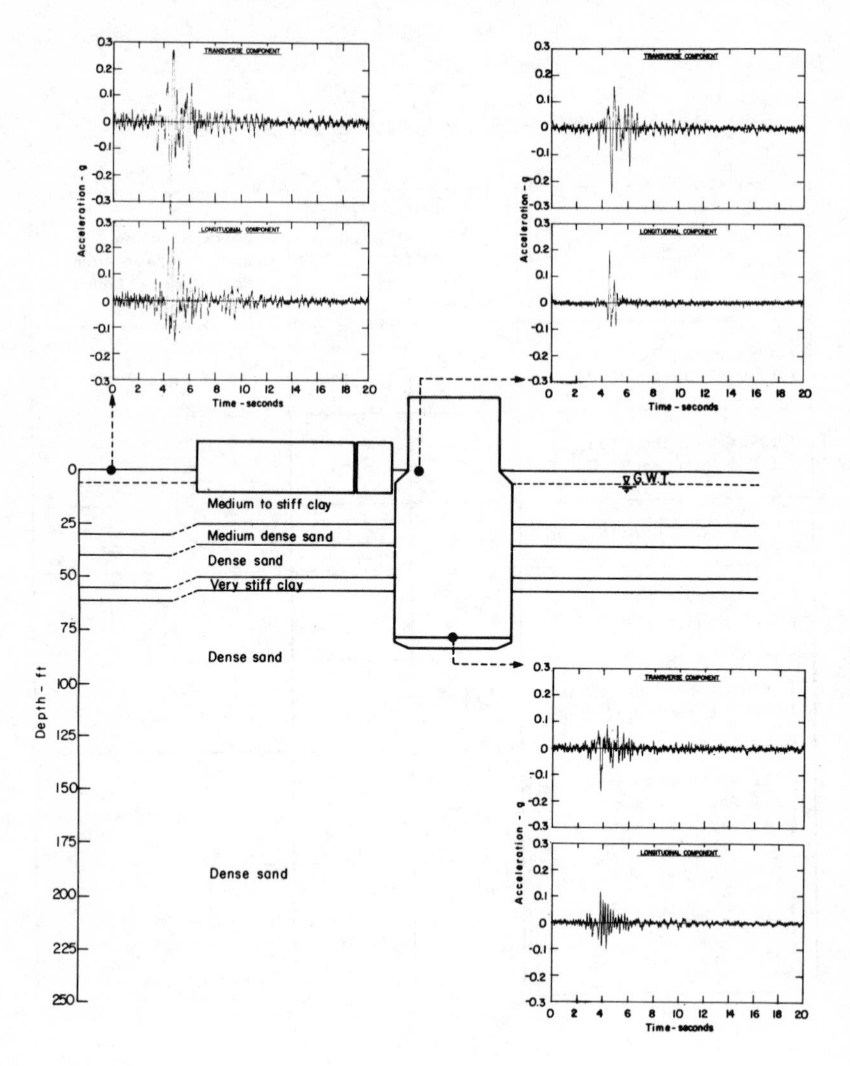

Fig. 20 GROUND MOTION RECORDS AT HUMBOLDT BAY NPS

validation that variations in ground motion with depth are not randomly variable, but characteristically decrease in the range of engineering interest, except for sites with unusual variations in soil characteristics with depth.

Variation of frequency characteristics with depth

Analytical considerations show not only a variation of peak horizontal accelerations with depth but also a deterministic variation of frequency content, and therefore of spectral shape at different depths below the ground surface. Specifically it is to be expected that at any depth z below the ground surface, frequencies of the order of $f = \overline{V}s/4z$ [Hz] will be suppressed due to ground surface reflection effects. (For Rayleigh waves the suppressed frequency would be approximately $\overline{V}_s/5z$ but it has already been shown that Rayleigh waves with frequencies above about 1 Hz could not persist in an extensive soil deposit due to the rapid attenuation of high frequencies in relatively short horizontal distances).

Corroborative evidence of this effect is provided by the data obtained from the Menlo Park array for the recorded motions shown in Figs. 16, 17 and 18. Acceleration response spectra for the motions recorded in these events are plotted in Figs. 21 to 23, and normalized spectra, obtained by dividing the spectral ordinates for any period by the spectral ordinate for the surface motions at that frequency are shown in Figs. 24 to 26. It may be seen that using this technique, the normalized surface spectrum becomes a broad band spectrum and the spectra at other depths are scaled proportionally. It may also be seen that for all three earthquakes, the normalized spectra for motions at a depth of 12 m show a marked suppression of frequencies (evidenced by a dip in the spectrum) corresponding to a period of 0.5 sec, which corresponds to the value $4z/\overline{V}_s$ for this deposit. Similarly the normalized spectra for a depth of 40 m show a suppression of frequencies corresponding to a period of 0.75 seconds, which corresponds to the value of $4z/\overline{V}_s$ for the same deposit. The deterministic value of the frequency suppression effect is clearly evident from this data.

Similar results are also obtained from an analysis of spectra for the motions recorded at Humboldt Bay Power Station (Fig. 20). The spectra for the transverse and longitudinal records of horizontal ground motions at the ground surface and at a depth of 80 ft are shown in

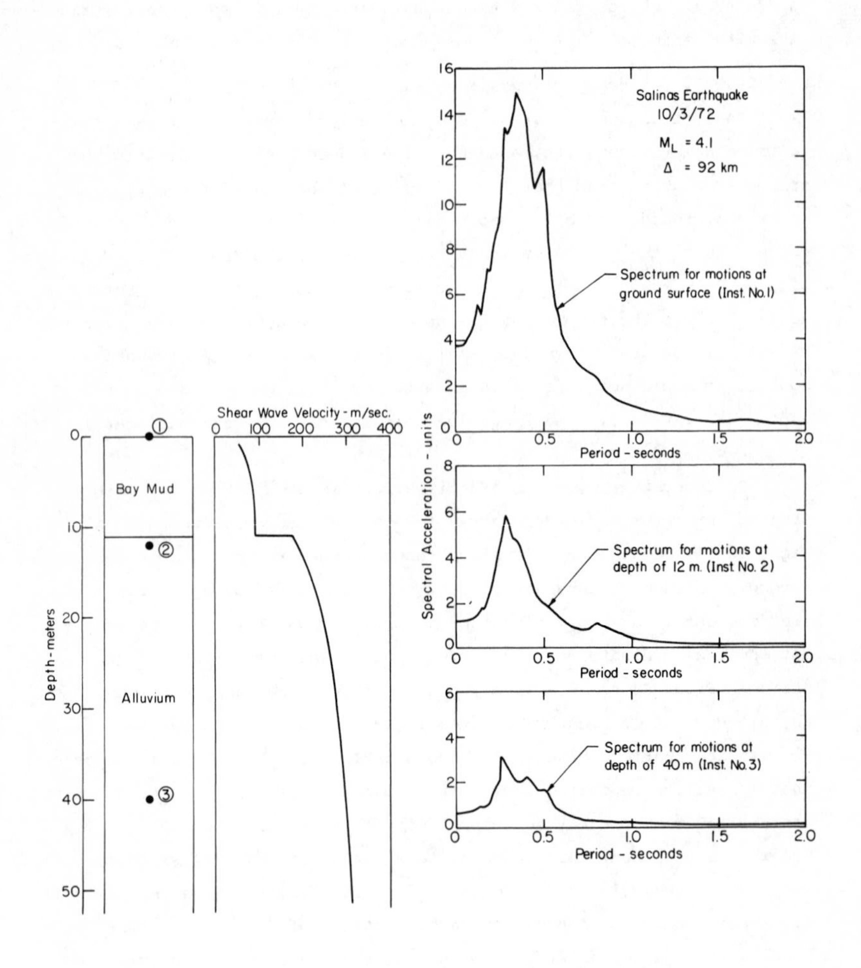

Fig. 21 SPECTRA FOR MOTIONS RECORDED IN SALINAS EARTHQUAKE, 1972

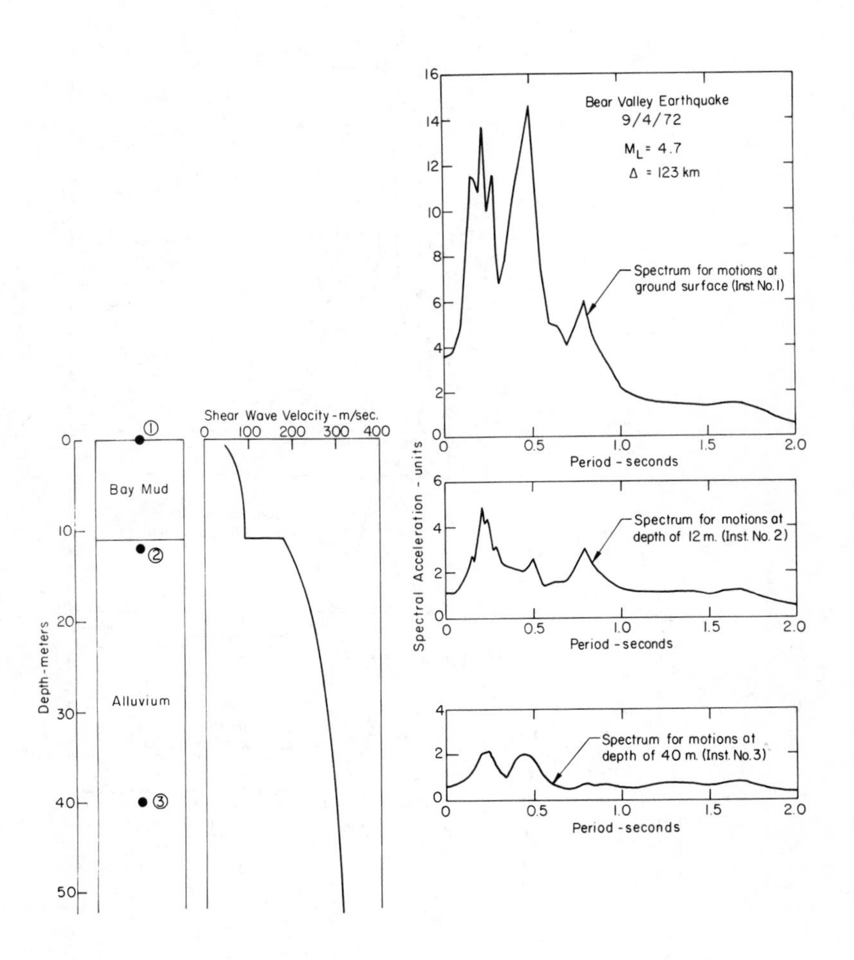

Fig. 22 SPECTRA FOR MOTIONS RECORDED IN BEAR VALLEY EARTHQUAKE, 1972

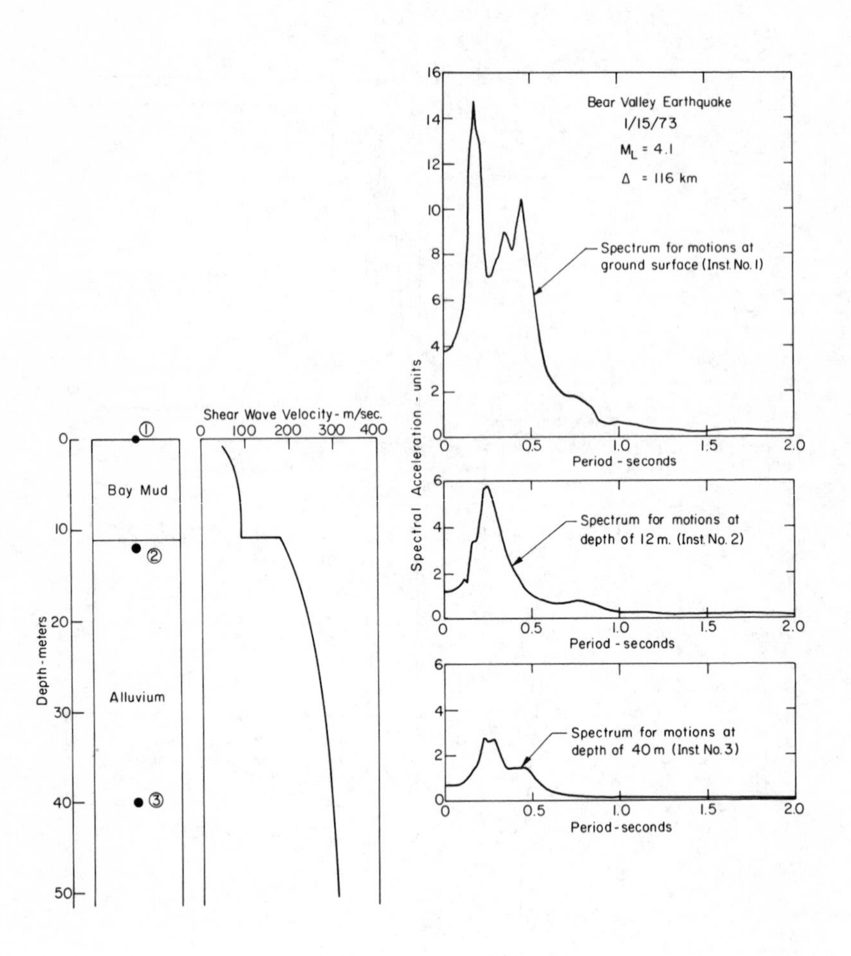

Fig. 23 SPECTRA FOR MOTIONS RECORDED IN BEAR VALLEY EARTHQUAKE, 1973

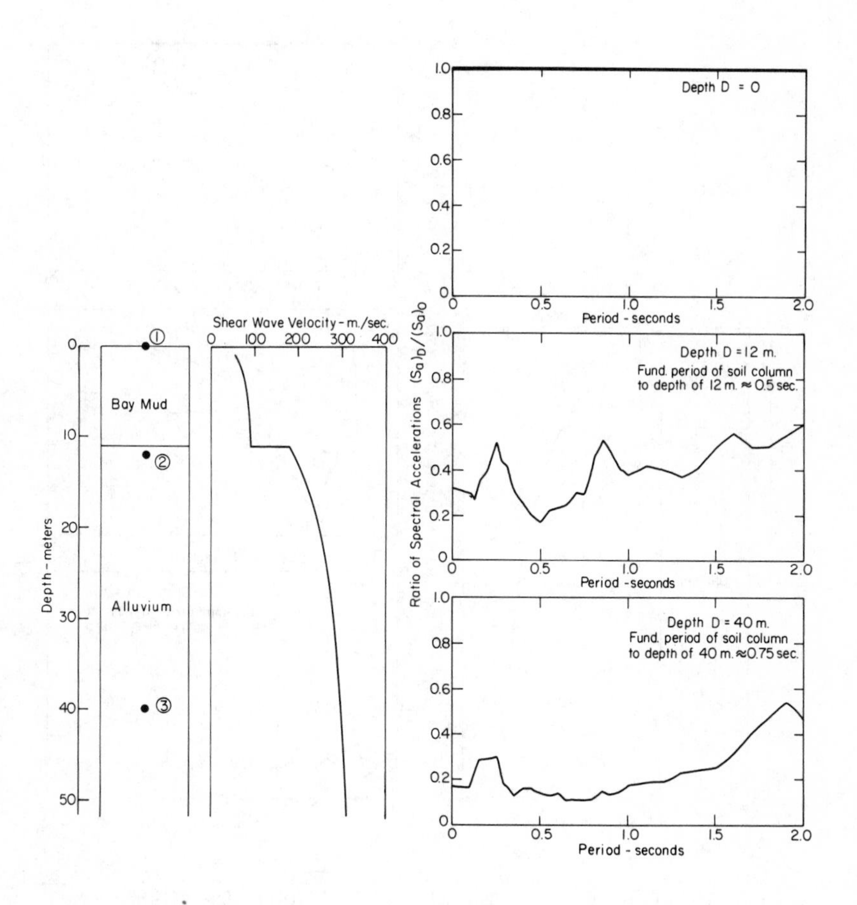

Fig. 24 RELATIVE SPECTRAL ACCELERATIONS AT DIFFERENT DEPTHS - SALINAS EARTHQUAKE, 1972

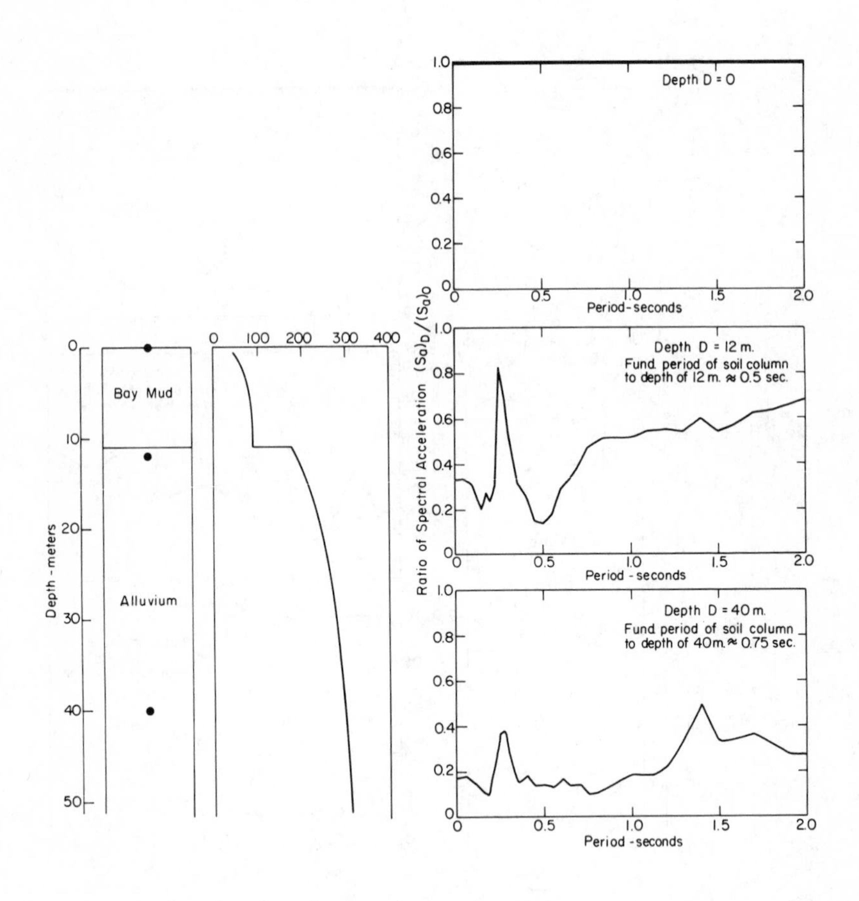

Fig. 25 RELATIVE SPECTRAL ACCELERATIONS AT DIFFERENT DEPTHS - BEAR VALLEY EARTHQUAKE, 1972

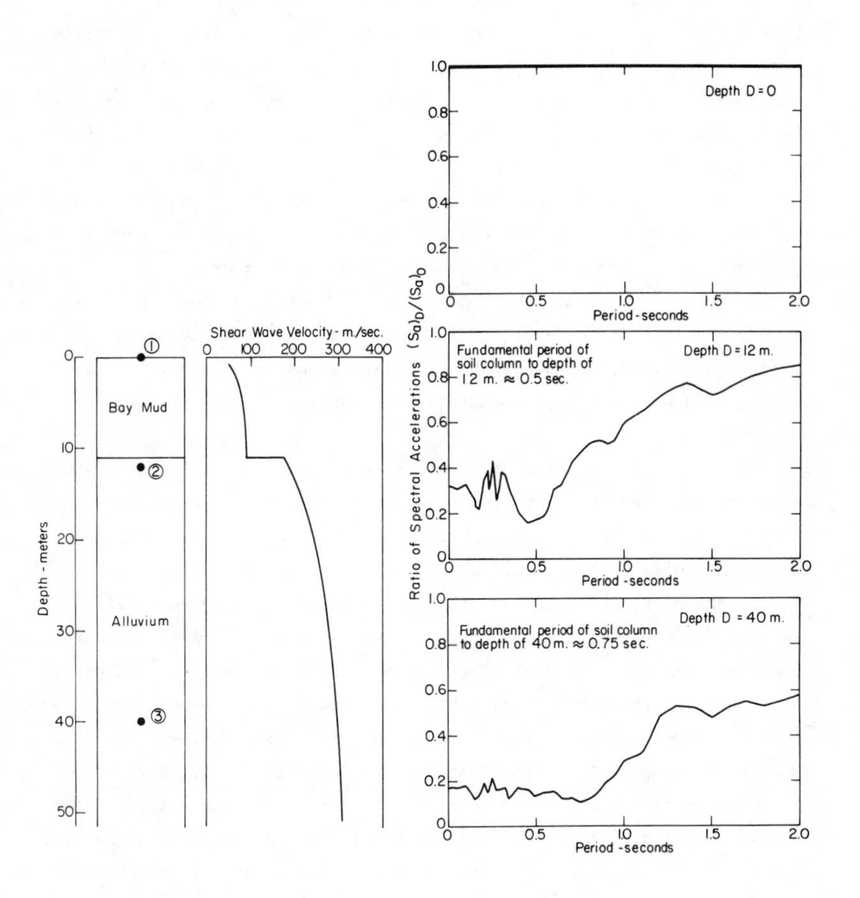

Fig. 26 RELATIVE SPECTRAL ACCELERATIONS AT DIFFERENT DEPTHS - BEAR VALLEY EARTHQUAKE, 1973

Figs. 27 and 28 and the normalized spectra for the same motions are shown in Figs. 29 and 30. Again it is apparent that there is a strong frequency suppression for both transverse and longitudinal motions at a period of about 0.5 sec, which corresponds closely to the computed value of $4z/\overline{V}_s$ for the soil deposit at this site. The great similarity in normalized spectra for transverse and longitudinal motions is shown more clearly in Fig. 31, where the spectra are plotted together and show almost identical characteristics, again illustrating the deterministic nature of this effect.

Hays et al. (1979) and Gazetas and Bianchini (1979) have recently reported data showing similar effects for motions recorded at depths below the ground surface. The data presented by Hays et al. is the average recorded at a distant site for eight nuclear detonations, which closely simulate earthquake effects at such distances. The average normalized spectra for the eight events are shown in Fig. 32, and the frequency suppression corresponding to the period $4z/\overline{V}_s$ is readily apparent.

The same is also true of the motions recorded at Ohgishima Station, Japan and analyzed by Gazetas and Bianchini. The site conditions are shown in Fig. 33 and the normalized spectra in Fig. 34. The frequency suppression effect at a period of 0.3 second at a depth of 15 m is readily apparent. Gazetas and Bianchini also made ground response analyses for the Ohgishima site using vertically propagating shear wave procedures and computed a similar suppression of frequencies at a period of 0.3 sec. by this procedure. They concluded that analyses of this type tend to underestimate the magnitude of the frequency suppression effect but that in general they provide a reasonable basis for evaluation.

In summary it seems apparent that the frequency suppression effect in soil deposits is not merely an analytical concept but that it is also apparent in recorded data. This agreement between analytical concepts and observational data clearly indicates the desirability of considering this phenomenon in evaluating site response or soil-structure interaction effects for embedded structures.

Summary and Conclusions

At the outset of this section it was shown that any analyses of soil-structure interaction must necessarily be based on a knowledge of

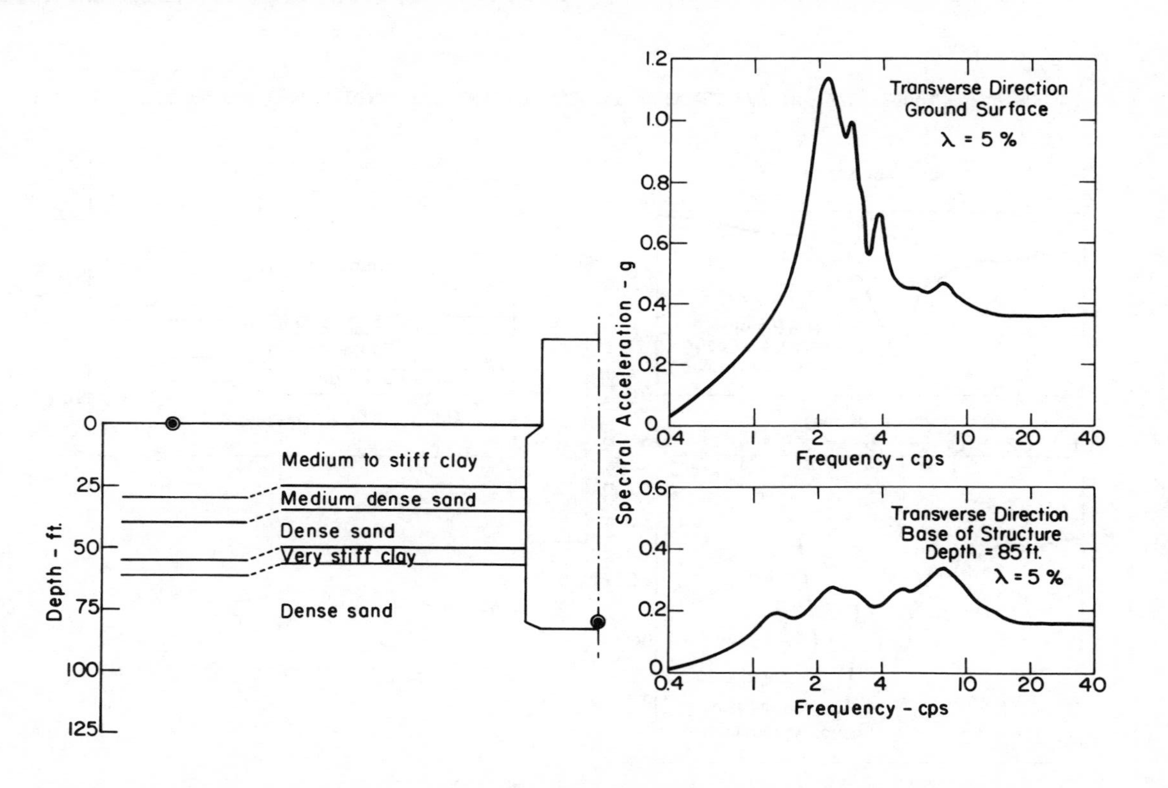

Fig. 27 SPECTRA FOR TRANSVERSE MOTIONS RECORDED AT HUMBOLDT BAY IN 1975 FERNDALE EARTHQUAKE

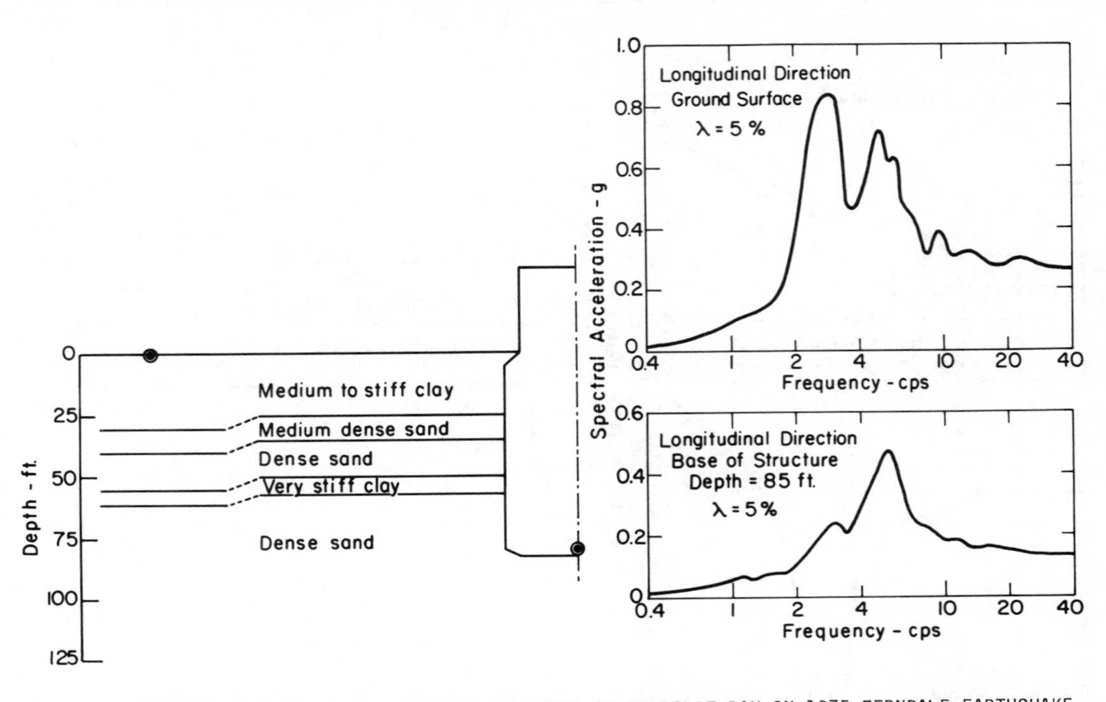

Fig. 28 SPECTRA FOR LONGITUDINAL MOTIONS RECORDED AT HUMBOLDT BAY IN 1975 FERNDALE EARTHQUAKE

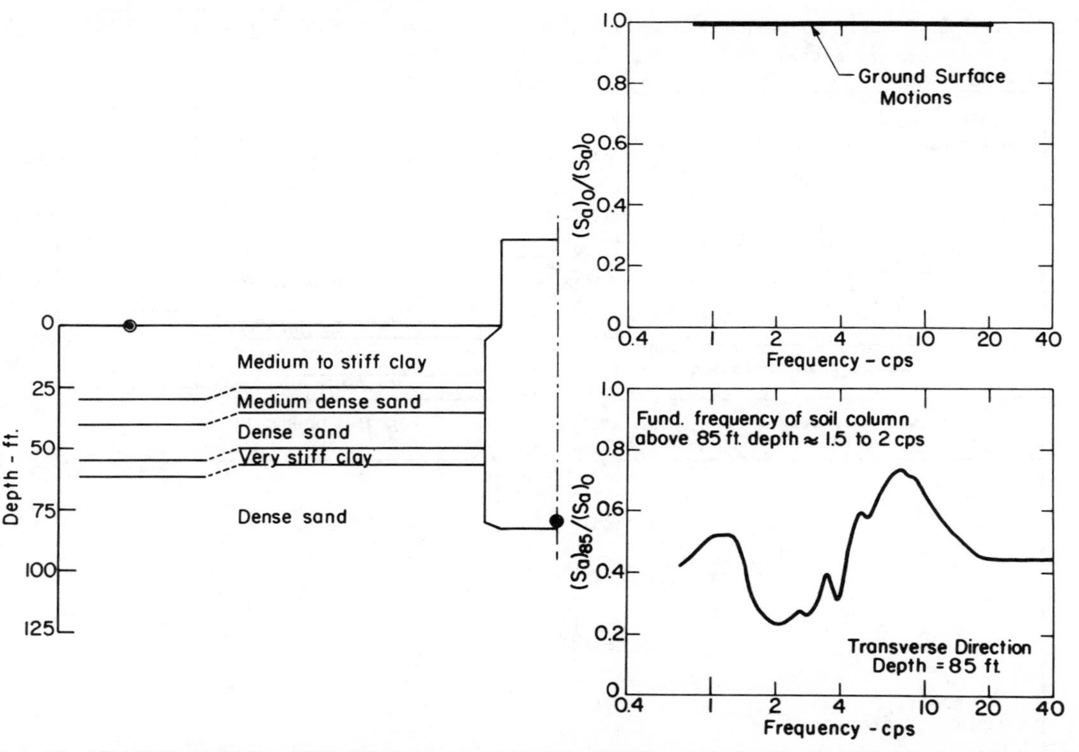

Fig. 29 RELATIVE SPECTRAL ACCELERATIONS AT DIFFERENT DEPTHS AT HUMBOLDT BAY (FERNDALE EARTHQUAKE, 1975 - TRANSVERSE DIRECTION)

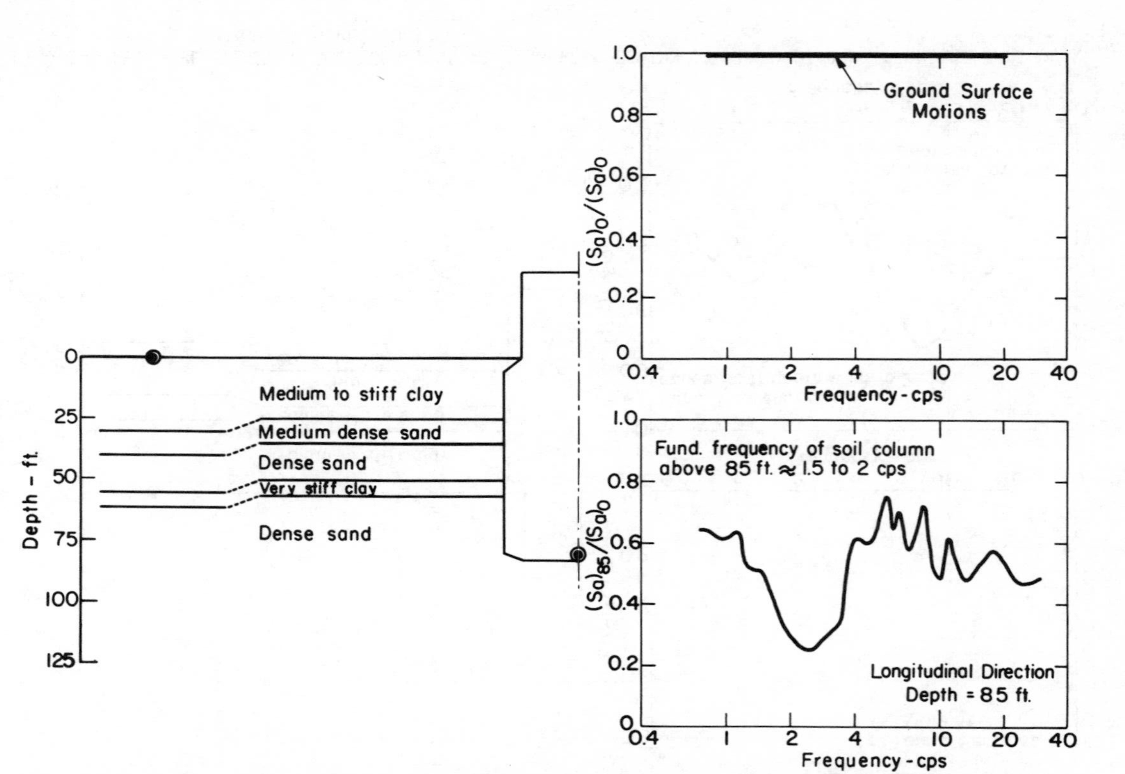

Fig. 30 RELATIVE SPECTRAL ACCELERATIONS AT DIFFERENT DEPTHS AT HUMBOLDT BAY (FERNDALE EARTHQUAKE, 1975 - LONGITUDINAL DIRECTION)

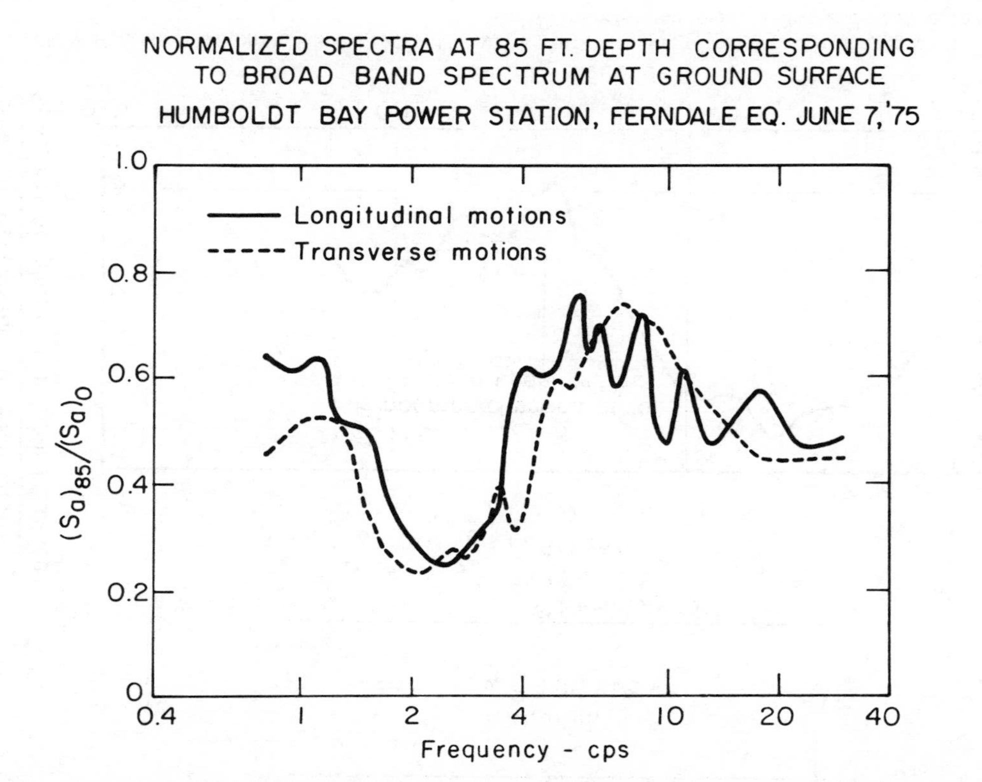

Fig. 31 COMPARISON OF NORMALIZED SPECTRA AT 85 FT DEPTH FOR TRANSVERSE AND LONGITUDINAL MOTIONS

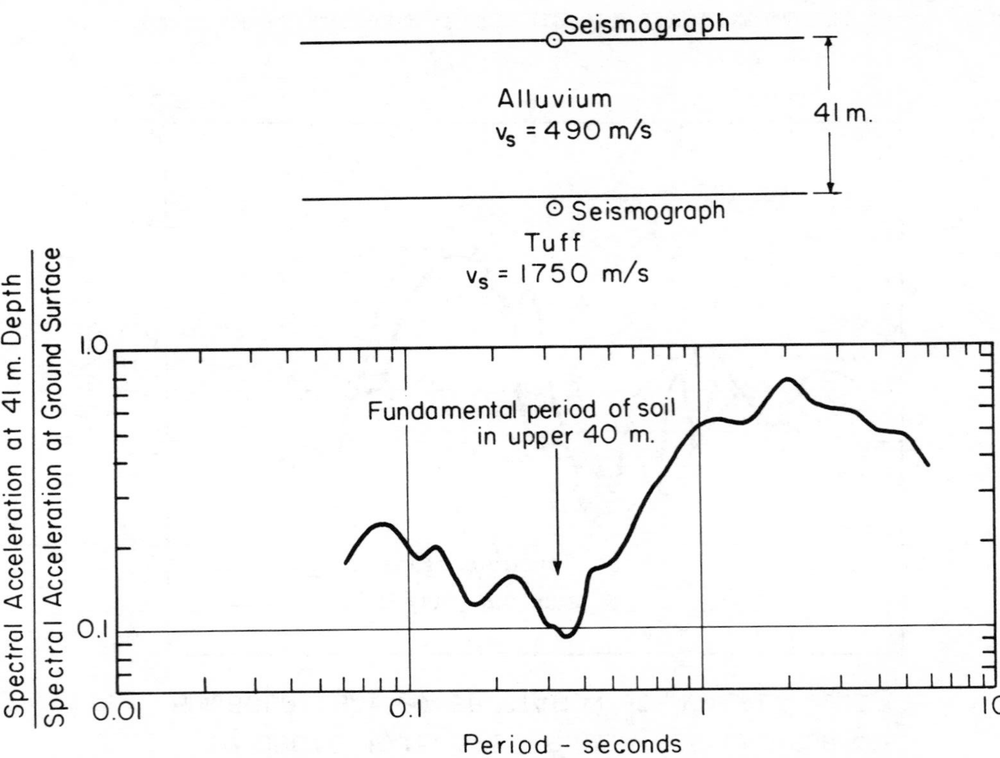

Fig. 32 VARIATIONS OF HORIZONTAL GROUND RESPONSE WITH DEPTH AT BEATTY, NEVADA — AVERAGE OF 10 RECORDS (After Murphy and West, 1974, and Hays et al, 1978)

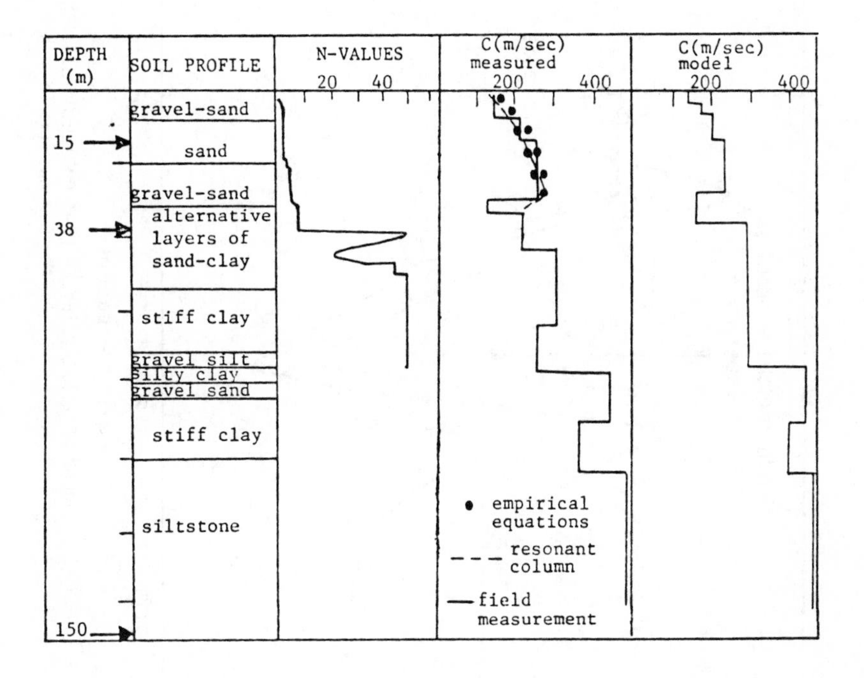

Fig. 33 SITE CONDITIONS AT OHGISHIMA STATION, TOKYO, JAPAN (After Gazetas and Bianchini, 1979)

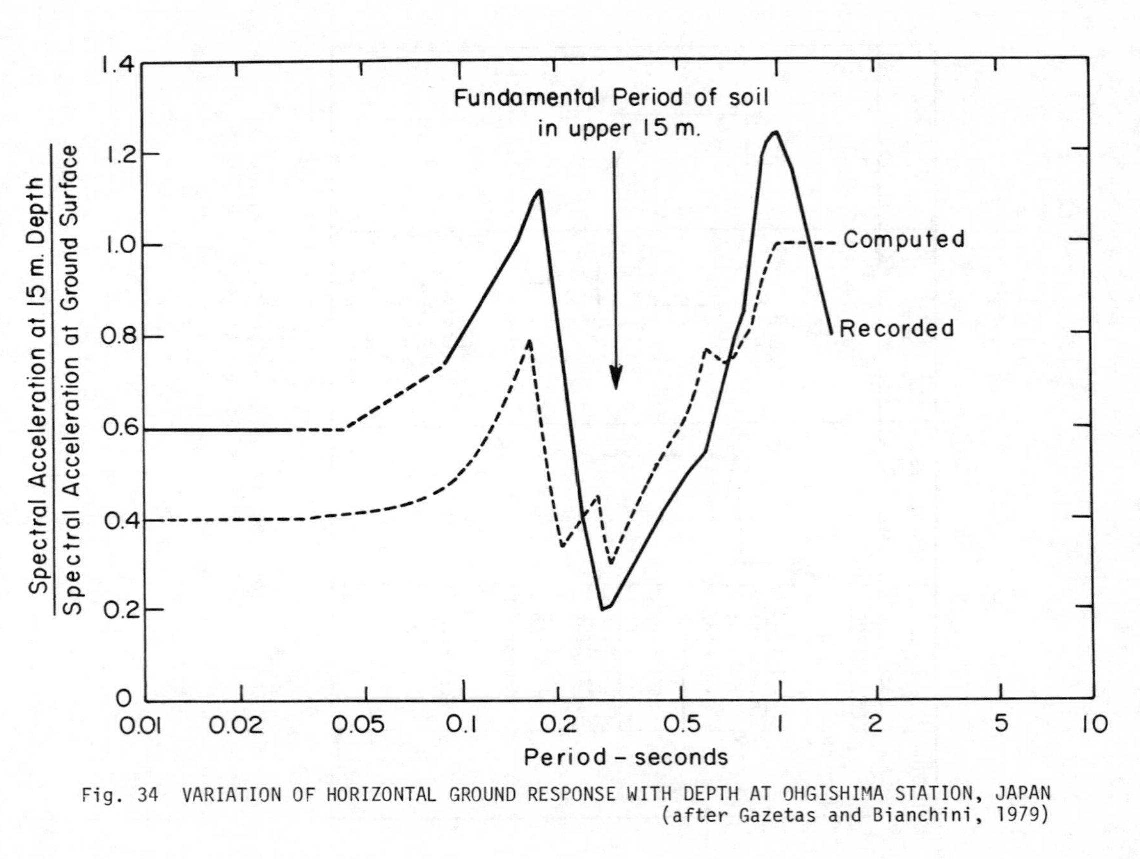

Fig. 34 VARIATION OF HORIZONTAL GROUND RESPONSE WITH DEPTH AT OHGISHIMA STATION, JAPAN (after Gazetas and Bianchini, 1979)

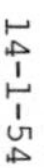

the seismic environment to which the structure will be subjected. This requires an understanding of the spatial distribution of motions in the ground within the depth of embedment.

In the light of the discussion of this subject presented in the preceding pages it seems reasonable to draw the following conclusions concerning the role of the seismic environment in soil-structure interaction analyses.

1. On rock sites structures are likely to be founded near the surface. For such sites, earthquake motions may consist of an unknown mixture of Rayleigh waves, Love waves and near-vertically propagating body waves. Because of the low damping in the rock, attenuation of Rayleigh and Love waves will be small within the general area of the site. It would be expected that the presence of such waves would tend to increase the rocking and torsional excitation on the base of a structure due to out of phase effects as the waves pass across the base. Thus structures located on rock should be analyzed for these motions to determine the potential severity of their contributions to the total response of the structure. However, since a substantial part of the response is likely to be due to the effects of vertically propagating body waves, the final evaluation may well show the influence of Rayleigh and Love waves to be small.

   In analyses using vertically propagating waves, however, it should be noted that because of the fact that these waves will in reality be inclined at different angles to the vertical and will be out-of-phase at different points on the base of the structure due to nonhomogeneities in the rock through which they must travel, some allowance could be made for the "base-slab averaging effect" which will cause the average motions developed in a stiff base slab to be somewhat less than those developed at individual points on the rock surface.

2. For soil sites, structures are likely to be embedded at some depth (say 20 to 80 ft) below the ground surface. The effects of fundamental Rayleigh and Love waves need not be considered at such sites in the design of nuclear plants because the high

frequency components of these waves (greater than 1 Hz) will have been damped out by the soil if it extends to any significant distance (say 1000 ft) around the location of the plant. Higher order Rayleigh modes can be simulated by inclined body waves. Thus the main source of excitation will be inclined body waves and for all practical purposes, these can be analyzed as if they propagated in a vertical direction. However in soil deposits there will be an important variation in motion characteristics with depth and this should be considered in the analysis if meaningful results are to be obtained. The assumption of uniform motions in the upper layers of a soil deposit is inconsistent with the physical nature of wave mechanics and observations in the field and can only lead to misleading results unless the specified control motion is intended to take the natural variations in motion characteristics into account in some way. Without knowing something about the variations in motion it is difficult to see how this can be done realistically without introducing an unwarranted degree of conservatism into the soil-structure analysis procedure.

It should not be construed from the above statements that the assumption of vertically propagating waves at soil sites is appropriate for all types of structures. The long-period components of horizontally propagating waves may be extremely important for the design of buried pipelines, tunnel linings and earth retaining structures. However, except for increased stresses in the walls of buried or embedded structures the change in the stress field due to these waves appear to have little effect on the overall horizontal motions of such structures.

The propagating nature of the displacement field may also induce additional displacements and stresses in long above-ground structures such as bridges, Bogdanoff et al. (1965), Johnson and Galletly (1972), Abdel-Ghaffar and Trifunac (1976); and rocking and torsional motions in long period single structures, Wong (1975), Scanlan (1976) and Wong and Luco (1976).

Finally, and perhaps most important, control motions should be chosen with due respect to site conditions and, if a broad band design spectrum is used the control point should be located at the ground surface or, alternatively, at an imaginary outcrop, where it could conceivably exist, and not at some arbitrary depth below the ground surface where the boundary conditions resulting simply from the existence of a ground surface preclude this possibility.

REFERENCES

1. "Statistical Analysis of Earthquake Ground Motion Parameters," Report No. NUREG/CR-1175 prepared for the U. S. Nuclear Regulatory Commission by Shannon & Wilson, Inc. and Agbabian Associates, December, 1979.

2. Abdel-Ghaffar, A. M. and Trifunac, M. D. (1976) "Antiplane Dynamic Soil/Bridge Interaction for Incident Plane SH-Waves (draft)," California Institute of Technology, Pasadena, California, 1976.

3. Bogdanoff, J. L., Goldberg, J. E. and Schiff, A. J. (1965) "The Effect of Ground Transmission Time on the Response of Long Structures," Bull. Seis. Soc. Am., Vol. 55, pp. 627-640, June, 1965.

4. Boncheva, H. (1977) "Soil Amplification Factor of Surface Waves," Proc. 6th World Conf. Earthq. Engrg., Vol. 6, pp. 6-189, New Delhi, 1977.

5. Chandra, U. (1972) "Angles of Incidence of S-Waves," Bull. Seism. Soc. Am., Vol. 62, No. 4, pp. 903-915, 1972.

6. Chen, J.-C. and Lysmer, J. (1979) Doctoral research at the University of California, Berkeley, 1979.

7. Desai, C. S. and Christian, J. T. (1976) "Numerical Methods in Geotechnical Engineering," McGraw-Hill Book Co., 783 pp., 1976.

8. Ewing, W. M., Jardetzky, W. S. and Press, F. (1957) "Elastic Waves in Layered Media," McGraw-Hill, 1957.

9. Faccioli, E. (1978) "Response Spectra for Soft Soil Sites," Proc. ASCE Specialty Conference, Earthquake Engineering and Soil Dynamics, Vol. 1, pp. 441-456, Pasadena, California, June 1978.

10. Finn, W.D.L., Lee, K. W. and Martin, G. R. (1977) "An Effective Stress Model for Liquefaction," Journal Geotech. Engrg. Div., ASCE, Vol. 103, No. GT6, pp. 517-533, June, 1977.

11. Gazetas, G. and Bianchini, G. (1979) "Field Evaluation of Body and Surface-Wave Soil-Amplification Theories," Proc. 2nd U. S. Nat. Conf. on Earthquake Engrg., August, 1979.

12. Ghaboussi, J. and Dikmen, S. U. (1978) "Liquefaction Analysis of Horizontally Layered Sands," Journal Geotech. Engrg. Div., ASCE, Vol. 104, No. GT3, pp. 341-356, March, 1978.

13. Gomez-Masso, Alberto, Lysmer, John, Chen, Jian-Chu and Seed, H. Bolton (1979) "Soil Structure Interaction in Different Seismic Environments," Report No. UCB/EERC-79/18, Earthquake Engineering Research Center, University of California, Berkeley, August, 1979.

14. Hannon, W. J. (1964) "An Application of the Haskell-Thomson Matrix Method to the Synthesis of the Surface Motion due to Dilatational Waves," Bull. Seis. Soc. Am., Vol. 54, pp. 2067-2079, December, 1964.

15. Haskell,N. A. (1953) "The Dispersion of Surface Waves in Multi-layered Media," Bull. Seis. Soc. Am., Vol. 43, No. 1, pp. 17-34, February, 1953.

16. Haskell, N. A. (1960) "Crustal Reflection of Plane SH Waves," Journal Geophysical Research, Vol. 65, No. 13, pp. 4147-4150, December, 1960.

17. Haskell, N. A. (1962) "Crustal Reflection of Plane P and SV Waves," Journal Geophysical Research, Vol. 67, No. 12, pp. 4751-4767, November, 1962.

18. Hayashi, S., Tsuchida, H. and Kurata, E. (1971) "Average Response Spectra for Various Subsoil Conditions," Third Joint Meeting, U.S. - Japan Panel on Wind and Seismic Effects, UJNR, Tokyo, May 2-12, 1971.

19. Hays, W. W., Rogers, A. M. and King, K. W. (1979) "Empirical Data About Local Ground Response," Proc. 2nd U.S. Nat. Conf. on Earthq. Engrg., Earthq. Engrg. Res. Inst., Stanford University, August, 1979.

20. Idriss, I. M. (1978) "Characteristics of Earthquake Motions--A State-of-the-Art Review," Proc. ASCE Geotechnical Engineering Division Specialty Conference on Earthquake Engineering and Soil Dynamics, Vol. III, Pasadena, California, June, 1978.

21. Idriss, I. M., Dobry, R., Doyle, E. H. and Singh, R. D. (1976) "Behavior of Soft Clays Under Earthquake Loading Conditions," Proc. Offshore Technology Conference, Paper No. OTC 2671, Dallas, May, 1976.

22. Johnson, N. E. and Galletley, R. D. (1972) "The Comparison of Response of a Highway Bridge to Uniform and Moving Ground Excitation," Shock and Vibration Bulletin, Vol. 42, p. 2, January, 1972.

23. Joyner, W. B. (1975) "A Method for Calculating Nonlinear Seismic Response in Two Dimensions," Bull. Seis. Soc. Am., Vol. 65, No. 5, pp. 1337-1357, October, 1975.

24. Joyner, W. B. (1977) "A FORTRAN Program for Calculating Nonlinear Seismic Ground Response," Open File Report No. 77-671, U. S. Geological Survey, Menlo Park, California, 1977.

25. Joyner, W. B. and Chen, A. T. F. (1975) "Calculation of Nonlinear Ground Response in Earthquakes," Bull. Seis. Soc. Am., Vol. 65, No. 5, pp. 1315-1336, October, 1975.

26. Joyner, W. B., Warrick, R. E. and Oliver, A. A. (1976) "Analysis of Seismograms from a Downhole Array in Sediments Near San Francisco Bay," Bulletin of the Seismological Society of America, Vol. 66, No. 3, pp. 937-958, June, 1976.

27. Kanai, K. (1952) "Relation Between the Nature of Surface Layer and the Amplitudes of Earthquake Motions," Bull. Earthquake Research Institute, Vol. 30, pp. 31-37, Tokyo, 1952.

28. Liang, G. C. and Duke, C. M. (1977) "Separation of Body and Surface Waves in Strong Motion Records," Proc. 6th World Conference on Earthquake Engineering, Vol. 2, pp. 215-220, New Delhi, India, January, 1977.

29. Liou, C. P., Streeter, V. L. and Richart, F. E., Jr. (1977) "Numerical Model for Liquefaction," Journal Geotech. Engrg. Div., ASCE, Vol. 103, No. GT6, pp. 589-606, June, 1977.

30. Lysmer, J. (1970) "Lumped Mass Method for Rayleigh Waves," Bull. Seis. Soc. Am., Vol. 60, No. 1, pp. 89-104, February, 1970.

31. Lysmer, J. and Kuhlemeyer, R. L. (1969) "Finite Dynamic Model for Infinite Media," Journal Engrg. Mech. Div., ASCE, Vol. 95, No. EM4, pp. 859-877, August, 1969.

32. Lysmer, J. and Waas, G. (1972) "Shear Waves in Plane Infinite Structures," J. Engrg. Mech. Div., ASCE, Vol. 98, No. EM1, pp. 85-105, February, 1972.

33. Martin, P. P. (1975) "Non-Linear Methods for Dynamic Analysis of Ground Response," Ph.D. Thesis, University of California, Berkeley, June, 1975.

34. Martin, P. P. and Seed, H. B. (1979) "Simplified Procedure for Effective Stress Analysis of Ground Response," Journal Geotech. Engrg. Div., ASCE, Vol. 105, No. GT6, pp. 739-758, June, 1979.

35. Nair, G. P. and Emery, J. J. (1975) "Spatial Variations in Seismic Motions," Second Canadian Conference on Earthquake Engineering, Preprint No. 5, Hamilton, Ontario, June, 1975

36. Ohsaki, Y. and Hagiwara, T. (1970) "On Effects of Soils and Foundations upon Earthquake Inputs to Buildings," Research Paper No. 41, Building Research Institute, Ministry of Construction, Japan.

37. Randall, M. J. (1971) "Revised Travel Time Table for S," Geophys. J., Vol. 22, pp. 229-234, 1971.

38. Richart, F. E., Jr., Hall, J. R., Jr. and Woods, R. D. (1970) "Vibrations of Soils and Foundations," Prentice-Hall, Inc., 1970.

39. Scanlan, R. H. (1976) "Seismic Wave Effects on Soil/Structure Interaction," Earthquake Engineering and Structural Dynamics, Vol. 4, pp. 379-388, June, 1976.

40. Schnabel, P. B., Lysmer, J. and Seed, H. B. (1972) "SHAKE - A Computer Program for Earthquake Response Analysis of Horizontally Layered Sites," Report No. EERC 72-12, Earthquake Engineering Research Center, University of California, Berkeley, December, 1972.

41. Seed, H. B. and Idriss, I. M. (1969) "The Influence of Soil Conditions on Ground Motions During Earthquakes," Journal Soil Mech. Found. Div., ASCE, Vol. 94, No. SM1, pp. 99-137, January, 1969.

42. Seed, H. B., Murarka, R., Lysmer, J. and Idriss, I. M. (1976b) "Relationships Between Maximum Acceleration, Maximum Velocity, Distance from Source and Local Site Conditions for Moderately Strong Earthquakes," Bull. Seis. Soc. Am., Vol. 66, No. 4, pp. 1323-1342, August, 1976.

43. Seed, H. B., Ugas, C. and Lysmer, J. (1976a) "Site Dependent Spectra for Earthquake Resistant Design," Bull. Seis. Soc. Am., Vol. 66, No. 1, pp. 221-243, February, 1976.

44. Silva, W. (1976) "Body Waves in a Layered Anelastic Solid," Bull. Seis. Soc. Am., Vol. 66, No. 5, pp. 1539-1554, October, 1976.

45. Streeter, V. L., Wylie, E. B. and Richart, F. E., Jr. (1974) "Soil Motion Computations by Characteristics Method," Journal Geotech. Engrg. Div., ASCE, Vol. 100, No. GT3, pp. 247-263, March, 1974.

46. Taylor, P. W. and Larkin, T. J. (1978) "Seismic Site Response of Nonlinear Soil Media," Journal Geotech. Engrg. Div., ASCE, Vol. 104, No. GT3, pp. 369-383, March, 1978.

47. Teng, T.-L. (1967) "Reflection and Transmission from a Plane Layered Core-Mantle Boundary," Bull. Seis. Soc. Am., Vol. 57, No. 3, pp. 477-499, June, 1967.

48. Thomson, W. T. (1950) "Transmission of Elastic Waves through a Stratified Solid Medium," Journal Appl. Physics, Vol. 21, February, 1950.

49. Toki, K. (1977) "Disintegration of Accelerograms into Surface and Body Waves," Proc. 6th World Conference on Earthquake Engineering, Vol. 2, pp. 209-214, New Delhi, India, January, 1977.

50. Trifunac, M. D. and Brune, J. N. (1970) "Complexity of Energy Release During the Imperial Valley, California Earthquake of 1940," Bull. Seis. Soc. Am., Vol. 60, No. 1, pp. 137-160, 1970.

51. Udaka, T. (1975) "Analysis of Response of Large Embankments to Travelling Base Motions," Ph.D. Dissertation, University of California, Berkeley, December, 1975.

52. Waas, G. (1972) "Analysis Method for Footing Vibrations through Layered Media," Ph.D. Dissertation, University of California, Berkeley, December, 1972; also published under the title "Earth Vibration Effects and Abatement for Military Facilities," Report 3, U. S. Army Engineer WES, Vicksburg, Miss., September, 1972.

53. Wong, H. L. (1975) "Dynamic Soil/Structure Interaction," Ph.D. Dissertation, California Institute of Technology, Pasadena, California, May, 1975.

54. Wong, H. L. and Luco, J. E. (1976) "Dynamic Response of Rigid Foundations of Arbitrary Shape," Earthquake Engineering and Structural Dynamics, Vol. 4, pp. 579-587, 1976.

55. Zienkiewicz, O. C., Chang, C. T. and Hinton, E. (1978) "Nonlinear Seismic Response and Liquefaction," Submitted for publication in the Int. J. Numerical and Analytical Methods in Geomechanics, 1978.

DEVELOPMENT OF SITE SPECIFIC RESPONSE SPECTRA

F. R. Hand[1], A. M. ASCE, G. V. Giese-Koch[2]

INTRODUCTION

Certain critical structures such as nuclear power plants, dams, emergency facilities, and hospitals, are designed to withstand the effects of earthquakes. To do this it is necessary to first determine the size (expressed as either magnitude or epicentral intensity) of the design earthquake, and then translate this event into a form suitable for incorporation into the design process. Such a form is often a response spectrum.

In the past generalized smoothed broad frequency band response spectra (1, 2, 13, and 14) have been developed for use at any site or for specific sites classified only as soil, deep soil, stiff, rock, etc. Once such a generalized spectrum shape is adopted, it is necessary to determine the appropriate "anchor point" acceleration (or other parameter) which is used to scale the general spectrum. This process has several shortcomings: (1) the generalized spectral shapes are based on a large range of earthquake sizes and epicentral distances which may not be representative of the tectonics of importance for the site and structure of current interest; (2) it neglects any particular site specific or unique characteristics; (3) of necessity the spectra are conservative or very conservative over a broad frequency range to cover a multitude of site, attenuation, and source characteristics; (4) a uniform measure of conservatism or risk is usually not present over the entire frequency range due to how the spectra were determined and scaled; and (5) no simple measure or expression of seismic risk is available.

---

[1]Senior Civil Engineer, Tennessee Valley Authority, Knoxville, TN
[2]Civil Engineer, Teneseee Valley Authority, Knoxville, TN

The construction of site specific response spectra and the calculation of uniform risk spectra would either avoid these shortcomings or qualitively address them. A procedure for calculating these is discussed.

DEVELOPMENT OF SITE SPECIFIC RESPONSE SPECTRA

In the Eastern United States a "tectonic province" approach is generally used to establish the earthquake design level. In this approach the largest (usually) historic earthquake is assumed to occur "at the site" of interest. In the following this method is used. Nevertheless, once the specific earthquake level has been established the remaining procedure is completely general and could be applied anywhere.

The tectonic provinces of the Southeastern United States are shown in figure 1. The site of interest is located in the Southern Appalachian Tectonic Province (SATP) and is shown in figure 1. The regional tectonics are used to define the size of the controlling earthquake event. This is a Modified Mercalli (MM) intensity VIII event and corresponds to the Giles County earthquake of May 31, 1897, the maximum historical intensity in the SATP. The epicenter location for this event is also shown in figure 1. Using the procedure developed by Nuttli, et al (15) the intensity is converted to a magnitude m_{bLg} + 5.8 event. For events in this range it is reasonable to approximate a magnitude m_{bLg} as a magnitude M_L. Thus, the controlling earthquake is assigned a magnitude M_L + 5.8.

Site specific response spectra are desired for this event occurring near a rock site. To develop these site specific design response spectra, a suite of existing strong motion records for earthquakes of appropriate conditions is investigated. In this investigation earthquakes of magnitude 5.3 to 6.3 (5.8 $\pm$ 0.5) recorded on competent rock within about 25 kilometers of the epicenter are considered.

Of the available world strong motion data, only thirteen existing records are found which meet these restrictions. These records are listed in table 1 (17, 18, and 19). Of these thirteen records, six are Western United States records and seven are Italy records. The magnitudes of these thirteen records range from 5.3 to 6.2 with an average of 5.7. The epicentral distances range from 7 to 27

TABLE 1

UNITED STATES AND ITALIAN EARTHQUAKES

USED IN THE STRONG MOTION ANALYSIS

Earthquake	Date/Time	Recording Station	Magnitude (M_L)	Epicentral Distance (km)	Instrument Orientation	Maximum Acceleration (g)
Helena, Montana	10-31-35/1138 MST	Carroll College	6.0	7	S00W S90W	.146 .145
San Francisco, California	3-22-57/1144 PST	Golden Gate Park	5.3	12	N10E S80E	.083 .105
Parkfield, California	6-27-66/2026 PST	Temblor	5.6	11	N65W S25W	.270 .348
Lytle Creek, California	9-12-70/0630 PST	Allen Ranch	5.4	24	S85E S05W	.071 .056
Lytle Creek, California	9-12-70/0630 PST	Devils Canyon	5.4	19	S00E S90E	.161 .165
Oroville, California	8-1-75/1320 PST	Oroville Dam	5.7	12	N53W N37E	.105 .113
Friuli, Italy	5-6-76/2000 GMT	Tolmezzo	6.2	27	N-S E-W	.346 .311
Friuli, Italy	5-9-76/0053 GMT	Tolmezzo	5.5	22	N-S E-W	.036 .032
Friuli, Italy	5-11-76/2244 GMT	Tolmezzo	5.3	13	N-S E-W	.027 .027
Friuli, Italy	9-11-76/1631 GMT	S. Rocco	5.5	16	N-S E-W	.040 .069
Friuli, Italy	9-11-76/1635 GMT	S. Rocco	5.9	14	N-S E-W	.089 .089
Friuli, Italy	9-15-76/0315 GMT	S. Rocco	6.1	9	N-S E-W	.066 .121
Friuli, Italy	9-15-76/0921 GMT	S. Rocco	6.0	20	N-S E-W	.142 .235

kilometers with an average of 15.8 kilometers. The peak accelerations range from 0.35 g to 0.03 g with an average of 0.13 g.

This suite of thirteen strong motion records is used to develop site specific response spectra. Response spectra for both horizontal components of these thirteen records for 4 and 7 percent of critical damping (the primary values of interest) are calculated for eighty different frequencies varying from 33 to 0.25 Hz. Site specific

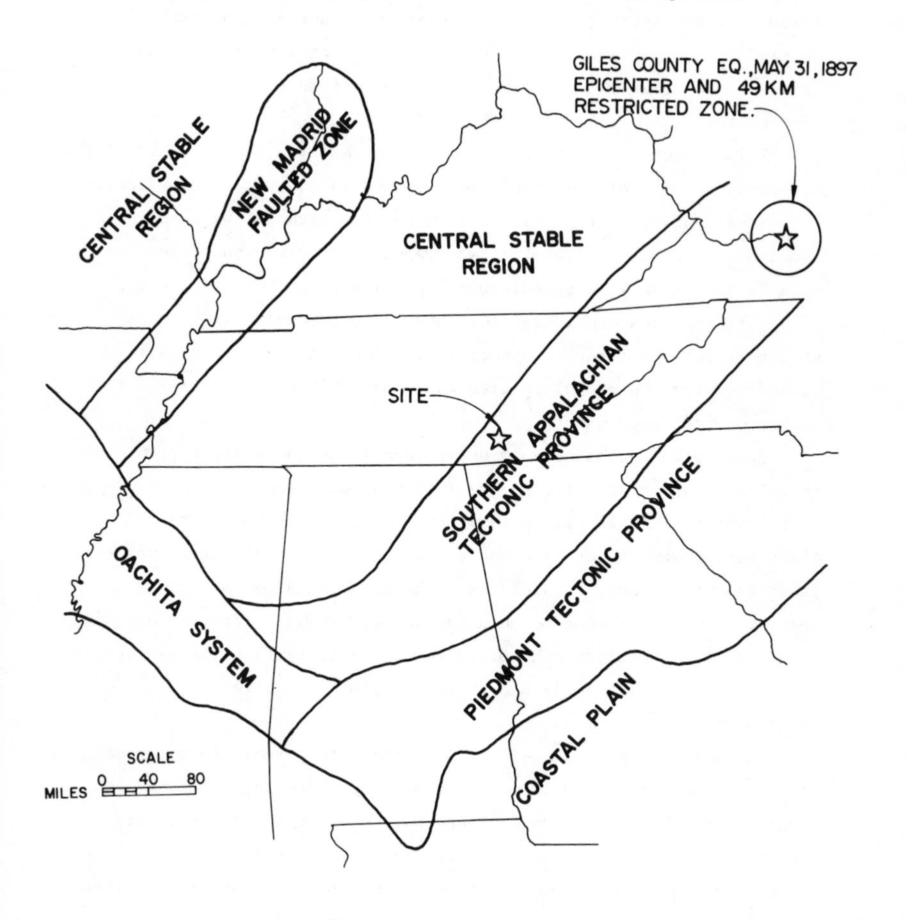

FIGURE 1 - MAP OF THE TENNESSEE VALLEY AREA SHOWING THE SEISMO-TECTONIC PROVINCES USED IN THE PROBABILITY STUDY

response spectra, both actual (unnormalized or absolute) and normalized, are developed from these individual spectra. The normalized spectra are obtained by scaling each of the twenty-six horizontal components to a peak acceleration of 1.0 g.

The statistical distribution of the actual and normalized response spectra is tested. Only normal (Gaussian) and lognormal distributions are considered. The data are compared visually to each assumed distribution for peak acceleration values and spectral values at periods 0.15, 0.40, and 4.0 seconds. This comparison indicates the data are more nearly lognormally distributed than normally distributed.

This tendency for the data to be lognormally distributed is also supported by a likelihood ratio test of the various spectral values. The likelihood ratio test was carried out for each of the eighty spectral frequencies. The test was repeated for frequency clusters with five or twenty frequencies in each cluster as well as for all eighty frequencies considered together. These results indicate a strong preference for the lognormal distribution, about a 2 to 1 preference for the actual spectra and about a 20 to 1 preference for the normalized spectra.

Thus, between the assumed normal and lognormal distribution, the actual data are better fit by the lognormal distribution. Also, for some lower percentile values (say 16th percentile), the normal distribution gave a negative entry indicating an acceleration value less than zero. Such a value has no physical meaning in this context. The occurrence of negative values has been reported also in reference 1. The occurrence of these possible negative values for lower percentile values is one shortcoming of the normal distribution.

The maximum, minimum, 16th, 50th, and 84th percentile response spectra for all thirteen records for 7 percent damping are shown in figure 2 for the actual spectra and in figure 3 for the normalized spectra (normalized to 0.10 g). From these figures approximate percent fractile levels may be selected to represent the desired site specific response spectrum. The spectra shown in these figures are based on the lognormal distribution.

Sensitivity Study

A sensitivity study is performed on the actual and normalized response spectra to determine the sensitivity of these spectra to include additional records in the data base. For both actual and normalized response spectra, six variations of additional records are considered. These six variations are the addition of: (1) two high

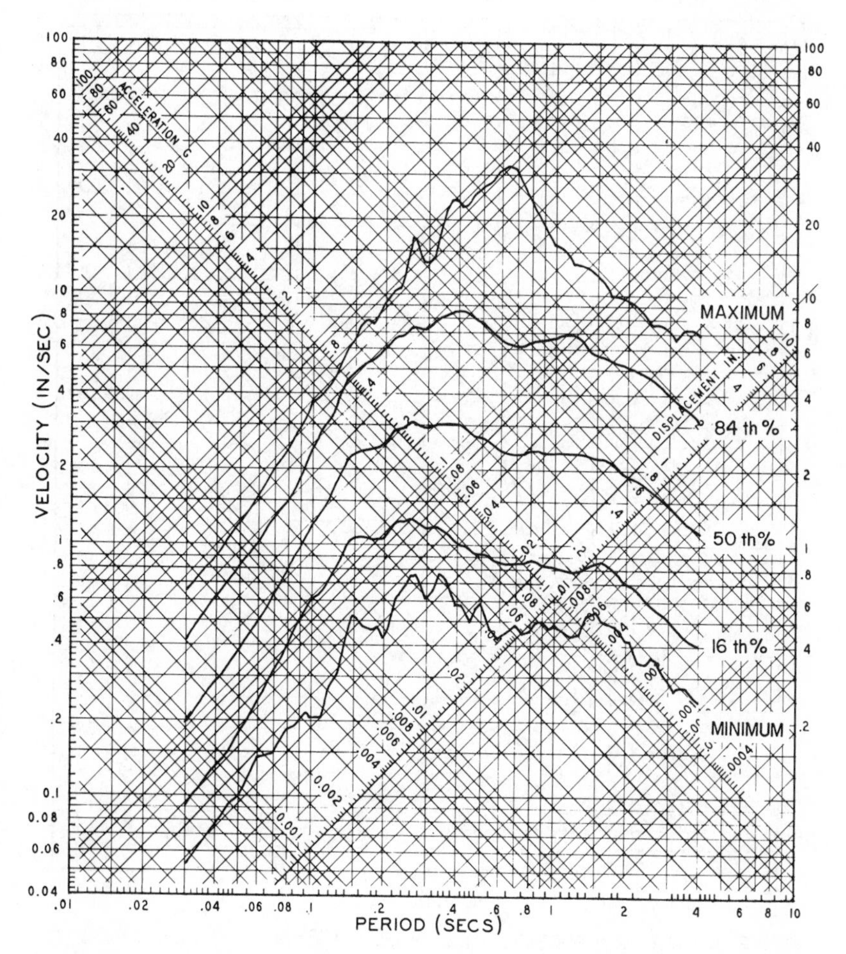

FIGURE 2 - MAXIMUM, MINIMUM, 16TH, 50TH, AND 84TH PERCENTILE RESPONSE SPECTRA FOR THIRTEEN UNITED STATES AND ITALY EARTHQUAKES - LOGNORMAL DISTRIBUTION - 7% DAMPING

pairs of records (two records of two components each), (2) four high pairs of records, (3) two low pairs of records, (4) four low pairs of records, (5) one high and one low pair of records, and (6) two high and two low pairs of records. In all cases the two horizontal components of the Tolmezzo, Italy, May 6, 1976, event (record No. 038) are used as the high pair of records. The peak accelerations are 0.346

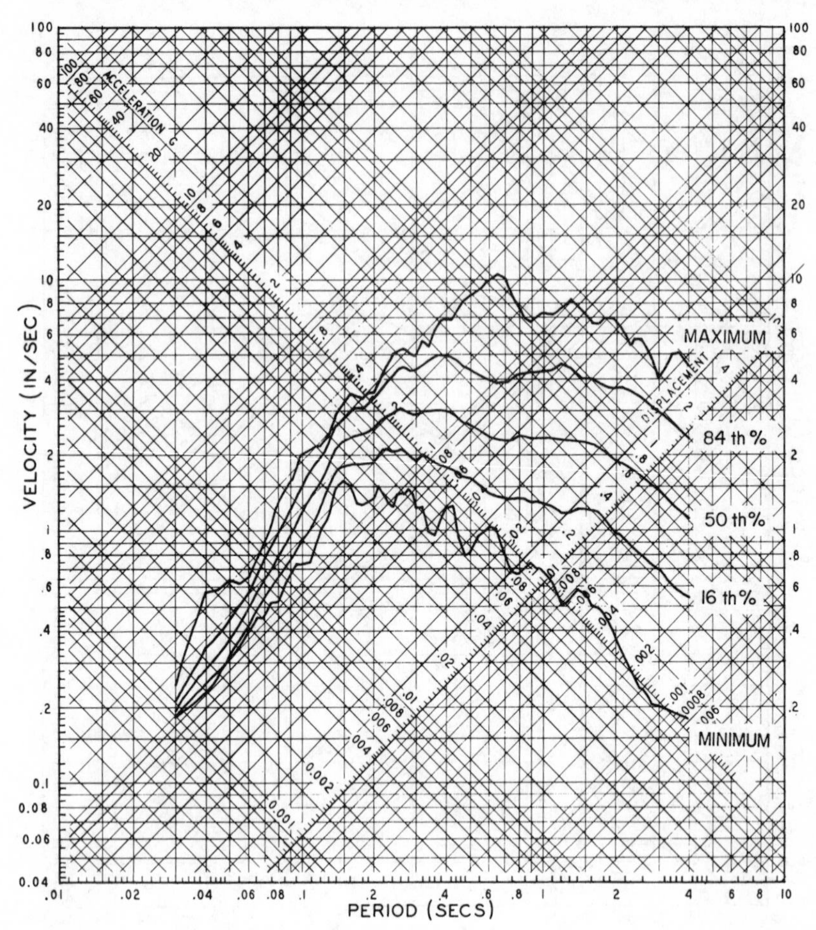

FIGURE 3 - MAXIMUM, MINIMUM, 16TH, 50TH, AND 84TH PERCENTILE NORMALIZED RESPONSE SPECTRA FOR THIRTEEN UNITED STATES AND ITALY EARTHQUAKES - LOGNORMAL DISTRIBUTION - 7% DAMPING

g and 0.311 g. As a pair these are the highest accelerations of all thirteen records. Temblor has a higher acceleration, 0.348 g, in one direction but a lower acceleration, 0.270 g, in the other direction. Similarly the Tolmezzo, Italy, May 11, 1976, event (record No. 063) is used as the low pair of records. It has the lowest acceleration, 0.027 g and 0.027 g, of all thirteen records.

Both high, low, and combined variations are considered to show the full impact of a parametric variation in the data base. It is reasonable to expect high and low values of acceleration for future earthquakes in the parameter range which we are considering. Examination of the recorded data during the Friuli earthquakes support this. For example, San Rocco recorded a magnitude 6.0 event at a distance of 20 kilometers with maximum values of 0.15 g and 0.24 g. A magnitude of 6.1 was also recorded at San Rocco at a distance of 9 kilometers which had maximum values of only 0.07 g and 0.12 g.

The 16th, 50th, and 84th percentile peak accelerations for the original data set and all six parametric variations of this data set are given in table 2. Figure 4 compares the 16th, 50th, and 84th percentile results for the actual response spectra for 7 percent

TABLE 2

VARIOUS PEAK ACCELERATION RESULTS FROM THE SENSITIVITY STUDY

	Peak Acceleration (g)		
Data Base Used	16th Percentile	50th Percentile	84th Percentile
Original 13 records	0.047	0.101	0.215
Original + 4 high pairs	0.058	0.133	0.306
Original + 2 high pairs	0.052	0.118	0.266
Original + 2 low pairs	0.037	0.084	0.195
Original + 4 low pairs	0.031	0.074	0.176
Original + 1 high and 1 low pair	0.043	0.100	0.232
Original + 2 high and 2 low pairs	0.040	0.099	0.244

damping for the original thirteen records, the original records plus four high pairs of records, and the original records plus four low pairs of records. The other four parametric variations are not shown since they fall between the extreme limits for four additional high or low pairs. Some noticeable effect is observed. Figure 5 gives the same comparison for the normalized response spectra for 7 percent

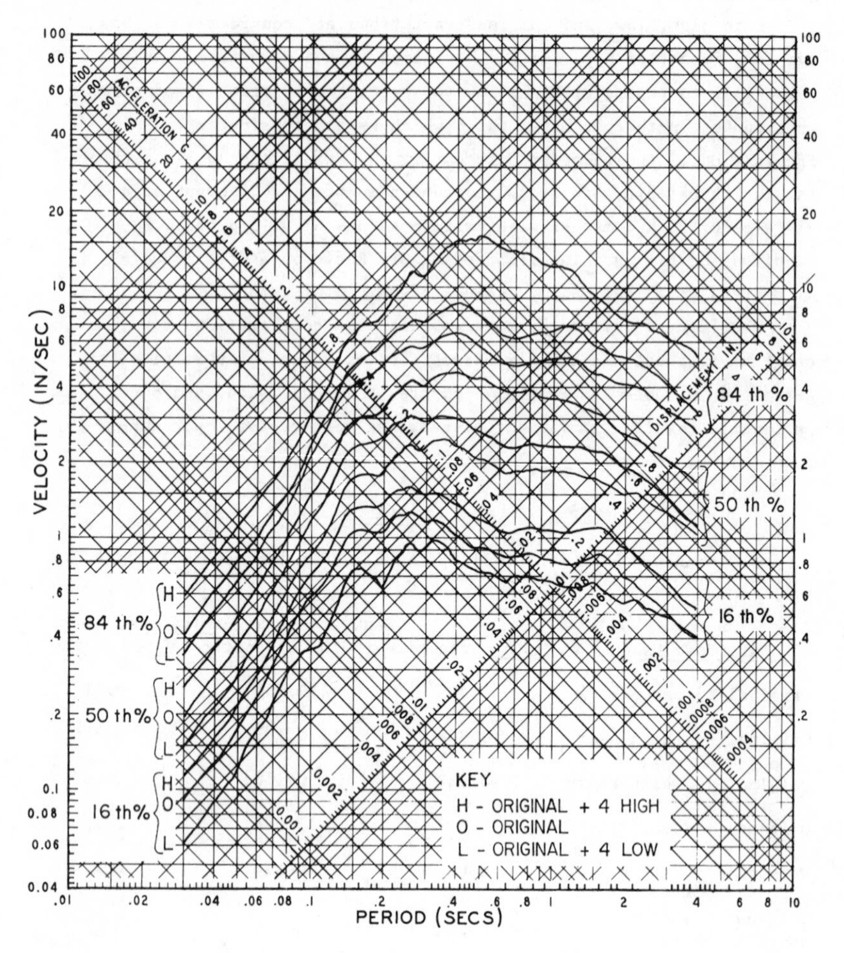

FIGURE 4 - SENSITIVITY STUDY - 16TH, 50TH, AND 84TH PERCENTILE RESPONSE SPECTRA FOR ORIGINAL 13 EARTHQUAKES, ORIGINAL PLUS 4 HIGH PAIRS, AND ORIGINAL PLUS 4 LOW PAIRS - LOGNORMAL DISTRIBUTION - 7% DAMPING

damping. In the higher frequency range (frequencies above 5 Hz) very little difference is noted between the parametric variations. In fact, these differences are so small that they are not discernible when plotted and shown on the same figure. Thus, for clarity of the plot, only the curves for the original thirteen records are shown in the higher frequency range. In the lower frequency range some

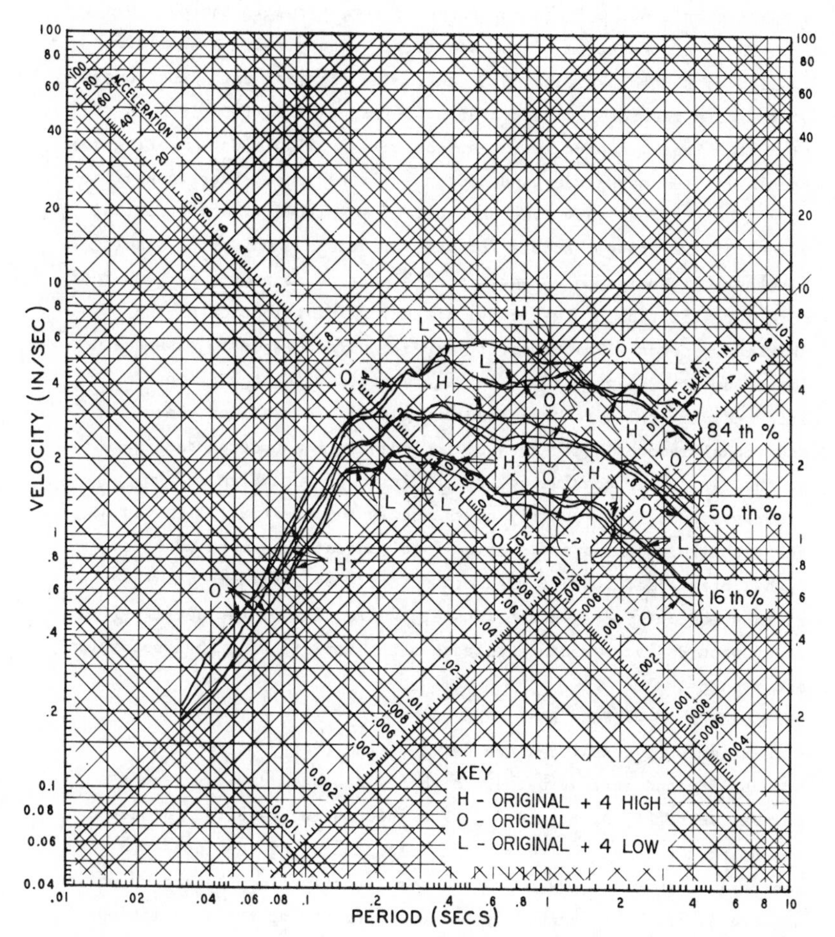

FIGURE 5 - SENSITIVITY STUDY - 16TH, 50TH, AND 84TH PERCENTILE NORMALIZED SPECTRA FOR ORIGINAL 13 EARTHQUAKES, ORIGINAL PLUS 4 HIGH PAIRS, AND ORIGINAL PLUS 4 LOW PAIRS - LOGNORMAL DISTRIBUTION - 7% DAMPING

deviation is observed and is shown in figure 5.

Based on the results of the complete sensitivity study and the results presented in figures 4 and 5, it is determined that (1) the actual response spectra are not overly sensitive to reasonable variations in the data base and (2) the normalized response spectra are very insensitive in the high frequency range and only moderately sensitive in the low frequency range to variations in the data base.

The interpretation of the sensitivity study results should be coupled with some practical judgment. The original data base consists of thirteen records recorded over a 43-year period, from 1935 to the present. These records represent the only available data meeting our specific site, magnitude, and distance limitations. Even though the peak accelerations vary over an order of magnitude, the statistical values are relatively stable for the variations considered. The hypothetical inclusion of additional high records should be tempered with a realization of the actual historical distribution of the thirteen records. Therefore, it appears unreasonable to assume a large number of high records may be recorded without additional intermediate or low records also being recorded. All of these additional records would then be included in the updated data base. In the variations considered, the addition of both high and low records results in virtually no change in the spectra. Although no quantitative measures are available to indicate how many additional high, intermediate, and low recordings will be available within a given number of years in the future, the results of the sensitivity study should give a measure to assess the impacts of such hypothetical occurrences.

DEVELOPMENT OF UNIFORM RISK SPECTRA

Seismic risk analyses are based on the theory of applied probability and statistics. This theory has been developed over several years (4, 5, 6, and 7). The information required to calculate the seismic risk at a site is (1) a description of the local and regional seismicity, (2) an attenuation function for the ground motion measure of interest, and (3) a means of performing the calculations. In the studies reported here, two earthquake models are used to describe the local and regional seismicity. Also, four attentuation functions are used. In all cases, the seismic risk computations are performed by a computer program developed principally by Cornell (4, 5, 6, and 7) and modified by McGuire (10 and 11). The end result of these studies is uniform risk spectra and comparison of these spectra with the various site specific spectra.

<u>Procedure For Calculating Probabilities</u>

The seismic risk analysis, as formulated by Cornell and McGuire, relies on the following basic assumptions:

1. The local and regional seismicity is described in the earthquake source model where the spatial, temporal, and size distributions the seismicity are defined. The spatial distribution is described by defining (a) the geographical extents of the tectonic province in which the site is located and sufficient adjacent provinces and/or (b) a system of capable faults or other earthquake sources. Within these regions the seismicity is further defined in terms of earthquake activity rate (earthquakes/year per defined surface area) and the relative distribution of earthquake sizes.
2. The relative frequency of earthquake magnitudes or epicentral intensities is of the form

   $$\log N_m \propto a - bM \qquad \text{or} \qquad \log N_I \propto a - bI_e \tag{1}$$

   where N is the expected number of events equal to or greater than magnitude M or epicentral intensity I_e in a given time period and a and b are parameters applicable to the source area examined. Implicit in this expression is the condition that $M \leq M_m$ or $I_e \leq I_m$ where M_m and I_m are the maximum (upper bound) magnitude or intensity for the region. The parameter b describes the relative distribution of small and large events; larger values of b imply relatively fewer large shocks, and vice versa. Since the program performs calculations using log base e (ln) rather than log base 10 (log), a parameter is also used where $\beta = b \ln 10 = 2.3b$.
3. The functional relationship between the desired ground motion parameter (peak ground acceleration, velocity, displacement, spectral acceleration, etc.) and the earthquake event is of the form

   $$Y = C_1' \, e^{C_2 M} R^{C_3} \varepsilon'$$

or taking the natural logarithm

$$\ln Y = C_1 + C_2 M + C_3 \ln R + \xi \qquad (2)$$

where

Y = The site ground motion parameter of interest,

R = The distance between the site and the event,

M = The size of the earthquake (magnitude or intensity),

C_1, C_2, C_3, C_1', = Constants applicable to the site ground motion parameter of interest and to the region of study ($C_1 = \ln C_1'$),

ξ' = A random lognormally distributed error term with median 1 and variance of $\ln \xi' = \sigma^2$, and

ξ = A random normally distributed error term with zero mean and standard error σ.

The above formulation may be transformed to calculate site intensity (I_s) by using the epicentral intensity (I_e) for M and replacing lnY by site intensity. The resulting expression is

$$I_s = C_1 + C_2 I_e + C_3 \ln R + \xi \qquad (3)$$

4. The annual risk associated with exceeding the desired ground motion parameter is calculated asssuming that earthquakes occur as Poisson arrivals; thus, the risk is determined from the total expected number of events causing exceedance by the expression

$$\text{Risk} = 1 - e^{(-\text{total expected number})} \qquad (4)$$

Predicting Probability of Spectral Acceleration Directly

The attenuation relationship expressed as equation 2 is for a general set of ground motion parameters. In seismic risk analyses in in the current literature, equation 2 is normally used to predict peak peak ground accelertion, velocity, or displacement. Similarly, equation 3 is normally used to predict site intensity; however, in the present work the desired end result is uniform risk spectra, not merely peak ground acceleration or site intensity. Thus, it is desired to extend the attenuation function used in the seismic risk analysis to encompass the entire transformation from earthquake size (be it described as magnitude or epicentral intensity) and distance to the spectral acceleration at a given frequency and percent of critical damping. It is possible to conveniently do this if the attenuation function thus derived is of the same general form as given in equation 2.

This concept of combining all transformations into one base expression (the attentuation function) is more desirable than isolating the individual parts. For example, if we proceed by the intensity route and calculate the individual parts the following steps are required: (1) determine the probability associated with a specified site intensity given the earthquake source model, (2) determine the probability associated with a specified peak ground acceleration given the site intensity, (3) determine the probability associated with a specified spectral acceleration (at a defined period and damping ratio) given the peak ground acceleration, (4) determine the total probability of exceeding the spectral acceleration by combining the probabilities due to steps 1, 2, and 3, and (5) take into account that the relationships used in steps 1, 2, and 3 are not deterministic but probabilistic with different measures of dispersion.

It is possible to integrate these five steps into one equation (the combined attenuation function) with one measure of dispersion. This allows the spectral acceleration (S_a) to be predicted directly from the earthquake size (M or I) at distance R. This greatly simplifies the theoretical formulation and considerably reduces the amount of numerical manipulation required.

Attenuation Models

Four different attenuation formulations are used in this study. These four formulations are described separately and are referred to as (1) Computer Science Corporation (CSC) formulation with unlimited intensity dispersion, (2) CSC formulation with limited intensity dispersion, (3) historical CSC formulation with unlimited intensity dispersion, and (4) historical CSC formulation with limited intensity dispersion.

CSC Formulation with Unlimited Intensity Dispersion - The general attenuation expression relating spectral acceleration (S_a) to the earthquake epicentral intensity (I_e) and distance (R) is derived from the following three expressions.

1. Bollinger's empirical expression is used to define the relationship between epicentral intensity and site intensity (3).

$$I_s = I_e + 2.87 - 2.88 \log R + \varepsilon_I \qquad R \geq 10 \text{ km} \qquad (5)$$

where

I_s = Site intensity (MMI),

I_e = Epicentral intensity (MMI),

R = Distance between epicenter and site (km), and

ε_I = A normally distributed error term with a zero mean and standard error of σ_I (σ_I = 1.2).

2. The CSC (Murphy and O'Brien, reference 12) empirical expression is used to define the relationship between the maximum horizontal acceleration and site intensity.

$$\log a_h = 0.24 I_s + 0.26 + \varepsilon_a \qquad (6)$$

where

a_h = Maximum peak horizontal ground acceleration (cm/s^2),

I_s = Site intensity (MMI), and

ε_a = A normally distributed error term with zero mean and standard error σ_a (σ_a = 0.34).

3. The site specific spectral amplifiction values are used to define the relationship between spectral acceleration and maximum ground acceleration (zero period spectral acceleration).

$$S_a = k \, a_{max} \, \varepsilon_{\ln k} \qquad (7)$$

where

S_a = Spectral acceleration at a specified frequency and damping ratio,

k = An empirically determined amplification factor dependent on the spectral period and damping ratio considered,

a_{max} = Maximum horizontal ground acceleration, and

$\varepsilon_{\ln k}$ = An error term assumed to be lognormally distributed such that $\ln \varepsilon_{\ln k}$ is normally distributed with zero mean and standard eror of $\sigma_{\ln k}$ where $\sigma_{\ln k}$ is dependent on frequency and damping ratio.

By successive substitution of equation 5 into 6 and the result into 7 the desired result is obtained.

$$\ln S_a = \ln k + 2.18 + 0.55\, I_e - 0.69 \ln R + 0.55 \xi_I + 2.30\, \xi_a + \ln \xi_{\ln k} \qquad (8)$$

This is in the form of equation 2 where

$C_1 = \ln k + 2.18$

$C_2 = 0.55$

$C_3 = -0.69$

S_a = Error term = $0.55\xi_I + 2.30\xi_a + \ln \xi_{\ln k}$

and where

I_e = Epicentral intensity (MMI),

R = Epicentral distance (km),

S_a = Spectral acceleration for a given frequency and damping ratio (cm/s^2), and

k = An empirically determined amplification factor dependent on spectral frequency and damping ratio.

The dispersion of the combined error term may be determined from the individual dispersions. Neglecting cross correlations, which is conservative (8 and 9), the dispersion may be combined approximately by the SRSS rule

$$\sigma_{\ln S_a} = \sqrt{(0.55\sigma_I)^2 + (2.30\sigma_a)^2 + \sigma_{\ln k}^2} \qquad (9)$$

where

$\sigma_I = 1.2$

$\sigma_a = 0.34$

$\sigma_{\ln k}$ = Varies with spectral frequency

and yields

$$\sigma_{\ln S_a} = \sqrt{1.048 + \sigma_{\ln k}^2}$$

The values of $\sigma_{\ln S_a}$ corresponding to individual values of $\sigma_{\ln k}$ for the various periods are listed in table 3.

CSC Formulation with Limited Intensity Dispersion - The attenuation relationship expressed in equation 8 does not specifically limit the upper dispersion intensity for events close to the site considered. This is an unrealistic and very conservative liberty. Based on historic data and the practice of assigning the maximum intensity as the epicentral intensity, it is not possible to have a site intensity at a given distance from the epicenter which is larger than the epicentral intensity. Nevertheless, the mathematics involved in calculating the seismic risk will permit this to occur.

To limit this dispersion on intensity the following procedure is used. Concentric circles about the site are defined with radii of 10, 22, and 49 kilometers. These radii represent in turn the limits of zero drop in intensity, a 1-unit drop in intensity, and a 2-unit drop in intensity as calculated from Bollinger's expression (equation 5). Within each of these circles or rings the dispersion on intensity is restricted by setting $3\sigma_I$ units equal to the average intensity drop within the zone. The $3\sigma_I$ level corresponds to 0.9987 confidence or 0.0013 exceedance probability. This approximation is conservative because it also reduces the calculated likelihood of intensities much less than the predicted intensity. The average intensity drop within each zone are 0, 0.5, and 1.5 and the resulting intensity dispersion σ_I are 0, 0.167, and 0.500, respectively. These three limited dispersions on intensity are then used in calculating the dispersion of the combined error term for these three zones by substituting the new σ_I at 10, 22, or 49 kilometers for the one originally used in

TABLE 3

NUMERICAL VALUES FOR THE VARIOUS ATTENUATION RELATIONSHIPS USED IN THE SEISMIC RISK ANALYSIS

Period (Seconds)	Program Constants					Standard Errors: Intensity Unlimited	Standard Errors: Intensity Limited		
T	k*	ln k	C_1	C_2	C_3	$\sigma_{\ln S_a}$	σ_{10}	σ_{22}	σ_{49}
			CSC Attenuation						
0.03	1.042	0.041	2.22	0.55	-0.69	1.026	0.785	0.790	0.832
0.06	1.297	0.260	2.24	0.55	-0.69	1.034	0.796	0.801	0.842
0.15	2.437	0.891	3.07	0.55	-0.69	1.047	0.813	0.818	0.858
0.40	1.196	0.179	2.36	0.55	-0.69	1.144	0.934	0.938	0.973
1.50	0.237	-1.440	0.74	0.55	-0.69	1.183	0.981	0.985	1.019
4.00	0.0450	-3.101	-0.92	0.55	-0.69	1.255	1.067	1.071	1.102
			Historical CSC Attenuation						
0.03	1.042	0.041	3.08	0.55	-0.94	0.836	0.693	0.696	0.720
0.06	1.297	0.260	3.30	0.55	-0.94	0.847	0.705	0.708	0.732
0.15	2.437	0.891	3.93	0.55	-0.94	0.862	0.725	0.727	0.750
0.40	1.196	0.179	3.22	0.55	-0.94	0.977	0.858	0.860	0.880
1.50	0.237	-1.440	1.60	0.55	-0.94	1.022	0.909	0.911	0.930
4.00	0.0450	-3.101	-0.061	0.55	-0.94	1.105	1.002	1.004	1.020

*k values are for 7 percent damping.

equation 9. The dispersions on the other terms are unchanged. The resulting $\ln S_a$ are shown in table 3. The attenuation relationship given by equation 8 remains otherwise unchanged.

Historical CSC Formulation with Unlimited Intensity Dispersion - The attenuation relationships developed in the previous two sections use equation 6 to transform site intensity to peak horizontal ground acceleration. A problem with this approach is that distance effects are not considered. For example, a site intensity VI due to an earthquake at a distance of 100 miles has a different site motion than a site intensity VI due to an earthquake at a distance of 20 miles. The distance independent equation 6 does not consider this.

Murphy and O'Brien (12) also determined a distance dependent acceleration/intensity correlation equation, namely

$$\log a_h = 0.17\ I_s + 0.07\ I_e - 0.45 \log R + 0.83 + \mathcal{E}_a \qquad (10)$$

where

I_s = Site intensity (MMI),

I_e = Epicentral intensity (MMI),

R = Epicentral distance (km),

a_h = Peak horizontal ground acceleration (cm/s^2), and

$\mathcal{E}_a$ = Error term with zero mean and standard error of σ_a (σ_a = 0.30).

This correlation expression is proposed for the assessment of historical earthquakes for which instrumentally determined magnitudes are not available; hence, we call this the historical CSC relationship. The composite attenuation function is determined by successive substitution of equation 5 into equation 10 and then into equation 7 to yield

$$\ln S_a = \ln k + 3.04 + 0.55\ I_e - 0.94 \ln R + 0.39\ \mathcal{E}_I + 2.30\ \mathcal{E}_a + \ln \mathcal{E}_{\ln k} \qquad (11)$$

where the symbols are defined the same as before. The dispersion on the combined error term is calculated as before and is listed in table 3.

Historical CSC Formulation with Limited Intensity Dispersion - The historical CSC attenuation relationship expressed in equation 11 does not specifically limit the upper dispersion on intensity as was previously discussed. Again the concentric circles with radii of 10,

22, and 49 kilometers are used to limit this dispersion. The resulting $\ln S_a$ are shown in table 3. The attenuation relationship (equation 11) remains otherwise unchanged.

The four attenuation formulations incorporate the site specific amplification values. These acceleration amplification values (k values) are determined from the 50 percentile normalized spectrum of a suite of twenty-six site specific earthquake records. An idealized form of this spectrum is used with control points at spectral periods of 0.03, 0.06, 0.15, 0.40, 1.5, and 4.0 seconds. The sensitivity of these amplification factors (k values) to variations in the suite of earthquakes is also checked. The k values are calculated for the six variations in the suite of earthquakes discussed in the sensitivity study. The k values are relatively insensitive to these variations. This fact implies that the uniform risk response spectra would not change significantly. Therefore, no further sensitivity analyses are performed.

Earthquake Source Models

The local and regional seismicity is described in the earthquake source model where the spatial, temporal, and size distributions of the seismicity are defined. Two basic earthquake source models are considered in this study with some minor variations also considered. The basic building blocks of these models are the tectonic provinces of the Southeastern United States. The two basic models are described below.

SATP Model - The earthquake source model consists of the tectonic provinces of the Southern United States as shown in figure 1. They are the Southern Appalachian Tectonic Province (SATP), Piedmont Tectonic Province, Ouachita Tectonic Province, Central Stable Region, and New Madrid Faulted Zone. The earthquake activity rates and the maximum historical intensity for each province are given in table 4. These values are determined from a historic earthquake list published by the U.S. Geologic Survey. A constant β value of 1.312 (b = 0.57) is used throughout. This value is reported by McGuire and Young (20) and is considered applicable for the Eastern United States. It also correlates well with the value calculated from the historical earthquake activity in the SATP. The maximum historical intensity in the SATP is a MMI VIII, the Giles County earthquake of May 31, 1897. A parametric variation is also considered where this intensity is a MMI IX.

SATP Model with Giles County Earthquake Restricted - This earthquake source model is the same as the SATP model except the Giles County earthquake of May 31, 1897, is restricted to a circle with radius 49 kilometers centered at the historical epicenter. As a result the maximum historical intensity in the remaining SATP is a MMI VII.

The Giles County earthquake is restricted based on the results of reference 16. This study strongly suggests the existence of a tectonic structural zone with which this earthquake was associated and to which a recurrence of an event of this magnitude would be restricted. The selection of the 49-kilometer radius is arbitrary. It restricts the earthquake to a region within a 2-unit intensity drop of the epicentral intensity (based on Bollinger's relationship). This region can be considered to approximate the intensity VI isoseismal.

RESULTS

A comparison of various spectral acceleration versus return period curves for a spectral period of 0.03 second is shown in figure 6. The SATP earthquake source model with and without the Giles County earthquake restricted is used. Three of the curves show results for CSC attenuation and compare the unlimited and limited intensity dispersion cases and the parametric variation of the SATP maximum historic intensity as a IX. The other two curves show the limited

TABLE 4

TECTONIC PROVINCE EARTHQUAKE DATA USED IN THIS STUDY

Tectonic Province	Maximum Historical Intensity (MMI)	Activity Rate (Earthquake/Year)	Recurrence Parameter β
Southern Appalachian	VIII	0.6020	1.312*
Piedmont	VII	0.1942	1.312
Ouachita	VII	0.2020	1.312
Central Stable Region	VII-VIII	1.1258	1.312
New Madrid Faulted Zone	XII	0.5780	1.312

*$\beta = b \ln 10$, => $b = 0.57$

intensity cases using historical CSC attenuation. Unless otherwise labeled, the curves are for the limited intensity case. Results similar to those shown in figure 6 are obtained for spectral periods of 0.06, 0.15, 0.40, 1.50, and 4.0 seconds. Uniform risk spectra are generated from these return period curves.

The earthquake activity of the New Madrid Faulted Zone is not considered in these models. In cases not presented this earthquake activity is considered. Detailed inspection of these probabilities indicate the New Madrid Faulted Zone is contributing up to 20 percent of the exceedance probability in the unlimited intensity dispersion case and up to 50 percent in the limited intensity dispersion case. These probabilities are for the exceedance of a given spectral acceleration. It is unreasonable, based on recorded acceleration attenuation with distance, to expect even an intensity XII in the New Madrid Faulted Zone to cause relatively high spectral or ground accelerations at distances of 300 kilometers. Therefore, the earthquake activity of the New Madrid Faulted Zone was not considered further.

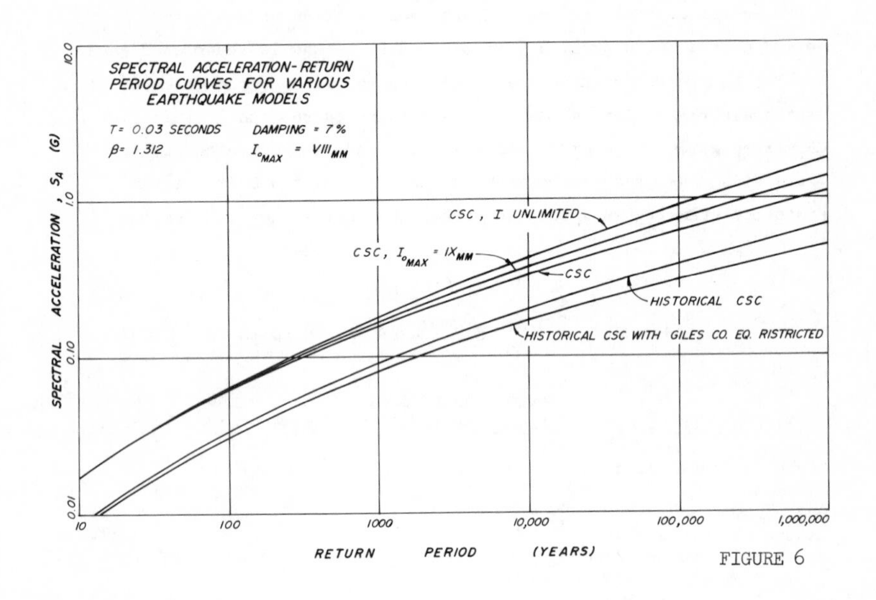

FIGURE 6

The sensitivity of the risk calculations to the specific site location was also tested by considering three other sites in the SATP. This study was made using the SATP earthquake source model when the upper limit on intensity dispersion is unlimited. The results show the return periods or probabilities of exceedance are basically the same at all sites in the SATP when the simple SATP earthquake source model is used. This generalization is not true for more refined models where the Giles County earthquake is restricted.

Figures 7 through 12 present the results of the probabilistic analysis as uniform risk spectra (risk is on an annual basis) and comparisons of these risk spectra to various site specific spectra. Uniform risk spectra and their comparisons are presented in figures 7 through 9. These three figures compare results for CSC attenuation when the maximum SATP intensity is VIII and IX and compare results for CSC and historical CSC attenuation with and without the Giles County earthquake restricted. As can be seen in figure 7 there is little difference between an upper bound intensity of VIII or IX in the SATP. Thus, comparisons of intensity IX uniform risk spectra with the site specific spectra are not made. This is contrasted with about a one-half order of magnitude difference in risk between CSC and historical CSC attenuation cases (figure 8) and up to an order of magnitude difference in risk between CSC attenuation and historical CSC attenuation with the Giles County earthquake restricted (figure 9).

Figures 10 through 12 compare the uniform risk spectra with the 50th and 84th percentile actual and 84th percentile normalized site specific spectra. These site specific spectra are designated as 50% A, 84% A, and 84% N, respectively, on the figures. The 84th percentile normalized spectrum is anchored to the 50th percentile (lognormal distribution) peak acceleration (0.10 g). Other anchor point values could be used. It is consistent to use a peak acceleration value defined from the site specific records rather than turn to some empirical relationship. This 50th percentile peak acceleration value is consistent with the nuclear industry and Nuclear Regulatory Commission (NRC) practice of using a "mean" acceleration value to anchor their general normalized spectra.

The probability of exceeding the different response spectra can be determined from figures 10 through 12. The relative differences of probabilities between the various response spectra are less than one order of magnitude for the same models. The probabilities of annual exceedance vary with the different models and ranges from 10^{-3} to 10^{-5}.

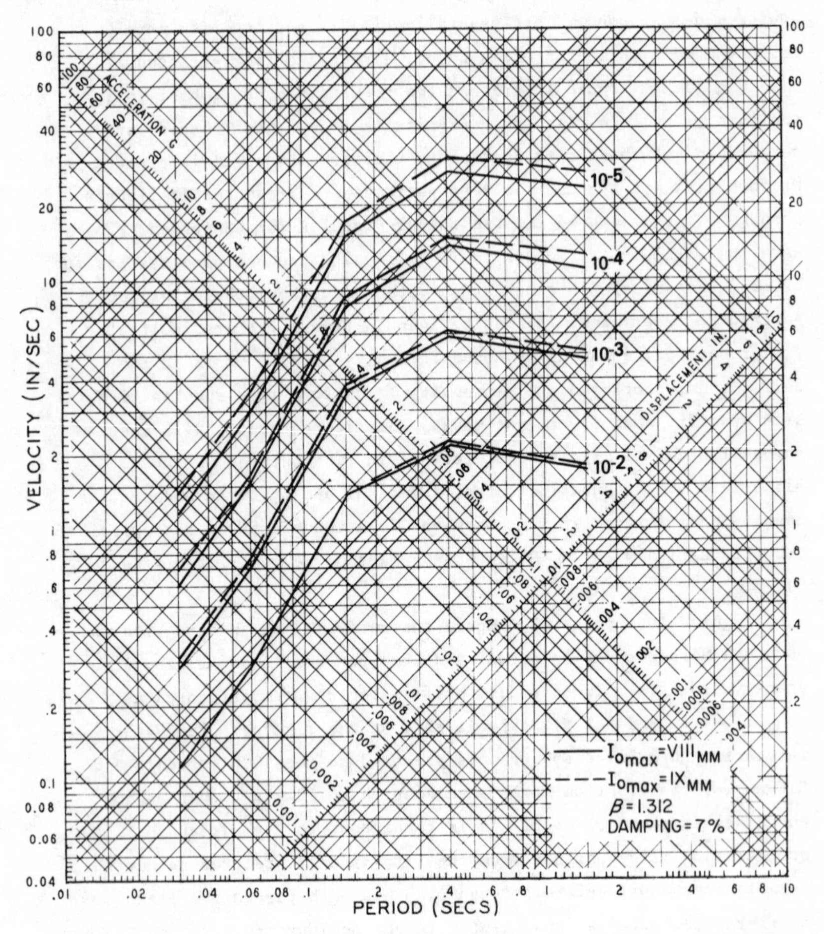

FIGURE 7 - COMPARISON OF UNIFORM RISK RESPONSE SPECTRA FOR UPPER BOUND INTENSITY OF VIII AND IX IN THE SATP.

SUMMARY AND CONCLUSIONS

Site specific response spectra and uniform risk spectra were developed for a rock site experiencing a near field magnitude M_L 5.8 (approximate) earthquake. Site specific earthquake records meeting certain restrictions of earthquake size, epicentral distance, and recording site conditions were used to generate these spectra. This

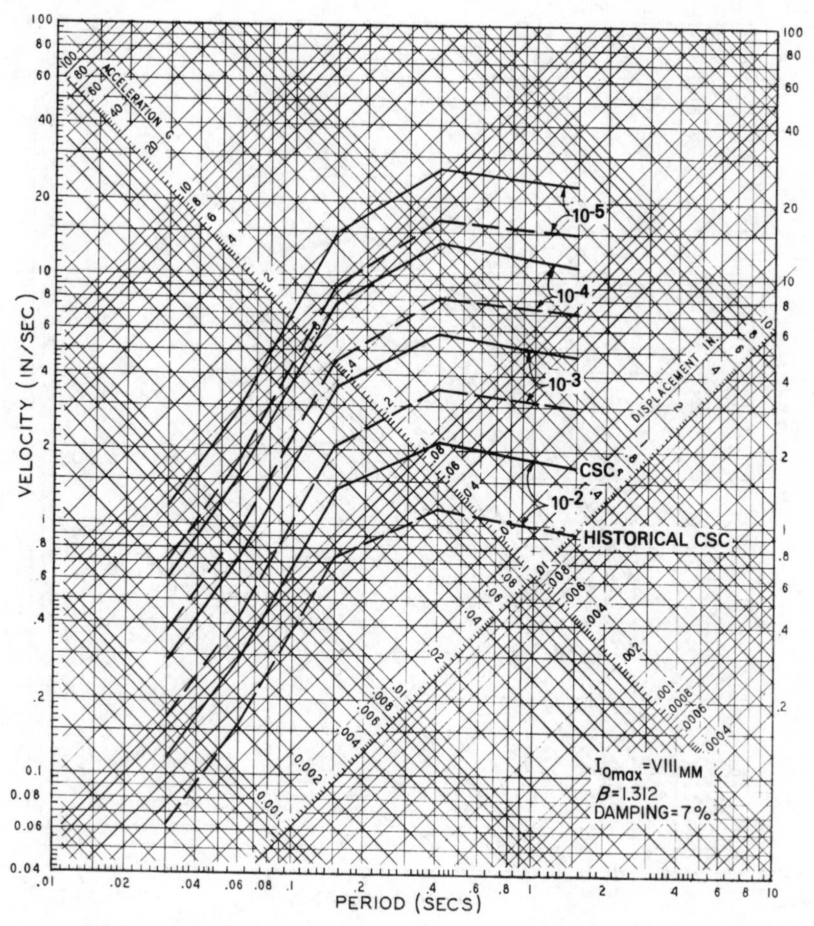

FIGURE 8 - COMPARISON OF UNIFORM RISK RESPONSE SPECTRA FOR CSC AND HISTORICAL CSC ATTENUATION.

process avoids many of the shortcomings of the current practice of using generalized broad frequency band spectra. The uniform risk spectra allow the probability of exceedance to be quickly determined for any selected site specific response spectrum or other spectrum one might wish to use. The probabilities of exceedance for the developed actual and normalized spectra obviously vary with the different

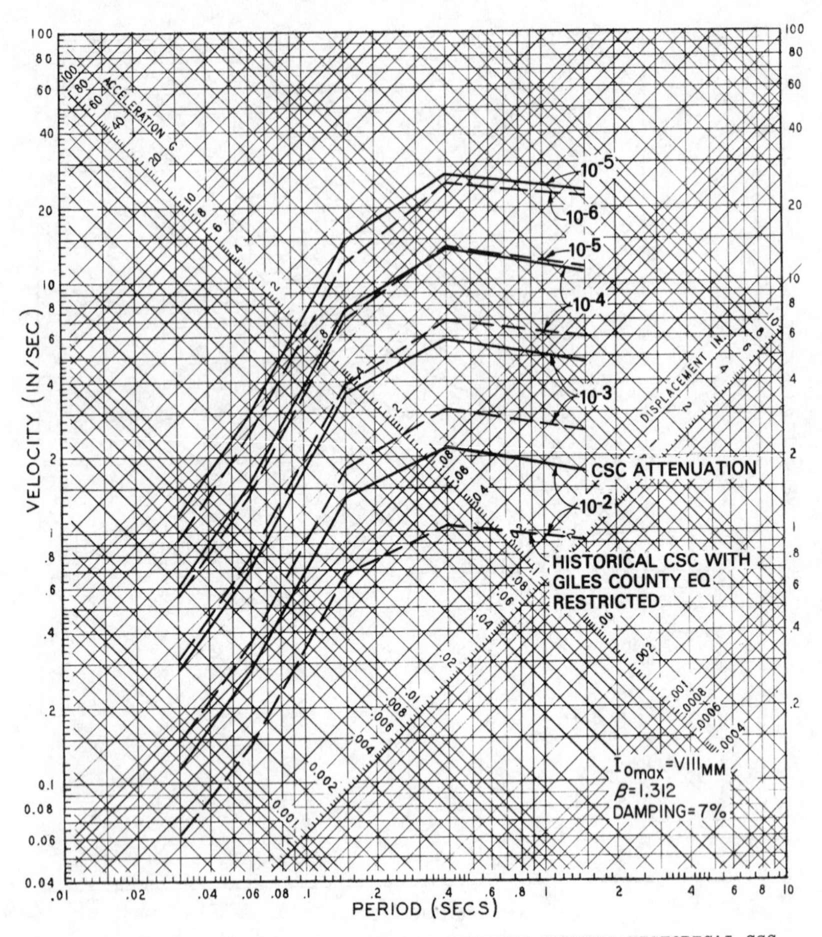

FIGURE 9 - COMPARISON OF UNIFORM RISK RESPONSE SPECTRA HISTORICAL CSC ATTENUATION WITH GILES COUNTY EARTHQUAKE RESTRICTED AND CSC ATTENUATION.

spectra and earthquake models used but range from 10^{-3} to 10^{-5} on an annual basis.

ACKNOWLEDGMENT

The authors thank Dr. C. A. Cornell, Professor of Civil Engineering, MIT, for his assistance in this work.

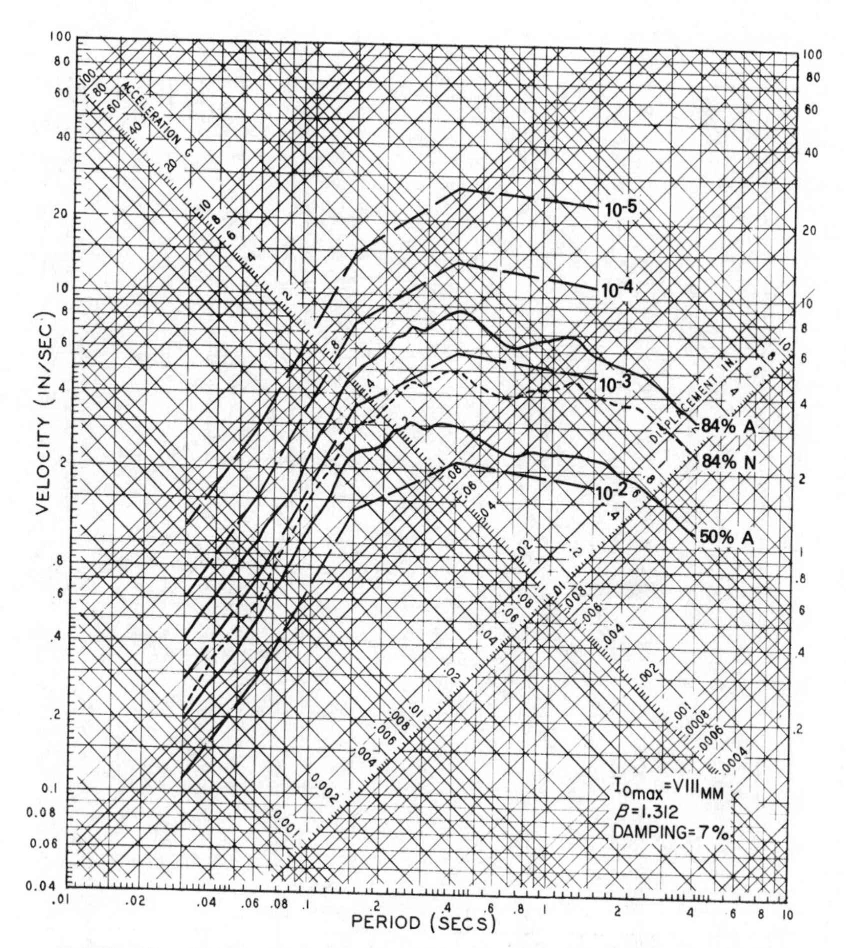

FIGURE 10 - COMPARISON OF UNIFORM RISK RESPONSE SPECTRA WITH VARIOUS SITE SPECIFIC SPECTRA - CSC ATTENUATION.

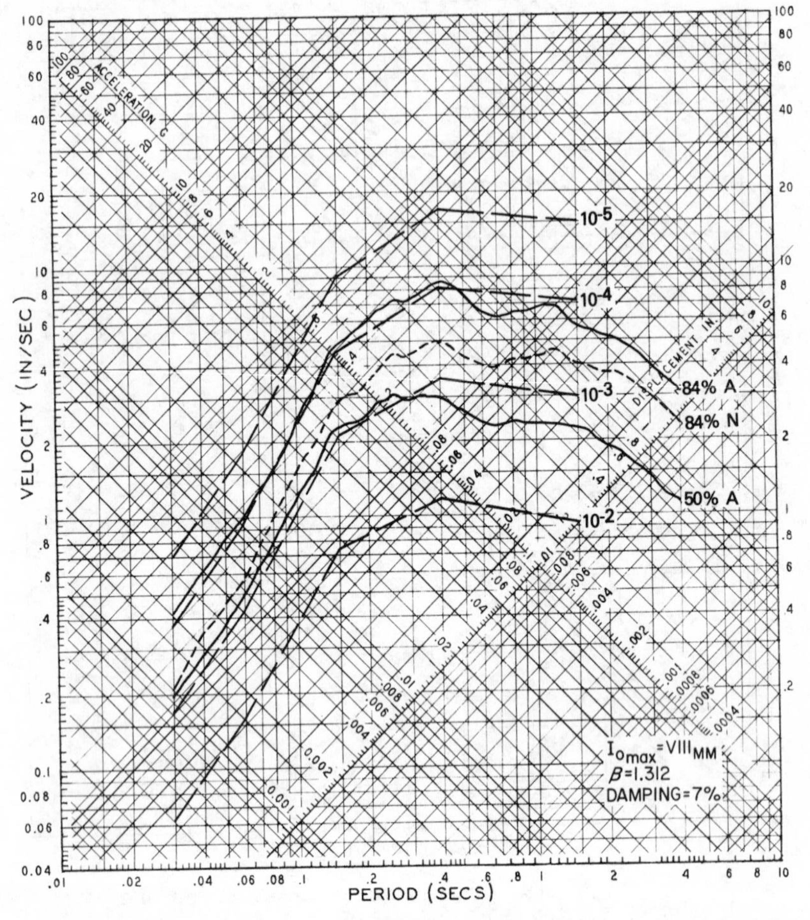

FIGURE 11 - COMPARISON OF UNIFORM RISK RESPONSE SPECTRA WITH VARIOUS SITE SPECIFIC SPECTRA - HISTORICAL CSC ATTENUATION

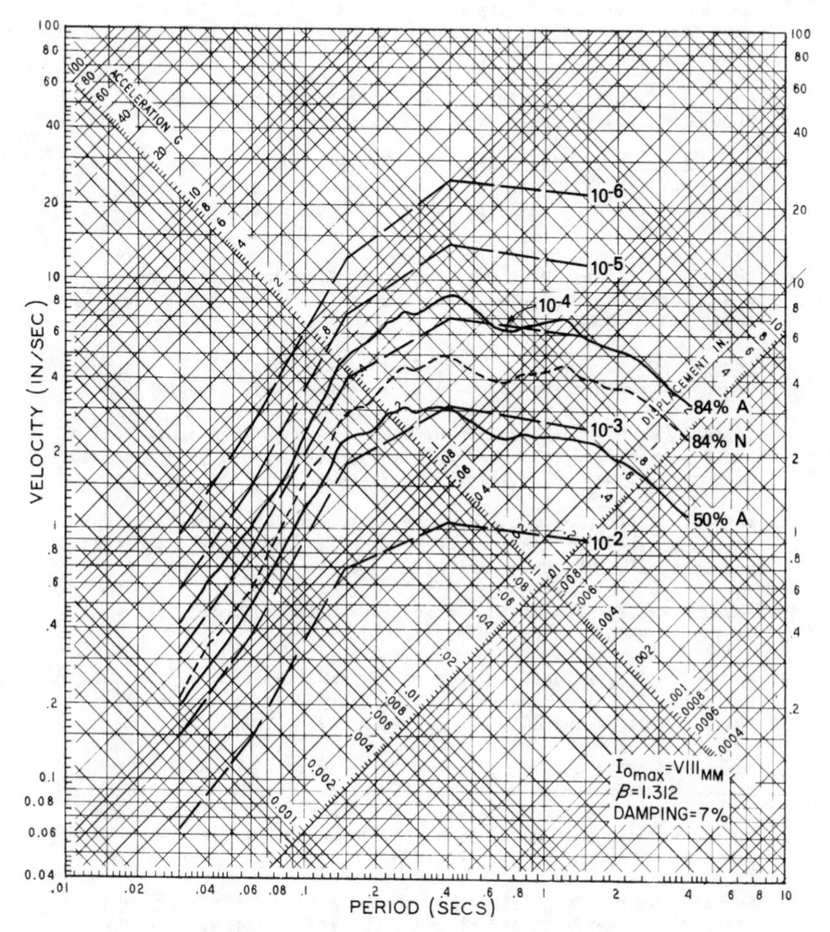

FIGURE 12 - COMPARISON OF UNIFORM RISK RESPONSE SPECTRA WITH VARIOUS SITE SPECIFIC SPECTRA - HISTORICAL CSC ATTENUATION - GILES COUNTY EARTHQUAKE RESTRICTED

References

1. Agbabian Associates, Correlation of Ground Response Spectra with Modified Mercalli Site Intensity, Topical Report for ERDA, June, 1977.

2. John A. Blume & Associates, "Recommendations for Shape of Earthquake Response Spectra," San Francisco, California, USAEC Contract No. AT (49-5)-3011, WASH-1254, February 1973.

3. Bollinger, G. A., Reinterpretation of the Intensiy Data for the 1886 Charleston, South Carolina, Earthquake USGS Professional Paper 1028-B, 1976.

4. Cornell, C. A., Engineering Seismic Risk Analysis, BSSA, Volume 58, No. 5, pages 1583-1606, 1968.

5. Cornell C. A., Probabilistic Analysis of Damage Structures Under Seismic Load, Chapter 27 in "Dynamic Waves in Civil Engineering," D. A. Howells, et al, Wiley Interscience, London, pages 473-488, 1971.

6. Cornell, C. A. and H. A. Merz, Seismic Risk Analysis Based on a Quadratic Magnitude-Frequency Law, BSSA, Volume 63, No. 6, pages 1999-2006, 1973.

7. Cornell, C. A. and H. A. Merz, Seismic Risk Analysis of Boston, Journal Structural Division, ASCE 10, pages 2027-2043, 1975.

8. Cornell, C. A., Hosshang Banon, and A. F. Shakal, Seismic Motion and Response Prediction Alternatives, Earthquake Engineering and Structural Dynamics, Volume 7, 1979, pages 295-315.

9. McGuire, R. K., Seismic Structural Response Risk Analysis, Incorporating Peak Response Regressions on Earthquake Magnitude and Distance, MIT, CE Report R74-51, 1974.

10. McGuire, R. K., FORTRAN Computer Program for Seismic Risk Analysis, USGS Open File Report 76-67, 1976.

11. McGuire R. K., Effects of Uncertainty in Seismicity of Estimates of Seismic Hazard for the East Coast of the United States, BSSA, Volume 67, No. 3, pages 827-848, 1977.

12. Murphy, J. R. and L. J. O'Brien, Analysis of a Worldwide Strong Motion Data Sample to Develop an Improved Correlation Between Peak Acceleration, Seismic Intensity, and Other Physical Parameters, Office of Standards Development, U.S. Nuclear Regulatory Commission Report NUREG 0402, 1978.

13. Newmark, N. M., Consulting Engineering Services, "A Study of Vertical and Horizontal Earthquake Spectra," Urbana, Illinois USAEC Contract No. AT (49-5)-2667, WASH-1255, April 1973.

14. Newmark, N. M., John A. Blume, and Kanwar K. Kapur, "Design Response Response Spectra for Nuclear Power Plants," ASCE Structural Engineering Engineering Meeting, San Francisco, April 1973.

15. Nuttli, O. W., Bollinger, G. A., and Griffiths, D. W., "On the Relation Between Modified Mercalli Intensity and Body-Wave Magnitude," Magnitude," Preprint submitted to the Bulletin of the Seismological Society of America , 1978.

16. Southern Appalachian Tectonic Study, TVA, January 1979.

17. Strong Motion Earthquake Accelerograms, Volume II, Earthquake Engineering Research Laboratory, California Institute of Technology, Pasadena, California, September 1971.

18. Strong Motion Earthquake Accelerograms, Digitized and Plotted Data, Part 1, Accelerograms 028 through 064, CNEN-ENEL, Rome, Italy, July, 1976.

19. Strong Motion Earthquake Accelerograms, Digitized and Plotted Data, Part 3, Accelerograms 120 through 177, CNEN-ENEL, Rome, Italy, November 1977.

20. Young, G. A., Problem Areas in the Application of Seismic Hazard Analysis Procedures, Energy Research and Development Administration Report SAU/1011-101, 1976.

HETERICK MEMORIAL LIBRARY
690.54 A111s v.2
ASCE Conference on/Second ASCE Conferenc
onuu
3 5111 00123 5187